APPLICATION AND CHARACTERIZATION OF SURFACTANTS

Edited by **Reza Najjar**

Application and Characterization of Surfactants
http://dx.doi.org/10.5772/65591
Edited by Reza Najjar

Contributors

Bogumil E. Brycki, Iwona Kowalczyk, Adrianna Szulc, Olga Kaczerewska, Marta Pakiet, Ana Maria Carmona-Ribeiro, Tomasz Kalak, Dorota Kregiel, Joanna Berlowska, Izabela Witonska, Hubert Antolak, Charalampos Proestos, Mirko Babic, Bolin Zhang, Ljiljana Babic, Darija Domazet Jurašin, Suzana Šegota, Vida Čadež, Atiđa Selmani, Maja Dutour Sikiric, Violeta Jakovljevic, Miroslav Vrvic, Eduardo Schulz, Erica Schulz, Pablo Schulz, Lidia Pinheiro, Celia Faustino, Abdelaziz El-Hoshoudy, Saad M. Desouky, Mohamed.H. Betiha, Ahmed Al-Sabagh, Katerina Lepkova, Deepak Dwivedi

Published by InTech

Janeza Trdine 9, 51000 Rijeka, Croatia

Publishing Process Manager Martina Usljebrka
Technical Editor SPi Global
Cover InTech Design team

First published June, 2017
Printed in Croatia
Legal deposit, Croatia: National and University Library in Zagreb

Additional hard copies can be obtained from orders@intechopen.com

Application and Characterization of Surfactants, Edited by Reza Najjar
p. cm.
Print ISBN 978-953-51-3325-4
Online ISBN 978-953-51-3326-1

Meet the editor

Reza Najjar (born in Marvdizaj/Soufian, Iran) studied Chemistry at the University of Tabriz (BSc, 1994), Sharif University of Technology (MSc, 1997), and Aachen University of Technology (RWTH), Germany (PhD, 2006). After completing a year of postdoctoral work at the University College Dublin (UCD), Ireland, he joined the Faculty of Chemistry at the University of Tabriz (August 2007) as an assistant professor, where currently he is working as an associate professor since January 2015. One of his research interests is to prepare polymeric surfactants and use them in various fields of applications.

Contents

Preface

The surfactants are among the materials that have a significant importance in everyday life of human. These materials are the main effective ingredients in systems that are based on the emulsions or microemulsions, where stabilization or solubilization of two or more components with opposite polar characteristics in a uniform system is required. Even the rapid growth in science and technology has opened new horizons in a very wide range, in which the surfactants play a major and vital role. Hence, the increasing number of applications as well as arising environmental issues has made this relatively old topic still a hot research theme in the various professions, ranging from pure physical chemists to oil engineers, pharmacologists, and biologists. The synthesis of novel and specially designed surfactants for certain applications, efforts to reduce their environmental impacts, preparation of the environmentally benign surfactants, or modifications of the methods to synthesize them from natural-based resources are still under the focus of many researchers. Furthermore, developing new ways or methods to use old techniques for studying the performance of these materials is of prime importance.

In the first section of this book, some of the applications of surfactants in various fields such as biology and petroleum industry, as well as their environmental effects, are described. The second section is devoted to the discussions on the experimental techniques used for characterization of the surfactants. In the first chapter, Dr. Kalak has described the environmental impacts of the use of some surfactants in the petroleum industry. Dr. Jurašin has summarized the recent progress in using mixtures of structurally different surfactants, especially catanionic surfactants and their various applications in nanotechnology and pharmaceutical formulations in Chapter 2. The use of hydrophobic polymers flooding in enhanced oil recovery (EOR) technologies is described by Dr. ElHoshoudy in Chapter 3. Gemini alkylammonium salts as Gemini surfactants that show unique surface and interfacial properties in aqueous solution and their applications in various fields is the subject of discussions made by Prof. Brycki in Chapter 4. Prof. Carmona in Chapter 5 has discussed dioctadecyldimethylammonium bromide as a quaternary ammonium surfactant with interesting properties and applications, with emphasis on its self-assembly in aqueous solutions to yield a wide range of novel uses in fields such as delivery of drugs, vaccines, etc. Steroidal and triterpenoid saponins are among the plant active compounds, which are traditionally used as natural detergents showing unique properties as foaming and emulsifying agents and can be used in many industrial applications, such as steroid hormones in the pharmaceutical industry to food additives. This issue is detailed by Prof. Kregiel in Chapter 6. Dr. Pinheiro has detailed in Chapter 7 the amino acid–based surfactants as surfactants prepared from natural resources that can be used easily in biomedical applications like drug and gene delivery. Dr. Jakovljevic (in Chapter 8) has presented the experimental evaluation of the ability of fungi species, isolated from municipal sewage and industrial wastewater, in removing of anionic surfactants from environment as well as their potential application in biotechnology. The application of several electrochemical techniques, such as cyclic voltammetry, chronoamperometry, chronocoulometry, and selective electrodes for studying the behavior of surfactants in solution as well as at the interfaces, has been discussed by Dr. Schulz in Chapter 9. The last chapter of the book is devoted to explain the use of SAXS and SANS techniques as powerful methods for studying surfactants and their behavior, by Dr. Lepkova.

Finally, I would like to express my gratitude to Ms. Martina Usljebrka, the publishing process manager, and InTech Open Access Publisher for their efforts in the publishing process. I would also like to thank my family and families of all other collaborators for their patience and acceptance of the lost evenings. I wish that this book can give an insight into the field.

Dr. Reza Najjar
Faculty of Chemistry
University of Tabriz
Tabriz, Iran

Section 1

Applications of Surfactants

Chapter 1

Environmental Impact of the Use of Surfactants and Oxygenates in the Petroleum Industry

Tomasz Kalak

Additional information is available at the end of the chapter

http://dx.doi.org/10.5772/intechopen.68683

Abstract

The role of surfactants and hydrophilic additives in gasoline fuel was demonstrated. The impact of anionic surfactant sodium bis-(2-ethylhexyl)sulfosuccinate (AOT) and hydrophilic oxygen containing additives, such as alcohols (methanol, ethanol, propan-2-ol, butanol, 2-methylpropanol) and methyl t-butyl ether (MTBE) on solubility of water, electrolytic conductivity in gasoline and interfacial tension in the water/gasoline system was studied. Small amounts of amphiphilic components improve the solubility of water in gasoline as a result of the occurrence of association phenomena with the formation of reverse micelles. The formation of surfactant aggregates and droplet clusters results in an increase in the solubility of water in gasoline, electrolytic conductivity, and a decrease in interfacial tension. The changes depend on concentration of the surfactant and type of applied biocomponents. Gasoline fuel in the form of microemulsion has a positive impact on the natural environment. The presence of water causes the almost complete combustion of hydrocarbons to the low toxic gases and the absence of carbon black among combustion products reduces fuel consumption, enhances engine power and decreases its temperature, reduces emissions of volatile organic compounds (VOCs), NO_x, SO_2, CO, and particulate matter. The alternative fuel may have a potential use in spark-ignition engines in the future.

Keywords: environmental protection, surfactants, fuel oxygenates, exhaust emissions, solubility of water, association phenomena, electrolytic conductivity, interfacial tension

1. Introduction

Energy consumption and the standard of living of a society are interrelated constantly growing. Nowadays, there are various sources of energy, such as solar, wind, geothermal, hydrogen, tidal, wave, hydroelectric, biomass, nuclear power, and fossil fuels (coal, oil, and natural gas). Among all the sources, crude oil still plays an important role in providing the energy supply of the whole world. It is the most readily available source of energy to humanity, but also a rich source of raw materials for a lot of chemical industries of any kind. The field of surface chemistry is linked to technological processes of crude oil, including the drilling, petroleum refining, and petrochemical processing, and also other related applications and industries. All the processes are associated with interfacial phenomena and surface chemical interactions, as well as have an impact on the environment.

Crude oil is processed into many products (**Figure 1**) and most of them are fuels used for transportation (**Figure 2**). Among the sources of energy, gasoline is the most commonly used fuel in the transportation industry (**Figure 3**). The global production of the fuel presents an upward trend and, in 2012, amounted to approximately 22,377,200 barrels per day [bbl/d]. Taking the regional production into account, the largest amount of gasoline is manufactured in North America (10,017,000 bbl/d in 2012). The next regions are Asia, Europe, South America and Africa. In Europe, gasoline production had continually increased up to 2006 (4,742,000 bbl/d), after the time it started to slowly decrease (**Figure 4**). The largest producer in the world is the United States, with a production of about 9,058,630 bbl/d. Other most productive countries are China, Japan, Russia, Canada, India, Germany, and others (**Figure 5**).

Gasoline is a petroleum-derived liquid that consists of mostly of organic hydrocarbons obtained by the fractional distillation of crude oil, such as paraffins (saturated and unsaturated), naphthenics, aromatics, and their derivatives. The fuel composition also includes other additives that help attain valuable physicochemical properties [1, 2]. The composition is continually improved by producers in order to achieve better performance and meet the requirements of today's advanced engine technology and environmental institutions. Vapor pressure, distillation curves, or octane rating are features closely associated with the fuel composition and the characteristics of its components. The appropriate additives should ensure antidetonation combustion, good and quick evaporation, high octane number, chemical

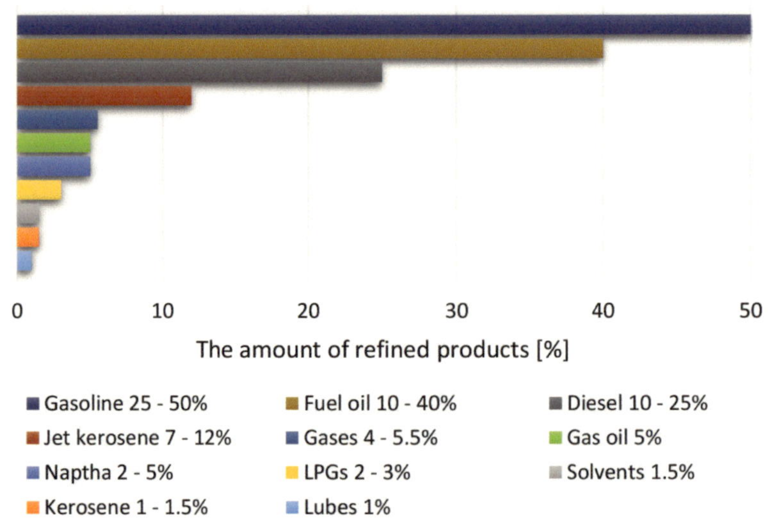

Figure 1. Global refined petroleum products attributable to one barrel [94]. *Source*: United States Energy Information Admini stration, 2017.

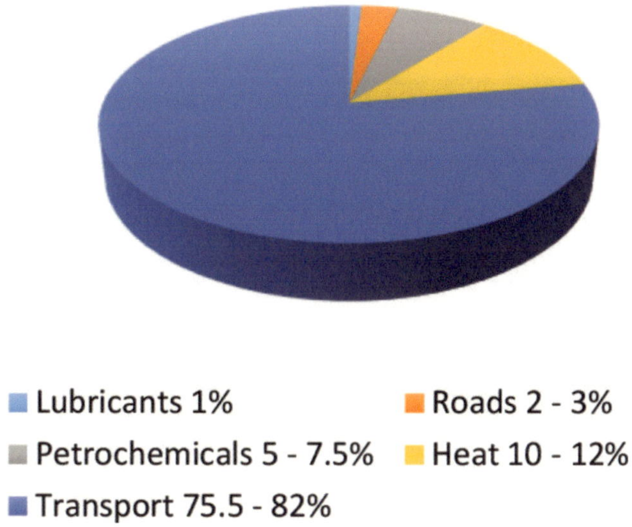

Figure 2. The global application of petroleum products attributable to one barrel [95]. *Source*: United States Energy Infor mation Administration, 2017.

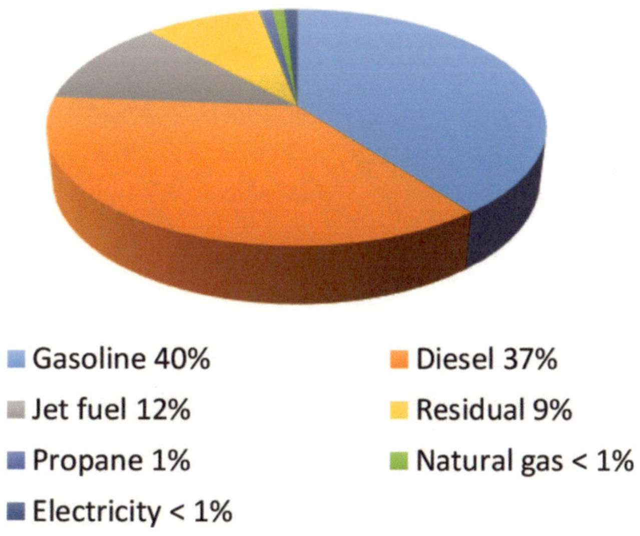

Figure 3. Various fuels used for global transportation in 2012 [96]. *Source*: U.S. Energy Information Administration, International Transportation Energy Demand Determinants (ITEDD-2015) model estimates.

stability, reducing emissions. Furthermore, fuel cannot be corrosive to metals and should not make deposits that interfere with the engine operation [3]. Most of additives belong to a few main functional groups, such as oxygenates (alcohols, ethers, esters, ketones, and others), complex binders, metalorganic compounds, heterorganic compounds, oxidizing organic compounds, petroleum fractions (aromatics, and light and heavy aliphatic hydrocarbons), surfactants, and polymers. Among the various types of components, high octane oxygen-containing

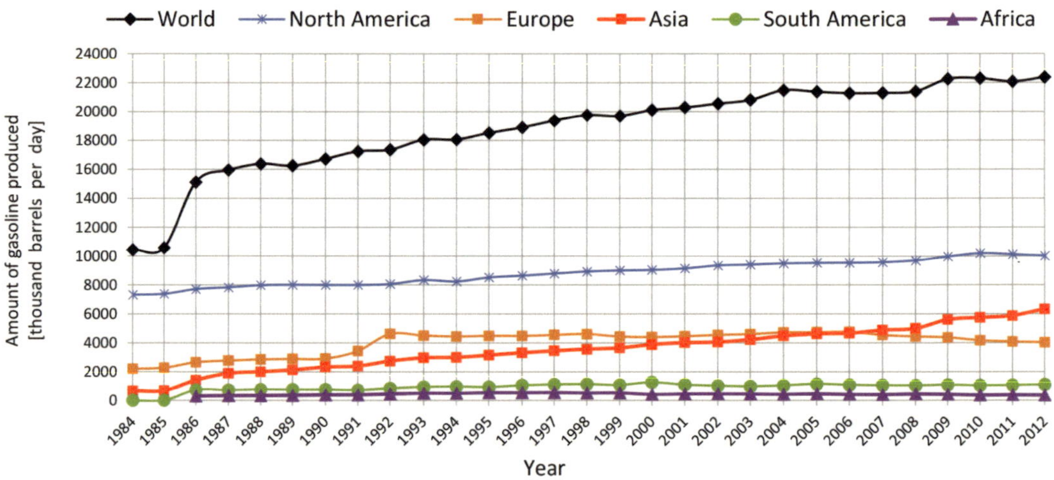

Figure 4. Global motor gasoline production by year [97]. *Source*: United States Energy Information Administration, 2016.

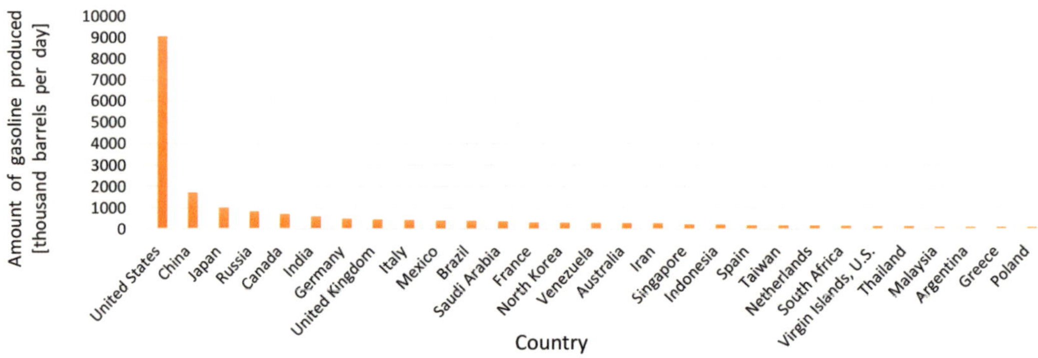

Figure 5. Motor gasoline production by country in 2012 [98]. *Source*: United States Energy Information Administration, 2016.

compounds, such as alcohols and ethers, are able to reduce pollutants from vehicle exhaust gases, increase the octane number [4], have antiknock properties, they can be obtained from renewable agricultural raw materials instead of fossil sources, they reduce carbon monoxide (CO), volatile organic compounds (VOCs), and unburned hydrocarbons emission [5–8].

A very important property of gasoline fuels is their hygroscopicity, which has a considerable effect on the reliability of vehicles and equipments. Operational experience showed that the permanent addition of a small amount of water to hydrocarbon fuel has a positive effect on a combustion process, provided that water in gasoline is in the form of an emulsion. Therefore, studies to determine physicochemical properties of microemulsions have significant practical importance. The emulsion, with the addition of 5 and 10% water, insignificantly increases engine torque, but a mixture with an addition of 15% water decreases engine torque. Water-in-gasoline emulsions (WiGEs) cause an increase in brake-specific fuel consumption (BSFC) and a decrease in exhaust temperature. Compared to basic gasoline, the WiGE fuel reduces NOx and CO and enhances O_2 emissions [9].

Owing to specific chemical structure, surfactants are commonly used as fuel additives with a range of various functions, such as reduction of surface tension [10], prevention of particle formation, removing deposits, dispersion of water, formation of protective layers on surfaces, and an increase in electric conductivity. The addition of surfactants into gasoline leads to a reduction of the amount of deposits formed in the injectors, intake valves, and combustion chambers of gasoline engines. Deposits cause various performance and emissions problems, so their continuous removal is needed. Surfactants dissolved in nonpolar solvents may undergo association with the formation of reverse micelles. In the aggregates, polar (hydrophilic) groups are directed to the center of the micelle and hydrocarbon chains toward the apolar (hydrophobic) phase [11].

Water-in-gasoline emulsion (WiGE) fuel with an addition of hydrophilic high-octane oxygen components has become the best alternative fuel to substitute gasoline fuel in spark-ignition engines. The growing interest to this type of fuel is due to simultaneous reduction of unburned hydrocarbons and CO, reduction of the formation of atmospheric ozone resulting from gasoline emissions, reduction of emissions of exhaust pollutants, such as volatile organic compounds (VOCs), NO_x, and particulate matters. This occurs as a result of the reduction in peak cylinder temperature and secondary atomization by a further breakup of gasoline spray due to microexplosion. Experimental investigation about the effect of various surfactants present in the WiGE fuel on engine performance and pollutant formation has not been fully known. Studies conducted in this field may constitute the basis for investigation the effects of blends of emulsified fuel with various surfactants and hydrophilic oxygen compounds on the combustion characteristics, emission formation processes, and engine behaviors also to determine the pollution formation suppression capability of the emulsified fuels by in-depth combustion characteristics analysis. It is also equally important to select the suitable emulsification method, optimized speed, agitation time, and suitable chemical stabilizers in order to achieve stable emulsions. It is reasonable to conduct intense studies in order to know the effect of water content on the engine combustion characteristics and to determine an optimum percentage of water content in the WiGE fuel. Systematic studies of the optimization of water content in the emulsion for best engine performance and emission by both experimental and numerical investigations are necessary so that it can give the best recommendations for the commercialization of the WiGE fuel as an alternative source of energy for the future spark-ignition engines.

The aim of this work is to study the effect of chosen hydrophilic additives (alcohols MeOH, EtOH, BuOH, IPA, IBA, and MTBE, 3% v/v) and the anionic surfactant (sulfosuccinic acid bis[2-ethylhexyl] ester (AOT) at various concentrations) upon the solubility of water in basic gasoline, electrolytic conductivity, and interfacial tension isotherms at water/gasoline interfaces.

2. Oxygenates used in gasoline fuel

Oxygenates are chemical substances that contain oxygen in their structure. There are several oxygenates that can be added into gasoline (**Table 2**) and they can be divided into several groups based on their functions in fuel. Antistatic additives are responsible for reducing the potential for static build up by improving electrical conductivity and charge dissipation. The electrolytic conductivity of basic gasoline is very low (25 pS/m). Static electricity can build up during pumping, filtering, and splash transfer operations within refineries and also at filling stations, so it can be a reason of static discharges presenting an obvious fire hazard due to low conductivity of gasoline. Grounding and bonding during liquid transfer is a need to protect against static discharge. In a container flow discharging back to the walls may happen, thus the rate at which it can discharge depends on the gasoline composition and properties. In the case of walls being conductive, the electric field achieved by the flow can induce a charge on the walls. The external part of the walls can achieve a charge equal to the charge of gasoline, and the internal part can achieve a charge that will be equal and opposite to that of fuel (**Figure 6**). In order to eliminate the possibility of an electrostatic discharge, various antistatic compounds are used for this purpose [15]. Metal deactivators' task is to extend the durability of fuel by reducing the effect of the catalytic metal to its oxidation. The inhibitory action of these additives involves the creation of inactive compounds with metal ions present in fuel. Metal ions bound in this way cannot catalyze the oxidation reaction any longer. The most active catalysts are copper and brass [16]. By the contact of hydrocarbons with oxygen at an elevated temperature, their oxidation to organic acids, resins, and other compounds usually occurs (**Figure 7**) [17]. Antioxidants interrupt the chain reaction of the oxidation at a stage of peroxides, delaying aging changes in fuel. The mechanism of action of antioxidants consists in inhibition or interruption of the chain oxidation process by decomposition of peroxides formed in radical reactions of the process. Antioxidants can also react with free radicals to give the stable compounds breaking chain reactions [18]. Anticorrosion additives protect metal from corrosion mostly caused by the acidic products of fuel oxidation. Due to the physical adsorption or chemical reaction, metal protective layers are formed (passivation). These layers are chemically

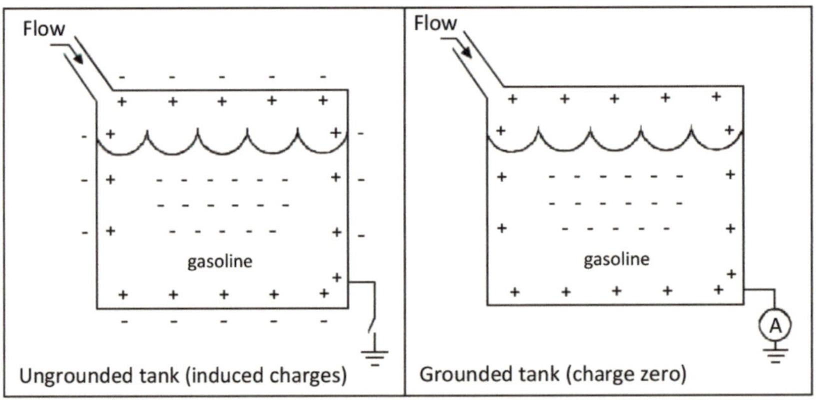

Figure 6. Fuel tank charging diagram [1, 14].

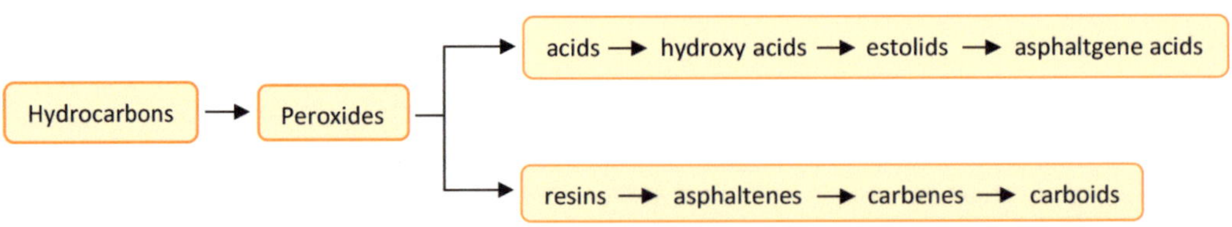

Figure 7. A scheme of hydrocarbons oxidation [17].

stable and resistant to damage caused by friction. Examples of anticorrosive compounds are the following: zinc dialkyldithiophosphates, dialkyldithiocarbamates, zinc alkylsuccinic acids and their monoesters, alkylsulfoamide acids, zinc and calcium salts, organic phosphorus compounds (phosphoric acid esters), and organic sulfur (sulfides) and amine compounds [19]. Dampness of gasoline at a temperature lower than 4°C causes the formation of ice crystals. Ice crystals are formed by the rapid evaporation of gasoline and are deposited on the surface of the shutter valve and its periphery, making it difficult for proper operation of the valve. While an engine is heated to a high temperature, ice melts, and water dripping from the surface of the aperture contributes to the restricted air flow. In order to eliminate crystallization of water, anticrystallization (anti-icing) substances are added that increase the solubility of water in fuel and reduce the temperature of crystallization of the aqueous solutions of the additive released from fuel. For this purpose, propan-2-ol, butanol, butan-2-ol, dimethylformamide (DMF) are used. Surface-active compounds are another group. They are adsorbed on the surface of ice-nucleating agents that prevent their growth and connection into deposited agglomerates [18]. The examples of components used in gasoline fuel are presented in **Table 1**.

Permanent removal of lead from gasoline was a cause of looking for other applicable substances, which are able to improve fuel properties. Therefore, refining technologies have been modernized in order to generate high-octane hydrocarbon (HC) compounds. In order to face the increasing

Group of chemicals	Chemical compound	CAS no.	Ref.
Aromatic amines	2,6-Dimethylanaline	87-62-7	[20]
	3,4-Dimethylaniline	95-64-7	[20]
	o-Toluidine	95-53-4	[20]
	o-Toluidine	95-53-4	[20]
	p-Toluidine	106-49-0	[20]
	Aniline	62-53-3	[20]
Aliphatic amines	Triethanolamine	102-71-6	[21]
	Diethanolamine	111-42-2	[21]
Phenols	2,6-Di-tert-butylphenol	128-39-2	[22]
	3,4,5-Trimethylphenol	527-54-8	[20]
	2,6-Dimethylphenol	576-26-1	[20]
	3,4-Dimethylphenol	95-65-8	[20]
	o-Cresol	95-48-7	[20]
	p-Cresol	106-44-5	[20]
	Phenol	108-95-2	[20]
Benzotriazoles	1-Methylbenzotriazole	13351-73-0	[20]
	Benzotriazole	95-14-7	[20]
Poly phenol	N,N-Disalicylidene-1,2-diaminopropane	94-91-7	[20]
Thiophenes	Benzothiophene	95-15-8	[23]
	Thiophene	110-02-1	[23]

Group of chemicals	Chemical compound	CAS no.	Ref.
Alcohols	2-Butoxy ethanol	111-76-2	[21]
	2-Ethyl 1-hexanol	104-76-7	[21]
	3-Methyl 1-butanol	123-41-3	[24]
	2-Methyl 1-butanol	137-32-6	[24]
	Isobutyl alcohol	78-83-1	[24]
	Tert-butyl alcohol	75-65-0	[1]
	2-Propanol	67-63-0	[21]
	1-Propanol	71-23-8	[24]
	Ethanol	64-17-5	[24]
	Methanol	67-56-1	[24]
	2-Methoxyethanol	109-86-4	[1]
	2-Ethoxyethanol	110-80-5	[1]
	Tetrahydrofurfuryl alcohol	97-99-4	[1]
	Tert-amyl alcohol	75-85-4	[1]
Ethers	Methyl tert-butyl ether (MTBE)	1634-04-4	[1]
	Ethyl tert-butyl ether (ETBE)	637-92-3	[1]
	Tert-amyl methyl ether (TAME)	994-05-8	[1]
	Tert-amyl ethyl ether (TAEE)	919-94-8	[25]
	Diisopropyl ether (DIPE)	108-20-3	[1]
	Tert-hexyl methyl ether (THEME)	38772-53-1	[26]
Ester	Ethyl acetate	141-78-6	[24]
Ester-acid	1,2-Bis(2-ethylhexyloxycarbonyl) ethanesulphonate potassium salt	7491-09-0	[21]
Neutral organics	1,1-Diethoxyethane	105-57-7	[24]
	2-Ethylhexyl nitrate	27247-96-7	[21]
	Tetrapropylenebutanedioic acid	27859-58-1	[21]
Undesignated	Dimethylformamide	68-12-2	[1]
	(Z)-4-Oxo-4-(tridecylamino)-2-butenoic acid	84583-68-6	[21]
	Polyolefin Mannich base	–	[21]
	1-Propene, 2-methyl-homopolymer, hydroformylation products, reaction products with ammonia	68891-84-9	[21]
	Di-sec-butyl-p-phenylenediamine	101-96-2	[27]

Table 1. Examples of additives used in gasoline fuel.

demands of environmental protection, oxygen-containing compounds, organic oxygen-containing compounds started to be used. The Environmental Protection Agency (EPA) allowed the addition of detergents to all types of motor gasoline in the United States in 1995 [28]. The minimum content of MTBE in gasoline is about 11% (v/v) in the United States, while in Europe the content is about 2.5% (v/v). In the European Union, according to Directive 2003/30/EC requirements, it is obliged to promote biofuels among the EU members and to recommend replacing conventional fuels by renewable energy sources (biofuels, etc.). The regulation initiated the reduction of greenhouse gas (GHG) emissions inter alia by allowing the use of ETBE and bioethanol. Based on the Directive 98/70/EC and the Directive 2009/30/EC, the limits of content of oxygenates are presented in **Table 2**.

Organic oxygen compounds, lead	Limits in the European Union		Limits in Poland	
	Minimum	Maximum	Minimum	Maximum
Oxygen content in gasoline [%, wt.]	—	3.7	—	2.7
Methanol [%, v/v] (required stabilizer)	—	3	—	3
Ethanol [%, v/v] (stabilizer may be needed)	—	10	—	5
Isopropanol [%, v/v]	—	12	—	10
Tert-butanol [%, v/v]	—	15	—	7
Isobutanol [%, v/v]	—	15	—	10
Ethers [%, v/v] (containing five or more carbon atoms per molecule)	—	22	—	15
Other oxygen compounds [%, v/v] (other mono-alcohols and ethers with a final boiling point no higher than that 210°C)	—	15	—	10
Lead content [g/l]	—	0.005	—	0.005

Table 2. Requirements for gasoline used in vehicles equipped with spark-ignition engines [12, 13].

3. The role of detergents in gasoline fuel

During combustion processes, fuel forms deposits in the combustion chamber, valves, piston rings, parts injectors, etc. Carbon deposits accumulating on valves can be a cause of their suspension on walls of the combustion chamber and piston head. They change the conditions of heat exchange and carbon deposits in the injector worsen the quality of fuel atomization. Deposits in the grooves of the volute on a piston may lead to their immobilization. This phenomenon deteriorates the conditions of air compression, facilitates the penetration of lubricating oil into a combustion chamber, may even lead to damage to the ring. The addition of detergents soluble in fuel reduces surface tension, but mainly removes all dirt and deposits from engine elements. Their function is to maintain engine cleanliness by counteracting the formation of sludge in the above-mentioned engine elements [17].

The mechanism of action of detergents includes such physicochemical processes as solubility and the stabilizing effect. Solubility is associated with the process of micelle formation, that is, of colloidal particles electrically charged and surrounded by a layer of associated solvent molecules. One theory explaining the mechanism of action of detergents brings to such processes as peptization and neutralization. Peptization is to move the pellet into sol or colloidal state under the influence of surfactants. Dirt particles of size from 10 to 150 nm may be subject to peptization. Larger particles are difficult to peptize. The particle size of the impurities is shown in **Figure 8**. Stages of an impact of surfactants on dirt particles are presented in **Figure 9**. Examples of detergents applied in gasoline fuel are shown in **Table 3**.

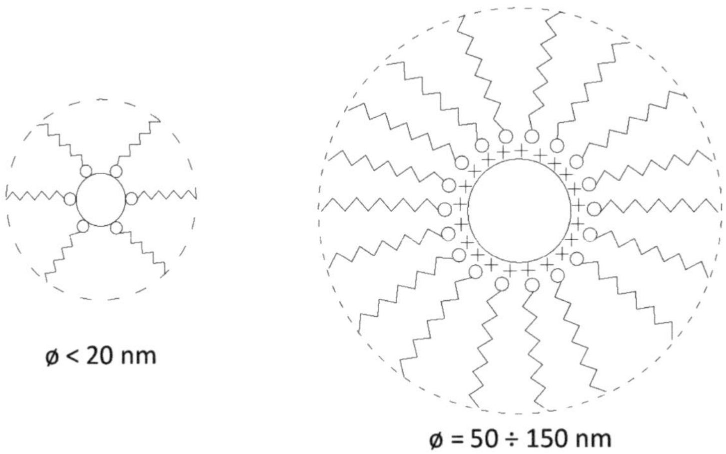

Figure 8. Schematic diagram of action of detergents [99].

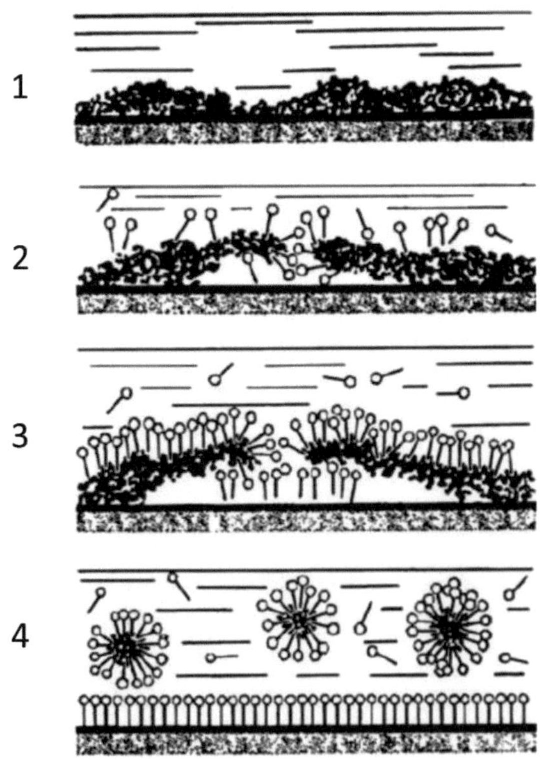

Figure 9. Steps of an influence of surfactants on dirt particles: 1, wetting and penetration; 2, adsorption; 3, emulsification and solubilization (rolling-up), dispersing; 4, emulsion [29].

SN	Examples of detergents with potential use in gasoline presented in patents and articles	Ref.
1	US 20040010966 A1, WO 2003074637. Gasoline detergents and intake valve deposit inhibitors consisting of high-molecular-weight hydrocarbyl amines in an alkoxylated alcohol carrier fluid. Gasoline detergents consist of a detergent additive, including a basic nitrogen atom that is substituted with a hydrocarbyl group, in a synthetic carrier oil component with the general structure R-O-(A-O)$_X$-H (R = linear or branched C_{6-18}-alkyl; A = C_{3-4}-alkylene; and x = 5–35). A preferred hydrocarbyl amine is polyisobutenylamine; preferred carrier oils are mono-C_{8-15}-alkyl-terminated polyoxyalkylenes, butoxylated tridecanol. The substances are added to remove and inhibit the formation of intake valve deposits.	[30]
2	WO 200302083. Diarylamine-functionalized ethylene-alkene-alkadiene copolymers as gasoline detergents. A preferred product is the reaction product of ethylene-propylene copolymer, maleic anhydride, and N-phenyl-(p-N'-phenyl)diamine. The detergent composition can also contain a second mixture of mixed fatty acid esters, a mono- or di(hydroxyalkyl amine) and a low-molecular weight ester. The compounds can be used as gasoline antifriction detergents.	[31]
3	US 6 454 818. Gasoline detergents and octane requirement reducing agents consist of aliph. amidoamine-terminated polyoxyalkylenes of general formula (R4)2N-R3-N((C:O)R2)-(CH2CH(R1)-O)xH, in which R1 = C1-12-aliph.; R2 = Me, Et, Pr, or Bu; R3 = C1-6-alkylene; R4 = C1-4-alkyl; and x = 5–30. To the above detergents, the gasoline can contain additional detergents, such as polyalkenyl amines, Mannich amines, alkylsuccinimides, poly(oxyalkylene)carbamates, and poly(alkenyl)-N-substituted carbamates. The compounds may be used as detergents and octane requirement reducing agents.	[32]
4	AN 2002:616810. Polyisobutylphenoxyethyl polyamines as gasoline detergents. A novel kind of detergents of polyisobutyl phenoxyethyl polyamines are synthesized using polyisobutylphenol (MW = 815,995), 1,2-dibromoethane, and polyamine with their structures identified as PIB--phenyl--OCH_2CH_2NH $CH_2CH_2NH_2$ and PIB--phenyl--$OCH_2CH_2NHCH_2CH_2NHCH_2CH_2NH_2$. The additives are utilized to inhibit deposits growth in modern gasoline engines to eliminate their adverse effects on vehicle performances in terms of drivability, power output, fuel economy, and much serious exhaust emissions.	[33]
5	US 2002066225. Mannich bases prepared from hydrogenated distilled nutshell liquid. Mannich bases functioning as deposit inhibitors and detergents in gasoline are prepared as the reaction products by reacting hydrogenated distilled cashew nut shell liquid (CNSL) with an amine having at least one active hydrogen, and an aldehyde in a CNSL-amine-aldehyde molar ratio of 1.0:1.0–1.2:1.0–1.2, at 70–175°C (preferably 90°C) for 6–12 h. Suitable amines are C_{1-20}-alkyl amines and -alkylamines (e.g., with C_{4-8}-alkyl side chains), and heterocyclic amines (e.g., morpholine, pyrrole, etc.). Suitable aldehydes are preferably aliphatic aldehydes, such as formaldehyde, acetaldehyde, and aldol (β-hydroxybutyraldehyde). The compounds may be used as gasoline detergents and deposit inhibitors.	[34]
6	A new type of adducts of polyisobutyl phenol-epoxy chloropropane-polyamine. The detergents are prepared using polyisobutylphenol, epichlorohydrin, and polyamines. The gasoline additives may inhibit deposit formation (e.g., intake valve deposits, combustion chamber deposits).	[35]
7	Poly(oxybutylene)amides. Effective compounds are amphiphilic copolymers consisting of a hydrophobic polyoxybutylene backbone and hydrophilic amide functionalities. The gasoline additives prevent automobile engine valve deposits.	[36]
8	Synthesis and *in situ* transformation of poly(oxybutylene)amides by butoxylation. A family of gasoline detergents was prepared by a two-step process: (1) preparation of amide initiators by reacting hydrophilic amines with Et acetate at high temperature; (2) consecutive butoxylation of the initiators with 1,2-epoxybutane (BO) to obtain oil-sol products. The additives improve engine performance of octane requirement.	[37]
9	Polyisobutenylsuccinimides may be used as detergents and dispersants. Reduction of forming of deposits in valve system.	[38]
10	WO 2000078898, US 6179885. Polyalkenylphenol-derived aromatic Mannich compounds may be used as diesel fuel and gasoline detergents.	[39]

SN	Examples of detergents with potential use in gasoline presented in patents and articles	Ref.
11	GB 2320719. The gasoline detergents are compounds of the formula Z.NH(CH_2CH_2NZ)$_{y.Z'}$ wherein Z is N, C_{8-16} straight of branched chain alkanoyl or RO-(CH_2CHR'O)$_X$OOC-Ph-CO with the proviso that at least one Z is RO-(CH_2CHR'O)$_X$OOC-Ph-CO and R is C_{2-16} straight of branched chain alkyl or dodecylphenyl; R′ is H, Me, or Et; x is an integer from 12 to 28; and y is an integer from 1 to 4.	[40]
12	WO 9736854, US 6053955. Polyoxyalkylene ether amino acid esters A polyalkylene glycol mono(alkylphenyl) ether p-$R_1C_6H_4(OCH_2CHR_2)_nOH$ ($R_1 = C_{4-25}$ alkyl; $R_2 = C_{1-3}$ alkyl; n = 5–50) is treated with cis-HO_2CCR_3:CR_4CO_2H (R_3, R_4 = H, C_{1-3} alkyl) or its anhydride to form a 1:1 (un)substituted maleate ester, which reacts in molar ratio ≤ (m + 2):1 with a polyamine $H_2NZ(NHZ)_mNH_2$ (Z = C_{2-6} alkylene; m = 0–6). The additives prevent or remove combustion chamber or fuel line deposits.	[41]
13	WO 9215656. Polyalkylenepolyamines, especially polyisbutylene polyamine. The additive has a structure $R_{12}NR_2(NR_2)_XNR_{32}$ [R_1 = H or polyolefin (of which ≥1 is a polyolefin); $R_2 = C_{1-8}$-alkylene; R_3 = H, C_{1-6}-alkyl; x = 0–5]. Gasoline additives for reducing valve sticking and to have detergency and deposition-inhibiting properties.	[42]
14	WO 9213047. Amine- and halogen-free gasoline detergents consisting of polypropylene glycol with hydrophobic end group.	[43]
15	GB 2247457. (Keto) diacid amides may be used as gasoline detergents being deposit inhibitors-octane requirement reducing additives.	[44]
16	US 5286265. Novel carbamates having the formula ROC(:O)NXY [X and Y are independently either H, a (hetero-substituted) hydrocarbyl group, or ZNHC(:O)OR (I), where Z is a divalent hydrocarbyl, a substituted hydrocarbyl, or (alkylene)m(NH)$_n$(alkylene)$_m$ in which n = 0–4, m = 1–4, and R is a (substituted) hydrocarbyl group, provided that if either one of X or Y is I, the other of X or Y is H]. The additives are able to remove deposits.	[45]
17	US 4729769. Reaction products of fatty acid esters and amines. Reaction products of C_{6-20} fatty acid ester with a mono- or di-(hydroxyhydrocarbyl)amine may be used as carburetor detergents.	[46]
18	US 4624682. Gasoline detergents are prepared by reacting an N-alkylalkylenediamine of formula $RNHR_1NH_2$ (R = C_{12-18} alkyl, $R_1 = C_{1-3}$ alkylene) with a bicyclic keto acid derived from a catalyzed rearrangement of a C_{6-10} cyclic alken-3-yl carboxylic acid anhydride in the presence of a Bronsted acid catalyst (e.g., Nafion H-501). The aim of the compounds is removing deposits.	[47]
19	EP 186473 A2. Lubricating oil detergents and fuel (especially gasoline) deposit inhibitors-detergents are prepared by the reaction (at 100–175°C≥) of ≥1 C_{10-20} fatty acids, ≥1 C_{12-26}-alkyl or -alkenylsuccinic acid or anhydride, and ≥1 polyalkylenepolyamine of formula $RNH(R_1NH)_XH$ (R = C_{1-5}-hydrocarbyl, $R_1 = C_{1-5}$-alkylene, x = 1–9). The compounds reduce carburetor deposits by 85% compared with the base fuel.	[48]
20	US 4508541. Vegetable oils [(esp. soybean oil, tall oil acids, or alkyl acids (esp. phenylstearic acid)] are reacted with polyamines (esp. tetraethylenepentamine) to form a product mixture for subsequent reaction with SO_2 to produce a product that has good detergent properties in fuels. The compounds can reduce deposits.	[49]
21	US 4505725. Fuel additives (detergents) obtained from borated, acid-treated mixtures of vegetable oil derived amides and esters. Reaction products of soybean oil with tetraethylenepentamine, sulfonated lubricating oil bright stock, and H_3BO_3.	[50]
22	US 4639255. Gasoline detergents (e.g., vegetable oil-polyamine reaction products) (and optionally hydrogenated polybutenes) are mixed with C_{18-32} paraffin waxes (m. 130–160°CF) or durene, foamed, and pelleted (or encapsulated) to provide deposit-control additives which float on the gasoline and readily dissolve. The additives, present at approximately 120 lb/1000 bbl unleaded gasoline, are sol. at extreme temperatures, do not change the gasoline octane rating, and do not promote gum formation or corrosion.	[51]
23	US 4400178. Polyamine carburetor dispersants. They are prepared by the Mannich reaction of primary or secondary amines with formaldehyde and 2-nitropropane followed by reduction of the nitro group.	[52]

SN	Examples of detergents with potential use in gasoline presented in patents and articles	Ref.
24	US 4394135. Gasoline detergents prepared by reacting tetraethylenepentamine with phenylstearic acid, isostearic acid, or tall-oil fatty acids. The additives reduce engine deposits by 94–95%.	[53]
25	US 4330303. The additives (carburetor detergents) are manufactured by condensation of 1:1 (molar) amounts of ethylenediamine and acid lactones prepared by acid-catalyzed lactonization of polyisobutenylsuccinic acid or anhydride.	[54]
26	Acylation of polyamines with fatty acids in the preparation of the detergent NAIK for gasoline.	[55]
27	US 4353711. The reaction products of glycidyl ethers, whose alkoxy portion contains 6–20 atoms C, with alkylenediamines. Excellent deposit reduction resulted in various parts of the engine and in carburetor.	[56]
28	US 4292046. The additives (carburetor detergents) are manufactured by reaction of acids (naphthenic acids or dimer acids) with 2-(2-aminoethylamino)ethanol or by reaction of the resulting imidazolines with (C_{18-24}-alkyl)succinic anhydrides or isostearic acid.	[57]
29	US 4269606. Fuel and lubricant additives (detergents) from acid treated mixtures of vegetable oil derived amides and esters. Vegetable oils such as corn oil, peanut oil, and soya oil are treated with polyamines to give mixtures containing amides, imides, half esters, and glycerol, treated with sulfonic acids give detergent for gasoline.	[58]
30	US 4249912. Amino amides, prepared by treating EDTA or NTA with fatty amines, phenylstearylamine, or oleylamine. The amino amides are combined with dodecylbenzenesulfonic acid to obtain fuel detergents. The additives can reduce carburetor deposits.	[59]
31	US 4251233. Liquid hydrocarbon-sol. rare-earth chelates prepared from the ligand 2,2,7-trimethyl-3,5-octanedione. The additives prevent formation of or remove carbonaceous combustion-chamber deposits.	[60]
32	US 4240804. Alkyl acrylate adducts of polyamines, ether amines and ether polyamines. A mixture of an adduct (I) of 2-ethylhexyl acrylate (II) and 4-aza-8-oxaeicosylamine (III) with III. The additives cause deposit reduction.	[61]
33	US 4247300. A detergent additive imidazoline prepared by treating carboxylic acid with polyamine. The imidazoline is further combined with a sulfonic acid. The additive, prepared by treating isostearic acid with tetraethylenepentamine and the formed imidazoline with dodecylbenzenesulfonic acid, is added to gasoline. The additives can reduce deposits in an internal combustion engine.	[62]
34	SU 755830. An additive of 10–20 wt% ethylene diisopropylxanthate in a gasoline additive containing 30–40 wt% aliphatic amine and a nitrated fraction of shale oil with a boiling point at 350–370°C. The deposits were decreased on the surface of engine parts by the gasoline additives.	[63]
35	US 4204841. A primary alkylaminoalkyl-substituted asparagine and an N-(primary alkyl)alkylene diamine. An additive contains 50% N,N'-bis(3-oleylaminopropyl)asparagine and 50% N-oleyl-1,3-propanediamine. The additive has excellent deposit prevention ability in a carburator and provides reduced corrosion.	[64]
36	US 4203730. Polyamine derivatives of oxidized olefinic substituted dicarboxylic acid compounds. The condensation of diethylenetriamine (I) with oxidized polybutenylsuccinic anhydrides (II) of different molecular weights. The detergents can reduce deposits.	[65]
37	EP 8591. Detergent additives from mixtures of vegetable oil derived amides and esters or acid treated mixtures. Vegetable oils were treated with polyethylenimine or tetraethylenepentamine and the products treated with alkylbenzenesulfonic acids or alkanesulfonic acids.	[66]
38	US 4240803. Reaction products of alkenylsuccinic anhydrides with tetraethylenepentamine or diethylenetriamine in which the alkenyl group is derived from mixed C16–28 olefins, being the bottoms from an olefin oligomerization. The succinic anhydrides were prepared by reaction of maleic anhydride with the mixed olefins at 200–210°C for 7 h and 235–240°C for 3 h. The compounds are carburetor detergents.	[67]

SN	Examples of detergents with potential use in gasoline presented in patents and articles	Ref.
39	US 4191537. Fuel compositions of poly(oxyalkylene) monoether (aminoethyl)carbamates. Poly(oxypropylene) mono-Bu ether (I) carbamtes, derivs. of H2NCH2CH2NH2 or polyethylenepolyamines, were prepared. Also prepared were poly(oxybutylene) mono(alkylphenyl) (2-aminoethyl)carbamates and poly(oxypropylene)-poly(oxybutylene) mono-Bu ether (2-aminoethyl)carbamate. Deposit-inhibiting dispersants (carburetor and intake-valve detergents) may be used in gasoline.	[68]
40	US 4179271. Amine oxide polymers. Gasoline having detergent properties contains 0.05–0.75 wt% of a tertiary amine oxide-containing polymer(I). Gasoline was prepared by mixing Neodol 25 L methacrylate 54.5, Alfol 1620 methacrylate 16.5, Bu methacrylate 20, and 4-vinylpyridine 9 wt%, polymerizing the mixture, and oxidizing the polymer by AcOH and H_2O_2).	[69]
41	US 4173456. Polyolefin-acylated poly(alkyleneamine) may be used as a component fuel additive to prevent deposits formation. Detergents containing triamide of tetraethylenepentamine and tall-oil fatty acids and polypropylene or polyisobutylene (the triamide 13.6, polypropylene (mol. wt. 800) 50, oxyalkylenated alkylphenol 1, corrosion inhibitor 1.1, and xylene (solvent) 34.3 wt% used at 25 lb/1000 bbl).	[70]
42	US 4132531. A detergent (to remove deposits) prepared by condensing 115 g 1-(2-aminoethyl) piperazine with 700 g polyisobutenylsuccinic acid-derived lactones (I) in 700 ml xylene.	[71]
43	US 4125382. Polyoxyalkylene ether demulsifiers. Alkylpolyamines as detergents and 5–30 ppm polyoxyalkylenes or their adducts with C_{8-18} epoxides as demulsifiers. For example, shaking 32 ml gasoline containing 500 ppm alkylpolyethylenepolyamine, 25 ppm acetal-coupled 28:72 polyethylene-polypropylene glycol (mol. wt. 2200), and 4 ppm 13:87 polyethylene-polypropylene glycol (mol. wt. 2800) and settling 2 h. The detergents can increase water tolerance.	[72]
44	US 4125383. Reaction products of a long-chain monocarboxylic acid, a polyamine, and a C_{12-18} isocyanate. For example, octadecyl isocyanate was reacted with triethylenetetramine and isostearic acid. Improved ashless gasoline detergents decrease the carburetor deposits by 70–80%.	[73]
45	US 4105417. Hydrocarbyl-substituted nitrogenous compounds (e.g., amides, carbamates, or ureas) are effective as gasoline detergents (to remove valve deposits) at a concentration of 50–1500 ppm. For example, 0.06 mol diethylenetriamine was added to 0.44 mol polyisobutenyl di-Et hydrazodicarboxylate (polyisobutenyl av. mol. wt. = 950) in 50 ml C_6H_6 to give polyisobutenyl cyclobiuret (I).	[74]
46	US 4059414. The detergents are prepared by treating long-chain monocarboxylic acids with trialkanolamines and sulfonic acids. For example, 10 lb of a detergent prepared by treating triethanolamine triisostearate (obtained by the reaction of triethanolamine and isostearic acid at 135–40°C for 6 h in the presence of p-$MeC_6H_4SO_3H$) with dodecylbenzenesulfonic acid. The detergents reduce gum deposits in an unleaded gasoline by 63%.	[75]
47	US 4054422, US 4121911. Mannich bases containing tertiary amines and fuel compositions containing said Mannich bases. Mannich condensation products have the formula $RNH(CH_2)_m[NR_1(CH_2)_m]_bNHR$ (R = $C_{20-1000}$ alkylhydroxybenzyl; m = 2 or 3; R1 = Me, Et; b = 1–5). A Mannich base was prepared by refluxing (11 h) a mixt. of 300 g of a 75 wt% solution in PhMe of polypropenylphenol (prepared. from PhOH and polypropylene of 840 mol. wt.), 14.5 g 3,3'-(methylimino)bis(propylamine) and 17 g of a 36 wt% aq. HCHO solution. The carburetor detergents can reduce gum deposits in unleaded gasoline.	[76]
48	US 4038043. Mixture of monoamine and polyamine (N,N-bis[2-hydroksy-4-(polipropyleno) benzylo]methylamine with triethyltetramine with N atoms and benzyl groups). A multifunctional gasoline additive that may be used as a carburetor detergent and at the same time minimizes intake valve deposits and quick-heat intake manifold deposits.	[77]
49	US 4038044. A combination of diamine and polyamine Mannich bases. The diamine Mannich base used was based on ethylenediamine with each N atom being substituted by an alkyl- and hydroxy-substituted benzyl group in which the alkyl substituent was derived from polypropylene of 840 mol. wt. The polyamine Mannich base was based on triethylenetetramine with the terminal N atoms each being substituted with one of the alkyl-and hydroxy-substituted benzyl groups described above. The multifunctional gasoline additive may be used as a carburetor detergent and can minimize intake valve deposits and quick-heat intake manifold deposits.	[78]

SN	Examples of detergents with potential use in gasoline presented in patents and articles	Ref.
50	US 4039300. Lubricating oils containing 60 wt% aromatics (average molecular weight 350–650) and detergents. The additives inhibit deposits in the carburetor, exhaust gas recycle system, and intake valves.	[79]
51	US 4024083. Substituted phenoxy propanol diamines and amino alcohol detergent additives for fuels to remove deposits. Detergents (I; R = alkyl with mol. wt. 200–1500, R_1 = H, C_{1-4} alkyl; Q C_{2-6} alkylene; X = NH, O; $x = 0, 1$; $y = 1, 2$, $x + y \leq 2$, and $z = 1–10$) for gasoline, diesel fuels, and lubricating oils. A detergent was obtained by treating polyisobutenylcatechol with epichlorohydrin in the presence of BF_3 etherate in xylene, stripping the solvent, treating the intermediate obtained with diethylenetriamine in xylene, washing with MeOH containing NaOH, and stripping off xylene.	[80]
52	Specific Mannich condensation products prepared from alkylphenols, HCHO, and an alkylenepolyamine. Control of deposits on intake valves and good detergency.	[81]
53	US 3951614, US 3785789. The gasoline detergents are reaction products of hydrocarbyl amines with polyhalides, polycarboxylic acids, or organic polyisocyanates.	[82]
54	US 3944397. Mannich condensation products as carburetor detergents. A benzyl polyamine of the formula $ZNH(C_nH_{2n}NR)_a(C_nH_{2n}NR)_bC_nH_{2n}NHZ$ (Z = alkyl- and hydroxy-substituted benzyl group wherein the alkyl group has 50–1000 C atoms and 60% of the alkyl group is para to the hydroxyl group, $n = 2–3$, R = H or Z, and the sum of a and b is 0–5).	[83]
55	US 3926578. Esters of 2-(alkylamino)propionic acid. For example, 62.2 g dodecylphenyl acrylate (prepared from dodecylphenol and acrylic acid) in 100 ml xylene was mixed with 54 g Armeen T (tallowamine) and heated 3 h at 120°C to give a product which, when added (7.5 lb/1000 gallons) to 100-octane gasoline. The compounds may be used as gasoline detergents and anticorrosive agents.	[84]
56	US 3923474. Alkylenediamineamides of fatty acids. A dual additive comprising $RNHCH_2CH_2CH_2NHCOR_1$ is improved by incorporation of a second additive $R_2OCH_2CH_2CH_2NHCH_2CH_2CH_2NHCOR_1$ in which R is a C_{10-20} aliphatic hydrocarbon radical, R_1 is a C_{9-19} aliphatic hydrocarbon radical and R_2 is C_{10-15} aliphatic hydrocarbon radical. These additives remain low in harmful deposits of varnish scale, which normally results from untreated fuel.	[85]
57	US 3907518. A combination of tert-alkyl primary amines, a surface-active NH_4 carboxylate salt-ethoxylated alkylphenol ester of a trimer or dimer acid, and a hydrocarbon-sol. polyisobutylene. The compounds may be used as a carburetor deposit inhibitor.	[86]
58	GB 1378709. Polyolefin carburetor detergent. The gasoline additive, $RN[C(CN):CH_2](CH_2)_3NMe_2$ (R = polyisobutenyl, mol. wt. 1300) was prepared from α,β-dibromopropionitrile by treatment with NaOH, N,N-dimethyl-1,3-propylenediamine, and polyisobutenyl chloride. The carburetor of an engine run on gasoline containing 100 ppm additive was cleaner than when the gasoline was used without additive.	[87]
59	US 3846089, US 3782912, US 3912771. Combinations of tert-alkyl primary amines, surface-active bis(alkylammonium) salts of ethoxylated alkylphenol esters of trimer acid, and dimer acid or trimer acid esters with mixtures of aliphatic and ethoxylated alkylphenols. For example, 845 g trimer acid (Emery 1834-18R), 316 g isodecyl alc., 338 g (octylphenoxy)poly(ethylene oxide), 200 ml toluene, and 1.0 g p-toluenesulfonic acid were heated to give diisodecyl (octylphenoxy) polyethylene glycol triester of trimer acid, of acid no. 1.0. Such mixed polyesters used in combination with rust inhibitors prevent rusting and pitting and provide carburetor and induction-system detergency.	[88]
60	GB 1368532. An alkylphenol, prepared by acid-catalyzed alkylation of PhOH with polybutene of average molecular weight 900–1100, with paraformaldehyde and $R_2N(CH_2)_3NH_2$ [R = Me, $HO(CH_2)_2$]. The additives clean carburetors, remove intake valve deposits, and reduce build-up of engine crankcase deposits.	[89]

Table 3. The role of chosen detergents applied in gasoline fuel: a review.

4. Experimental procedure

4.1. Materials

Basic gasoline, obtained from the petrochemical industry, was used in all experiments. The composition is shown in **Table 4**.

The following compounds were used in experiments:

- methanol (MeOH), p.a., POCH S.A., Gliwice, Poland,
- ethanol (EtOH), 96%, p.a., Sigma-Aldrich, Germany,
- propan-2-ol (isopropanol, IPA), p.a., POCH S.A., Gliwice, Poland,
- 2-methylpropanol (isobutanol, IBA), p.a., POCH S.A., Gliwice, Poland,
- butanol (BuOH), POCH S.A., p.a., Gliwice, Poland,
- methyl t-butyl ether (MTBE), 99%, p.a., Sigma-Aldrich, Germany,
- sodium bis-(2-ethylhexyl)sulfosuccinate (AOT), 99%, p.a., Sigma-Aldrich, Germany,
- Hydranal Composite 5, Sigma-Aldrich, Germany,
- deionized water.

4.2. Advantages and disadvantages of the use of sodium bis-(2-ethylhexyl)sulfosuccinate (AOT)

Sodium bis-(2-ethylhexyl)sulfosuccinate (AOT) is an anionic surfactant with a sulfone group directly connected to a hydrophobic group ($C_{20}H_{37}NaO_7S$, molecular weight 444.56 g/mol, CAS Number 577-11-7). Water solubility in hydrocarbon fraction can be increased with an addition of AOT due to the formation of ternary microemulsion system consisting of water, the organic phase, and AOT. Microemulsions can form different types of structures as discontinuous spherical water droplets, interconnected channels of water, and so on. Their size can be controlled by the water content (wo), that is, the molar ratio of water to AOT (wo = $[H_2O]/[AOT]$). Microemulsions may be created as "oil-in-water" (o/w) or "water-in-oil" (w/o) depending on the nature of the solvent. Due to high interfacial activity and good hydrophilic properties, the surfactant AOT is able to form reversed micelles in the hydrocarbon phase, which absorbs large amounts of water. Reverse micelles occur in the situation when $w_o = [H_2O]/[AOT] < 10–15$. AOT can be hydrolyzed in the presence of an acid or a base, which results in formation of 2-ethylhexyl alcohol and sulfosuccinate anion (**Figure 10**) [90].

The surfactant AOT contains seven oxygen atoms in its structure, which presence has a positive effect. The compound contained in fuel introduces additional oxygen into the system, which plays an important role in combustion processes. Air-fuel ratio is of great importance and essential measure for antipollution and performance reasons. Air-fuel ratio is the amount of air needed to burn fuel in the engine and in other words, it is mass ratio of air to fuel present in the combustion chamber. Combustion efficiency depends on the right amount of air,

http://dx.doi.org/10.5772/intechopen.68683

Number of carbon atoms	The components of basic gasoline [%, v/v]						Total
	n-Alkanes	Iso-alkanes	Olefins	Naphthenes	Aromatics	Oxygenates	
C1	—	—	—	—	—	—	0.000
C2	—	—	—	—	—	—	0.000
C3	0.055	—	—	—	—	—	0.055
C4	1.204	2.633	4.820	—	—	—	8.658
C5	3.218	10.137	3.869	0.56	—	—	17.783
C6	0.764	12.351	1.427	1.781	0.484	—	16.806
C7	1.308	4.695	0.949	1.913	6.590	—	15.454
C8	0.655	6.887	0.142	2.381	7.833	—	17.897
C9	0.286	1.665	0.011	0.847	11.612	—	14.421
C10	0.125	0.841	0.074	0.093	5.495	—	6.629
C11	0.047	0.680	—	0.042	0.215	—	0.984
C12	0.045	0.071	—	0.035	0.747	—	0.898
C13	0.058	—	—	—	—	—	0.058
C14	0.004	—	—	—	—	—	0.004
Total	7.768	39.960	11.292	7.652	32.975	0.000	99.647
Total heavies							0.034
Total unknown compounds							0.320
Grand total							100.000

Table 4. The components of basic gasoline used for the research [1].

which is reflected on the engine power. Air contains about 21% oxygen, 79% nitrogen, and smaller amounts of other elements. When fuel burns in the presence of O_2 and N_2, it is converted to carbon dioxide, water, nitrogen, and heat according to Eq. (1):

$$CH_4 + 2O_2 + 7.53N_2 \rightarrow CO_2 + 2H_2O + 7.53N_2 + \Delta H \tag{1}$$

The exhaust gases from internal combustion engines mainly consist of the products of complete combustion, small amounts of the oxidation products of sulfur and nitrogen, and components derived from the fuel and various lubricants. The composition of gases is shown in **Table 5** [91].

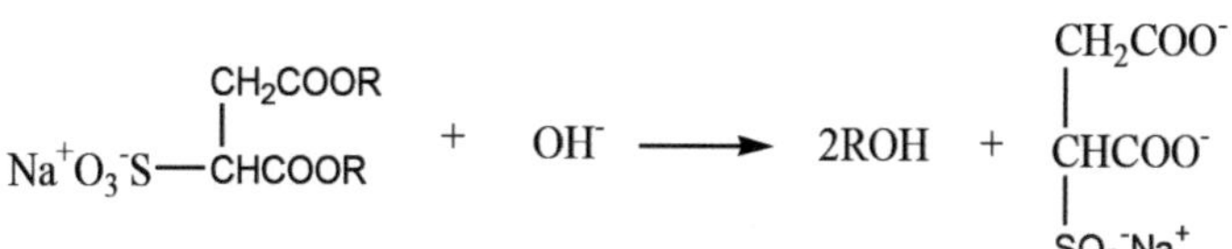

AOT

Figure 10. Alkaline hydrolysis of AOT (R—ethylhexyl group) [90].

Major components (>1%)	Minor components (<1%)
Water, H_2O (c)	Oxides of sulfur, SO_2, SO_3 (c)
Carbon dioxide, CO_2 (c)	Oxides of nitrogen, NO, NO_2 (c)
Nitrogen, N_2 (c)	Aldehydes, C_nH_mCHO (c)
Oxygen, O_2 (c)	Organic acids, C_nH_mCOOH (c)
Carbon monoxide, CO (a)	Alcohols, C_nH_mOH (c)
Hydrogen, H_2 (a)	Hydrocarbons C_nH_m (c)
	Carbon monoxide, CO (b)
	Hydrogen, H_2 (b)
	Smoke (c)

(a) Spark-ignition engine, (b) diesel engine, (c) both engines.

Table 5. Components of internal combustion engine exhaust gases [91].

The disadvantage of the surfactant AOT is the presence of one sulfur atom in its structure. There is a tendency to eliminate sulfur from fuel composition in order to reduce its content in emitted gases after combustion processes in engines. In the air, SO_2 is present in the largest quantities, but other sulfur oxides (SO_x) are found in the atmosphere at much lower concentrations. SO_2 influences human health when it is breathed in, at concentrations above 1000 μg/m^3, measured as a 10-min average. The gas irritates the nose, throat, and airways to cause wheezing, coughing, shortness of breath, and a tight feeling around the chest. The large amounts of SO_x in the atmosphere can harm all types of plants by damaging foliage and decreasing growth. Sulfur oxides are responsible for contributing to acid, which can harm sensitive ecosystems. Therefore, the concentration of sulfur oxides in air is constantly monitored in order to react appropriately in the case of a high concentration.

4.3. Apparatus and procedures

Samples of basic gasoline (25 ml) containing 3% (v/v) of a hydrophilic additive were mechanically shaken with 1% of deionized water for 2 h at 4000 revolutions per minute and left to phase separation for 24 h. The content of water in saturated gasoline samples was determined using the Karl Fischer method. The potentiometer 702 SM Titrino (Metrohm, Switzerland) was used for titration using Hydranal Composite 5 (Sigma-Aldrich, Germany). The basic gasoline was modified with the anionic surfactant AOT at various concentrations and hydrophilic additives. After saturation with deionized water, the content of water was determined.

Electrolytic conductivity of modified gasoline samples was determined using pH/conductivity meter CPC-551. The K12 tensiometer with a platinum ring (Krüss, Germany) was used to measure the interfacial tension (water/gasoline). After preparation of the systems (15 ml of water and 9 ml of modified gasoline) for measurements, interfacial tension was measured using the Du Noüy ring method at room temperature. All experiments mentioned above were made in triplicate for each method.

5. Results and discussion

5.1. The influence of biocomponents on water solubility in gasoline

The composition of gasoline, type, and concentration of hydrophilic oxygen-containing additives (i.e., alcohols, ethers), amphiphiles (i.e., surfactants), and other functional components affect the solubility of water in the fuel. Preliminary studies demonstrated that the content of water in basic gasoline saturated with 1% of deionized water was about 0.01% (v/v). The solubility of water in basic gasoline modified with the anionic surfactant AOT and chosen hydrophilic alcohols and ether (3%, v/v) is shown in **Figure 13**. The addition of AOT causes significant changes and depends considerably on its concentration. The multifunctional surfactant increases the solubility of water up to about 1%. The sudden increase is observed at very low AOT concentration equal to 6.25×10^{-4} mol/l. This phenomenon can be explained by the fact that the surfactant AOT initiates structural changes and it is able to increase the solubility of water in the fuel through the formation of reverse micelles. The relative standard deviation of the measurements is presented in **Table 6**.

The experimental data shown in **Figure 13** demonstrate the dependence that the solubility of water in gasoline increases with the growing number of carbon atoms in the alcohol molecule. The greatest values of water content are observed in the case of an addition of AOT and mixtures of AOT and MTBE in a range of concentrations from 10^{-5} to 10^{-3} mol/l. Samples of fuel containing AOT and MTBE were very cloudy, which can be probably the result of reverse micelles formation. The hydrophilic part of the surfactant AOT creates the micelle cores, which are filled with deionized water and thus, a quick rise of solubility of water is noticed. The association phenomena are dependent on a type of hydrophilic components, which are able to act as cosurfactants, which is shown in the schematic diagram in **Figures 11** and **12**. The components present in reverse micelles lead to an increase in micelle's size and water solubility, and they promote the charge of structure with formation of microemulsion.

Figure 13 shows the higher surfactant AOT concentration, the higher amount of water in the modified fuel. In the presence of MTBE, the content of water is higher compared to samples including various alcohols (3%, v/v). At first, it may be explained that hydrophilic components enhance polarity of gasoline mixture and cause an increase in the solubility of water. Larger amounts of the additives may not act as cosurfactants and furthermore they can delay association of surfactant AOT into reverse micelles with water pools. Second, alcohols methanol and ethanol contain short hydrocarbon chains, therefore due to their low molecular weights they cannot join the micelles. As a result of that, water solubility in gasoline consequently achieves low level and the situation is improved only by an increase in the concentration of

Molar concentration AOT [mol/l]	2.25×10^{-2}	1.13×10^{-2}	5.63×10^{-3}	1.88×10^{-3}	6.25×10^{-4}	2.09×10^{-4}	6.95×10^{-5}
Relative standard deviation for AOT (RSD)	±0.048	±0.051	±0.049	±0.086	±0.154	±0.161	±0.225

Table 6. Relative standard deviation values for content of water measurements. Source: own research.

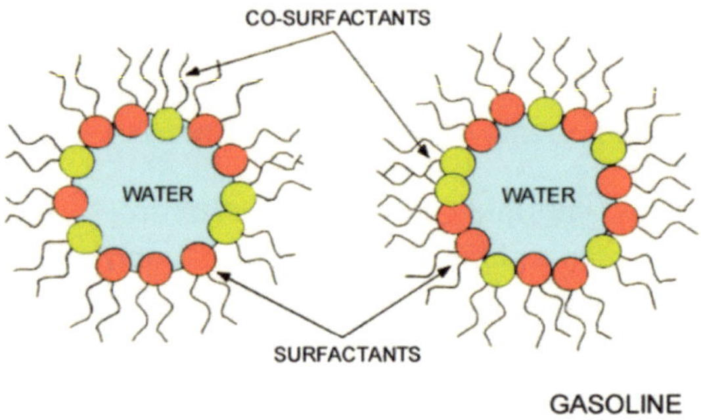

Figure 11. Reverse micelles formed in gasoline modified with hydrophilic additives and surfactants [1].

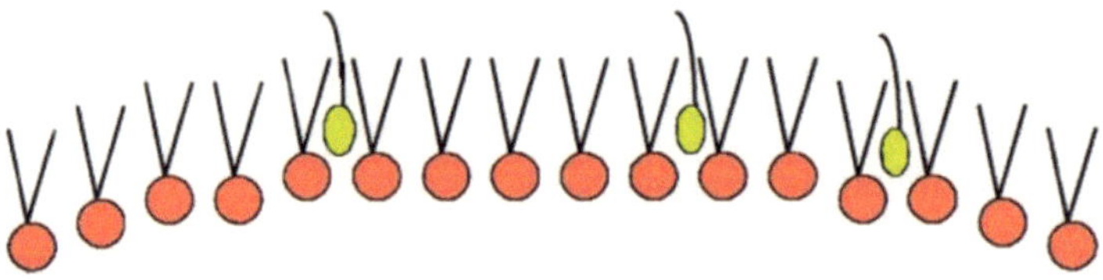

Figure 12. A scheme of solubilization site for alcohols [1].

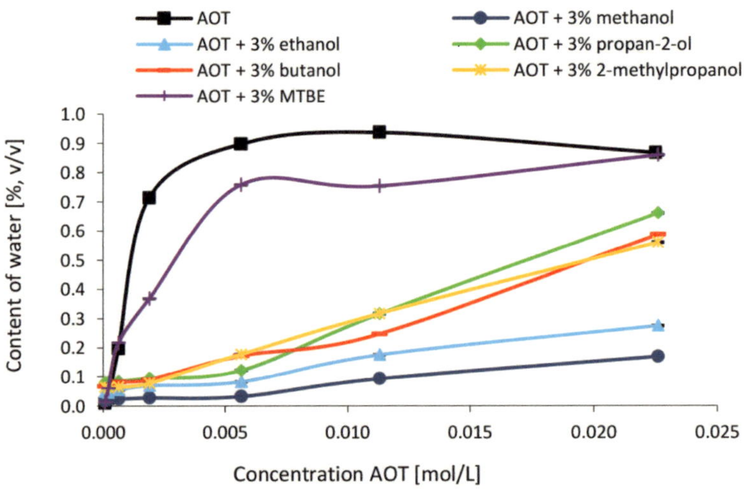

Figure 13. Solubility of water in gasoline containing hydrophilic additives (3%, v/v) and AOT.

surfactant AOT. The ether MTBE has a higher molecular weight and a branched hydrocarbon chain; therefore, in the presence of a small amount of AOT it is possible to enhance the water content significantly. In comparison to methanol and ethanol, such alcohols as propan-2-ol (IPA), 2-methylpropanol (IBA), and butanol (BuOH) have ability to be highly included in micelles, because of their higher molecular weights and amphiphilic properties. Solubility of water caused by the tested alcohols is low and very similar to basic gasoline. Only an addition of MTBE significantly improves water solubility (**Figure 13**). The impact of the additives changes in the following order: AOT > AOT/MTBE > AOT/IPA < AOT/IBA > AOT/BuOH >

AOT/EtOH > AOT/MeOH. Similar results were obtained in the case of an addition of AOT, MTBE, and the alcohols in the amount of 2% (v/v) in our previous studies [1].

5.2. Conducting properties improvement of modified gasoline

Initial studies demonstrated that electrolytic conductivity of basic gasoline and basic gasoline previously saturated with an addition of deionized water (1%, v/v) was 0 μS/cm. **Figure 14** shows the presence of only AOT caused a sudden and fast increase in electrolytic conductivity even at a concentration of 2×10^{-4} mol/l. The relative standard deviation of measurements is shown in **Table 7**. Yet, the conductivity achieved the value 0.16 μS/cm and next remained on a constant level above the concentration of 1.88×10^{-3} mol/l. The highest values were obtained after modification with MTBE (3%, v/v). While the conductivity achieved the level of 0.19 μS/cm, it did not change with the increasing concentration of AOT. The lowest values were observed in the case of an addition of 3% MeOH (0–0.02 μS/cm) and EtOH (0–0.05 μS/cm). Nonetheless, higher branched alcohols (IPA, IBA, and BuOH) with a higher molecular weight significantly caused an increase in conductivity. The anionic surfactant AOT in the presence of MTBE and water generated the highest values due to the formation of reverse micelles. It is reported in the literature that electrolytic conductivity is very sensitive to the microemulsion system structure [92]. The occurrence of conductivity percolation is revealed due to an increase in the droplet size, interactions and the exchange rate of substances between droplets. The percolation threshold coincides with the formation of the first clusters of droplets [93]. The change of electrolytic conductivity demonstrates the alteration of the reverse micellar microstructure and after that the percolation

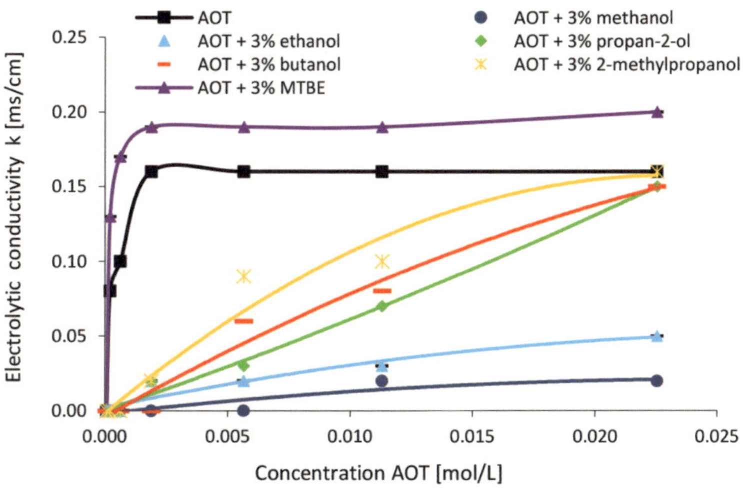

Figure 14. Conductivity of gasoline containing hydrophilic additives (3%, v/v) and AOT.

Molar concentration AOT [mol/l]	2.25×10^{-2}	1.13×10^{-2}	5.63×10^{-3}	1.88×10^{-3}	6.26×10^{-4}	2.09×10^{-4}	6.95×10^{-5}
Relative standard deviation for AOT (RSD)	±0.084	±0.178	±0.180	±0.567	±0.221	±0.178	±0.189

Table 7. Relative standard deviation values for electrolytic conductivity measurements. Source: own research.

transition occurs. It is reported that conductivity is firmly related to droplet diameter, however, a temperature, the presence of external entity, or the composition of the microemulsion system also have an influence on the conducting properties of reverse micelles. Microemulsion is able to transport charges and affects the changes in the electrolytic conductivity [1].

5.3. The effects of additives on interfacial tension

In preliminary studies, it was indicated that interfacial tension at the interface of basic gasoline/deionized water was 27.16 mN/m. The basic gasoline saturated with 1% of deionized water demonstrated the value equal to 25.12 mN/m. **Figure 15** shows the influence of various additives on interfacial tension at the gasoline/water interface. At the abscissae of **Figure 15**, a common logarithm (log to base 10) of the molar concentration c [mol/l] of the surfactant AOT was used to present the interfacial tension isotherms in a clearer way. The relative standard deviation of measurements is shown in **Table 8**. The decrease in interfacial tension depends

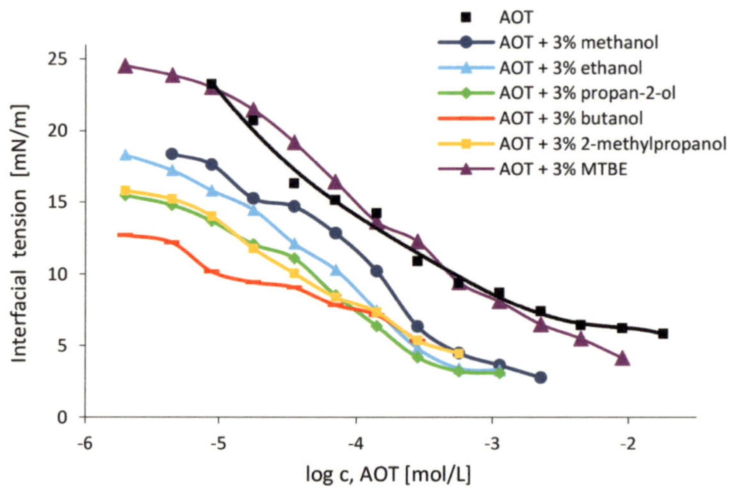

Figure 15. Interfacial tension isotherms at the modified gasoline/water interface.

Molar concentration AOT [mol/l]	**1.49×10^{-2}**	**7.43×10^{-3}**	**3.72×10^{-3}**	**1.86×10^{-3}**	**9.29×10^{-4}**	**4.65×10^{-4}**
Relative standard deviation for AOT (RSD)	0.011	0.008	0.026	0.028	0.031	0.018
Molar concentration AOT [mol/l]	2.32×10^{-4}	1.16×10^{-4}	5.81×10^{-5}	2.9×10^{-5}	1.45×10^{-5}	7.26×10^{-6}
Relative standard deviation for AOT (RSD)	0.032	0.037	0.023	0.041	0.015	0.016

Table 8. Relative standard deviation values for interfacial tension measurements. Source: own research.

on concentration of the surfactant AOT and the type of an additive. Interfacial tension isotherms of gasoline samples with AOT (23.2–5.8 mN/m) and with AOT and 3% MTBE (24.5–4.12 mN/m) have a similar course. Alcohols BuOH (12.7–5.3 mN/m), IPA (15.5–3.0 mN/m), and IBA (15.8–4,4 mN/m) showed the greatest surface activity. The effect of examined gasoline additives can be presented in the following order: AOT/BuOH > AOT/IPA > AOT/IBA > AOT/EtOH > AOT/MeOH > AOT/MTBE > AOT.

6. Conclusions

The multifunctional anionic surfactant AOT causes an increase in the solubility of water and electrolytic conductivity in gasoline. The obtained properties are the result of the association phenomenon of the surfactant and formation of reverse micelles comprising water pools in the hydrophilic micelle cores. Alcohols containing higher number of carbon atoms in their molecule lead to an increase in the solubility of water. Yet, the compounds may reduce the positive effect of the surfactant AOT on water solubility in gasoline. The results showed that some examined additives may act as cosurfactants.

Alcohols with highly branched hydrocarbon chains (isopropanol, isobutanol, and butanol) essentially increase the electrolytic conductivity. Modification of gasoline with the surfactant AOT and ether MTBE indicated the highest increase in electrolytic conductivity because of the formation of reverse micelles, which are able to transport charges.

The effect of the addition of AOT is the decrease in the interfacial tension at the water/gasoline interface. The decrease depends on the surfactant concentration and type of hydrophilic additives. The lowest values were observed in the presence of butanol, isopropanol, and isobutanol. The examined components have an influence on the interfacial tension, electrolytic conductivity, and the solubility of water in the same order: butanol > 2-methylpropanol > propan-2-ol > ethanol > methanol > MTBE. The research results demonstrated strong relationship between the length of the hydrocarbon chain, the molecular weight of hydrophilic components, and the tested properties of gasoline.

The conducted studies are innovative and can significantly contribute to an increase in knowledge and research of new water-in-gasoline emulsion (WiGE) fuel. The fuel with the addition of hydrophilic oxygen components in the presence of small amounts of surfactants and water may have unique properties. Oxygen compounds have a lot of useful properties, including antiknock properties, enhancing octane number, and they can be produced from renewable agricultural raw materials. Gasoline as an emulsion may have a beneficial effect on the combustion process, and the result is the almost complete combustion of hydrocarbons to the low toxic gases and the absence of carbon black among combustion products. The presence of water in gasoline reduces fuel consumption, increases engine power, decreases the temperature of its work, thus reducing emissions of volatile organic compounds, NO_x, SO_2, CO, and particulate matter. The use of water in fuel can be a unique chance for development of global economy in terms of energy production. The research may contribute to the commercialization of new environmentally friendly fuel that may provide an alternative source of energy for spark-ignition engines in the future.

Author details

Tomasz Kalak

Address all correspondence to: tomasz.kalak@ue.poznan.pl

Department of Commodity Science and Ecology of Industrial Products, Faculty of Commodity Science, Poznań University of Economics and Business, Poznań, Poland

References

[1] Kalak T, Strus B. Influence of selected surfactants and high-octane oxygen components on water content, electrolytic conductivity in gasoline, and interfacial tension in the water/gasoline system. Energy & Fuels. 2014;**28**:1926-1939. DOI: 10.1021/ef4018338

[2] Marshal EL, Owen K. Motor Gasoline. Cambridge, UK: The Royal Society of Chemistry; 1995. DOI: 10.1021/ja965502k

[3] Weston KC. Energy Conversion. Chapter 3: Fuels and combustion. PSW Publishers: St. Paul, MN, 1992: 85-132. ISBN: 978-0534938611

[4] Nadim F, Zack P, Haag GE, Liu S. United State experience with gasoline additives. Energy Policy. 2001;**29**(1):1-5. DOI: 10.1016/S0301-4215(00)00099-9

[5] Taljaard HC, Jaardaan CFP, Botha JJ. The effect of content in different oxygenates gasoline blends on performance and emission in a single cylinder, spark ignition engine. SAE 91037; 1991

[6] Pumphrey JA, Brand JI, Scheller WA. Vapour pressure measurements and predictions for alcohols-gasoline blends. Fuel. 2000;**79**(11):1405-1411. DOI: 10.1016/S0016-2361(99)00 284-7

[7] Perry R, Gee IL. Vehicle emissions in relation to fuel composition. Science of the Total Environment. 1995;**169**(1-3):149-156

[8] Lin ChY, Wang KH. Effects of an oxygenated additive on the emulsification characteristics of two- and three-phase diesel emulsions. Fuel. 2004;**83**:507-515. DOI: 10.1016/j.fuel.2003.08.014

[9] Nguyen Q-A, Wu Y-Y. Experimental investigations of using water-gasoline emulsions as a NOx treatment and its effects on performance and emissions of lean-burn spark-ignition engine. In: Proceedings of the International Conference on Power Engineering-09; Nov. 16–20, 2009; Kobe, Japan

[10] Kalak T, Cierpiszewski R. Correlation between surface properties of laundry detergents and mixed surfactants solutions with their washing abilities. Textile Research Journal. 2015;**85**(18):1884-1906. DOI: 10.1177/0040517515578329

[11] Stępień Z, Oleksiak S. Deposit forming tendency in spark ignition engines and evaluation of gasoline detergent additives effectiveness. Journal of KONES Powertrain and Transport. 2009;**16**(2):421-431

[12] The Regulation of the Minister of Economy of 9 December 2008 on quality requirements for liquid fuels (Journal of Laws No. 221, item. 1441), Poland

[13] Directive 2009/30/EC of The European Parliament and of The Council of 23 April 2009 amending Directive 98/70/EC as regards the specification of petrol, diesel and gas-oil and introducing a mechanism to monitor and reduce greenhouse gas emissions and amending Council Directive 1999/32/EC as regards the specification of fuel used by inland waterway vessels and repealing Directive 93/12/EEC, the European Union

[14] Phair B, Bensch L, Duchowski J, Khazan M, Tsalyuk V. Overcoming the electrostatic discharge in hydraulic, lubricating and fuel-filtration applications by incorporating novel synthetic filter media. Tribology Transactions. 2012; 48(3):343-351. DOI: 10.1080/05698190591002195

[15] Anderson EK, Carlucci AP, De Risi A, Kyritsis DC. Electrostatic effects on gasoline direct injection in atmospheric ambiance. Energy Conversion and Management. 2007;**48**:2762-2768. DOI: 10.1615/AtomizSpr.v17.i4.10

[16] Zwierzycki W. Paliwa i oleje silnikowe. Instytut Technologii Eksploatacji w Radomiu; Gorlice: Rafineria Nafty Glimar, Radom; 1997

[17] Zwierzycki W. Paliwa, oleje, motoryzacyjne płyny eksploatacyjne. Instytut Technologii Eksploatacji w Radomiu; Gorlice: Rafineria Nafty Glimar, Radom; 1998

[18] Kajdas C. Petrochemical Technology, Part I, II. Warszawa: Wydawnictwo Politechniki Warszawskiej; 1977

[19] Podniało A. Paliwa, oleje i smary w ekologicznej eksploatacji. Warszawa: Wydawnictwo Naukowo-Techniczne; 2002

[20] Schmidt TC, Kleinert P, Stengel C, Goss Kai-Uew, Haderlein SB. Polar fuel constituents: Compound identification and equilibrium partitioning between nonaqueous phase liquids and water. Environmental Science and Technology. 2002;**36**:4074-4080. DOI: 10.1021/es010307d

[21] Videncenter for Jordforurening (VFJ). Fuel additives: A risk screening of additives to gasoline and diesel. Teknik og Administration Nr. 3; 2006

[22] Landells RG. Precombustion Gasoline Additives. In: Motor Gasoline. Marshall EL, Owen K, editors Cambridge, UK: RSC; 1995; 170-200

[23] Quimby BD, Giarrocco V, Sullivan JJ, McCleary KA, Fast analysis of oxygen and sulfur compounds in gasoline. Journal of High Resolution Chromatography. 1992;**15**:705-709. DOI: 10.1002/jhrc.1240151103

[24] Weaver JW, Skaggs SA, Spidle DL, Stone GC. Composition and Behavior of Fuel Ethanol. EPA/600/R-09/037; 2009

[25] Varisli D, Dogu T. Simultaneous production of tert-amyl ethyl ether and tert-amyl alcohol from isoamylene–ethanol–water mixtures in a batch–reactive distillation column. Industrial & Engineering Chemistry Research. 2005;**44**(14):5227-5232. DOI: 10.1021/ie049241w

[26] Snelling J, Barnett MO, Zhao D, Arey JS. Methyl tertiary hexyl ether and methyl tertiary octyl ether as gasoline oxygenates: Assessing risks from atmospheric dispersion and deposition. Journal of the Air and Waste Management Association. 2006;**56**(10):1484-1492. DOI: 10.1080/10473289.2006.10464552

[27] Owen K. Gasoline and Diesel Fuel Additives. Wiley; New York, USA, 1989

[28] Driveability and Performance of Reformulated and Oxygenated Gasoline. DAI Informational Document No. 970302. Downstream Alternatives, Inc.; March 1997

[29] Zoller U, editor. Handbook of Detergents, Part E: Applications. CRC Press, Taylor & Francis Group; Boca Raton, USA, 2009

[30] Schwahn Ha, Posselt D. Gasoline detergents and intake valve deposit inhibitors consisting of high-molecular-weight hydrocarbyl amines in an alkoxylated alcohol carrier fluid. Germany: BASF Aktiengesellschaft; 2003. PCT Int. Appl., Patent US 20040010966 A1

[31] Derosa TF. Kaufman BJ, Deblase FJ, Rawdon MG, Hayden TE, Ketcham JR, Diarylamine-functionalized ethylene-alkene-alkadiene copolymers as gasoline antifriction detergents. USA: Texaco Development Corporation; 2003. PCT Int. Appl., Patent

[32] Lin J-J, Ku W-S, Lee C-N, Ho Y-S, Lin K-H, Wei K-M. Amidoamine-terminated polyoxyalkylenes as gasoline detergents and octane requirement reducing agents. Taiwan: Chinese Petroleum Corp.; 2002. U.S., Patent US 6 454 818

[33] Guo H, Liu Z, Li S, Xu S., Research on novel polyisobutylphenoxyethyl polyamines as gasoline detergents. Xian Research Institute of Hi-Tech, P.R. China, Xian City, Peop. Rep. China, American Chemical Society. Fuel Chemistry Division Preprints. 2002;**47**(2):57

[34] Puri SK, Gupta AA, Rajesh M, Prasad R, Sharma VK, Vanamamalai M, Singh, V, Misra AK, Raje NR, Srivastava SP, Bhatnagar AK. Mannich bases prepared from hydrogenated distilled nutshell liquid as gasoline detergents and deposit inhibitors. India. 2002. U.S. Pat. Appl. Publ., Patent US 2002066225

[35] Guo H, Liu Z. Synthesis and performance evaluation of a new type of adducts of polyisobutyl phenol-epoxy chloropropane-polyamine as gasoline detergents. American Chemical Society, Division of Petroleum Chemistry. 2002; **47**(2); 139-143

[36] Lin J-J, Wu J-J, Shau S-M, Ho Y-S. Thermal stability and combustion behaviors of poly(oxybutylene)amides. Department of Chemical Engineering, National Chung-Hsing University, Taichung, Taiwan. Polymer Journal (Tokyo, Japan). 2002;**34**(2):72-80. DOI: 10.1295/polymj.34.72

[37] Lin J-J, Wu J-J, Ho Y-S. Synthesis and in situ transformation of poly(oxybutylene) amides by butoxylation. Department of Chemical Engineering, National Chung-Hsing University, Taichung, Taiwan. Journal of Applied Polymer Science. 2001;**82**(2):435-445. DOI: 10.1002/app.1869

[38] Aleman-Vazquez LO, Villagomez-Ibarra JR, Polyisobutenylsuccinimides as detergents and dispersants in fuel: Infrared spectroscopy application. Gerencia de Productos Quimicos,

eje Central Lazaro Cardenas No. 152 col. San Bartolo Atepehuacan, Instituto Mexicano del Petroleo, Mexico City, Mexico. Fuel. 2001;**80**(7):965-968. DOI: 10.1016/S0016-2361(00)00167-8

[39] McAtee RJ. Polyalkenylphenol-derived aromatic Mannich compounds as diesel fuel and gasoline detergents. USA: The Lubrizol Corporation; 2000. PCT Int. Appl., Patent WO 2000078898, US 6179885

[40] Palmer Jr. CF. Gasoline Detergents. UK: Associated Octel Co. Ltd.; 1998., Brit. UK Pat. Appl., Patent GB 2320719

[41] Lee BH, Kim WS, Lim SC, Kwon HY, Yang DJ, Yukong Ltd., S. Korea, Lee BH, Kim WS, Lim SC, Kwon HY, Yang, DJ. Polyoxyalkylene ether amino acid esters, their preparation and use. 1997. PCT Int. Appl., Patent WO 9736854, US 6053955

[42] Block D. Polyalkylenepolyamines, especially polyisbutylene polyamine, as gasoline detergents-deposit inhibitors, for prevention of sticking of inlet valves. USA: Exxon Chemical Patents, Inc.; 1992. PCT Int. Appl., Patent WO 9215656

[43] McLeary DB, Mitchell JA. Amine- and halogen-free gasoline detergents consisting of polypropylene glycol with hydrophobic end group. UK: British Petroleum Co. PLC; 1992. PCT Int. Appl., Patent WO 9213047

[44] Wang PC. (Keto) Diacid amides as gasoline detergents-deposit inhibitors-octane requirement reducing additive. Shell Internationale Research Maatschappij B. V., Neth., Brit. 1992. UK Pat. Appl., Patent GB 2247457

[45] A'Court R, Fox WJ, Hamlin JE, O'Connor SP. Novel carbamates, their production and use as fuels additives. UK: BP Chemicals Ltd.; 1988. PCT Int. Appl., Patent US 5286265

[46] Schlicht RC, Levin MD, Herbstman S, Sung RL. Gasoline compositions containing reaction products of fatty acid esters and amines as carburetor detergents. USA: Texaco Inc.; 1988. U.S., Patent US 4729769

[47] Sung RL, Kaufman BJ. Motor fuel detergent additives. USA: Texaco Inc.; 1986. Patent US 4624682

[48] Andress HJ, Ashjian H, Gawel HA. Compounds containing amide linkages from mono- and polycarboxylic acids in the same molecule and lubricants and fuels containing them. USA: Mobil Oil Corp.; 1986. Eur Pat Appl, Patent EP 0186473 A2

[49] Kaufman BJ, Sung RL. Mannich reaction product and a motor fuel composition containing it. USA: Texaco Inc.; 1985. Patent US 4508541

[50] Schuettenberg AD. Fuel additives from borated, acid-treated mixtures of vegetable oil derived amides and esters. USA: Phillips Petroleum Co.; 1985. Patent US 4505725

[51] Schuettenberg AD, Gragson JT. Solid additives for fuel. USA: Phillips Petroleum Co.; 1984. Patent US 4639255

[52] Hoke DI. Liquid fuels containing polyamine dispersants. USA: Lubrizol Corp.; 1983. Patent US 4400178

[53] Andress HJ. Liquid hydrocarbon fuel composition. USA: Mobil Oil Corp.; 1983. US 4394135

[54] Cummings WM, Powell JC. Detergent fuel composition. USA: Texaco Inc.; 1982. US 4330303

[55] Nikitina EA, Emel'yanov VE, Rudyak KB, Lyamenkova TA. Study of the Detergent Properties of Gasoline Additives. USSR: Sbornik Nauchnykh Trudov-Vsesoyuznyi Nauchno-Issledovatel'skii Institut po Pererabotke Nefti; 1980. p. 37

[56] Childs ME. Multifunctional gasoline additives. USA: UOP Inc.; 1981. Patent US 4353711

[57] Piotrowski AB. Detergent compositions. USA: Mobil Oil Corp.; 1981. US 4292046

[58] Bonazza BR, Devault AN. Fuel and lubricant additives from acid treated mixtures of vegetable oil derived amides and esters. USA: Phillips Petroleum Co.; 1981. Patent US 4269606

[59] Bonazza BR, Holtz HD. Amino amide fuel detergents. USA: Phillips Petroleum Co.; 1981. Patent US 4249912

[60] Sievers RE, Wenzel TJ. Liquid hydrocarbon-soluble rare earth chelates prepared from the ligand 2,2,7-trimethyl-3,5-octanedione and fuels containing them. USA: University Patents, Inc.; 1981. Patent US 4251233

[61] Shields TC. Alkyl acrylate adducts of polyamines, ether amines and ether polyamines. USA: Sherex Chemical Co., Inc.; 1980. Patent US 4240804

[62] Bonazza BR, Holtz HD. Imidazoline fuel detergents. USA: Phillips Petroleum Co.; 1981. Patent US 4247300

[63] Bedrik BG. Multifunctional gasoline additive. USSR; 1980. Patent SU 755830

[64] Biasotti JB, Dille K, Dorn P, Herbstman S. Detergent gasoline composition. USA: Texaco Inc.; 1980. Patent US 4204841

[65] Hanson JB. Polyamine derivatives of oxidized olefinic substituted dicarboxylic acid compounds. USA: Standard Oil Co. (Indiana); 1980. Patent US 4203730

[66] Bonazza BR, DeVault AN. Detergent additives from mixtures of vegetable oil derived amides and esters or acid treated mixtures thereof and their use in fuels and lubricants. USA: Phillips Petroleum Co.; 1980. Eur. Pat. Appl., Patent EP 8591

[67] Andress Jr HJ. Fuel containing detergent. USA: Mobil Oil Corp.; 1980., Eur. Pat. Appl., Patent US 4240803

[68] Lewis RA, Honnen LR. Fuel compositions of poly(oxyalkylene) monoether (aminoethyl) carbamates. USA: Chevron Research Co.; 1980. Patent US 4191537

[69] Vartanian PF, Ippolito AL. Amine oxide polymers and their use as fuel detergents. USA: Texaco Inc.; 1979. Patent US 4179271

[70] Scheule HJ, Starn Jr RE. Polyolefin-acylated poly(alkyleneamine) two component fuel additive. USA: du Pont de Nemours, E. I., and Co.; 1979. Patent US 4173456

[71] Cummings WM, Powell JC. Detergent additive and motor fuel composition. USA: Texaco Inc.; 1979. Patent US 4132531

[72] O'Brien EF, Langdon WK. Fuels containing polyoxyalkylene ether demulsifiers. USA: BASF Wyandotte Corp.; 1978. Patent US 4125382

[73] Holtz HD. Ashless fuel detergent additives. USA: Phillips Petroleum Co.; 1978. Patent US 4125383

[74] Coon MD. Fuel additive. USA: 1978. Patent US 4105417.

[75] Holtz HD, Bonazza BR. Ashless fuel detergent additives. USA: Phillips Petroleum Co.; 1977. Patent US 4059414

[76] Garth BH. Mannich bases containing tertiary amines and fuel compositions containing said Mannich bases. USA: du Pont de Nemours, E. I., and Co.; 1977. Patent US 4054422, US 4121911

[77] Garth BH. Gasoline additive compositions comprising a combination of monoamine and polyamine Mannich bases. USA: du Pont de Nemours, E. I., and Co.; 1977. Patent US 4038043

[78] Garth BH. Gasoline additive compositions comprising a combination of diamine and polyamine Mannich bases. USA: du Pont de Nemours, E. I., and Co.; 1977., Patent US 4038044

[79] Chloupek FJ, Chao TS. Gasoline fuel composition. USA: Atlantic Richfield Co.; 1977. Patent US 4039300

[80] Kablaoui MS, Biasotti JB. Substituted phenoxy propanol diamines and amino alcohol detergent additives for fuels and mineral oils. USA: Texaco Inc.; 1977. Patent US 4024083

[81] Anon. Mannich condensation products, Org. Chem. Dep., E. I. du Pont de Nemours and Co., Inc.; 1976. Wilmington, DE, USA, Research Disclosure

[82] Honnen LR, Coon MD. Fuel detergents. USA: Chevron Research Co.; 1976. Patent US 3951614, US 3785789

[83] Gardiner RA, Garth BH. Motor fuel composition containing Mannich condensation products. USA: du Pont de Nemours, E. I., and Co.; 1976. Patent US 3944397

[84] Cummings WN. Hydrocarbyl amine-substituted propionic ester and motor fuel composition containing same. USA: Texaco Inc.; 1975. Patent US 3926578

[85] Yount JB. Alkylenediamineamides of fatty acids as gasoline additives. USA: ICI United States, Inc.; 1975. Patent US 3923474

[86] Machleder WH, Kuhn RR. Detergent motor fuel. USA: Rohm and Haas Co.; 1975. Patent US 3907518

[87] Forbes ES, Reid AJD. Polyolefin carburetor detergent. British Petroleum Co. Ltd., Britain; 1974. Patent GB 1378709.

[88] Machleder WH, Kuhn R. Motor fuel composition. Rohm and Haas Co.; 1974. Patent US 3846089, US 3782912, US 3912771

[89] Worrell CJ. Gasoline detergent. USA, Brit.: Ethyl Corp; 1974. Patent GB 1368532

[90] De T, Maitra A. Solution behaviour of aerosol OT in non-polar solvents. Advanced Colloid and Interface Science. 1995;**59**:95-193. DOI: 10.1016/0001-8686(95)80005-N

[91] Elliott MA, Nebel GJ, Rounds FG. The composition of exhaust gases from diesel, gasoline and propane powered motor coaches. Journal of the Air Pollution Control Association. 1955;**5**(2):103-108. DOI: 10.1080/00966665.1955.10467686

[92] Giustini M, Palazzo G, Colafemmina G, Monica Marcello Giomini MD, Ceglie C. Microstructure and dynamics of the water-in-oil CTAB/n-pentanol/n-hexane/water microemulsion: A spectroscopic and conductivity study. Journal of Physical Chemistry. 1996;**100**:3190-3198. DOI: 10.1021/jp952263z

[93] Kallay N, Chittofrati A. Conductivity of microemulsions: Refinement of charge fluctuation model. Journal of Physical Chemistry. 1990;**94**:4755-4756. DOI: 10.1021/j100374a070

[94] Global refined petroleum products. United States Energy Information Administration [Internet]. 2017. Available from: https://www.eia.gov/ [Accessed: 2017-01-10]

[95] The global application of petroleum products. United States Energy Information Administration [Internet]. 2017. Available from: https://www.eia.gov/ [Accessed: 2017-01-10]

[96] World transportation consumption by fuel, 2012. U.S. Energy Information Administration, International Transportation Energy Demand Determinants (ITEDD-2015) model estimates [Internet]. 2017. Available from: https://www.eia.gov/todayinenergy/detail.php?id=23832 [Accessed: 2017-01-10]

[97] Global motor gasoline production. United States Energy Information Administration [Internet]. 2016. Available from: https://www.eia.gov/ [Accessed: 2016-12-08]

[98] Motor gasoline production by country. United States Energy Information Administration [Internet]. 2016. Available from: https://www.eia.gov/ [Accessed: 2016-12-08]

[99] Kalak T. The role of selected additives in gasoline fuel [thesis]. Poznań, Poland: Poznań University of Technology; 2004.

Chapter 2

Recent Advances in Catanionic Mixtures

Darija Domazet Jurašin, Suzana Šegota, Vida Čadež,
Atiđa Selmani and Maja Dutour Sikirć

Additional information is available at the end of the chapter

http://dx.doi.org/10.5772/67998

Abstract

Most surfactant mixtures display synergistic physicochemical properties, which have led to their extensive application in various technologies. Aqueous mixtures of two oppositely charged surfactants, so-called catanionic surfactant mixtures, exhibit the strongest synergistic effect, which is manifested as high surface activity, enhanced adsorption and a low critical aggregation concentration. In addition, catanionic systems display rich phase behavior and a range of nano and microstructures, including small spherical micelles, rod-like micelles as well as open and closed bilayers (vesicles). The spontaneous formation of catanionic vesicles is of special interest due to their various applications in nanotechnology and pharmaceutical formulations. In this chapter, the properties of catanionic mixtures of amphiphilic molecules with advantageous properties are discussed. Since numerous papers dealing with catanionic mixtures of monomeric surfactants already exist, the aim of this chapter is to summarize recent progress in mixtures of structurally different surfactants. At the end of the chapter, special emphasis is placed on applications of catanionic mixtures.

Keywords: surfactants, catanionic mixtures, vesicles, phase behavior, application

1. Introduction

Due to their amphiphilic structure, surfactants exhibit unique physicochemical properties both in solutions and in solid state. Mixtures of two or more different surfactants often show improved properties compared to individual surfactant solutions. As a result, in household and industrial applications, surfactant mixtures are usually used [1, 2]. Aqueous mixtures of two oppositely charged surfactants, that is, catanionic surfactant mixtures, exhibit the strongest synergistic effect, which is manifested as high surface activity, enhanced adsorption and low critical aggregation concentration [3–7]. In addition, catanionic mixtures display rich phase behavior governed by electrostatic and hydrophobic interactions, steric effects (geometric packing constraints) and hydrogen bonding. Therefore, such systems offer numerous possibilities in controlling molecular self-assembly by adjusting bulk properties and using appropriate surfactant molecules. Consequently, they are of special interest not only from fundamental point of view but also because of a wide range of industrial applications.

In this chapter, catanionic mixtures of amphiphilic molecules with advantageous properties are discussed. Since numerous papers dealing with catanionic mixtures of monomeric surfactants already exist, the aim of this chapter is to summarize recent year's progress in mixtures of structurally different surfactants. At the end of the chapter, special emphasis is placed on applications of catanionic mixtures. It should be pointed out that the field of catanionic mixtures investigation is vast and still expanding, so the present review can be neither fully comprehensive nor final.

2. Properties of catanionic mixtures

Catanionic mixtures can be prepared by mixing a cationic surfactant with an anionic one. Because two oppositely charged surfactants are present in the mixture, catanionic mixtures possess unique features, which can be summarized as follows [2–8]:

(1) strong electrostatic attractions between oppositely charged headgroups and ion pairing (**Figure 1**),

(2) pronounced synergism and solution behavior that considerably deviate from ideal mixing, that is, interfacial and aggregation properties of such systems are enhanced compared to those of single surfactants,

(3) strong dependence of the physicochemical properties, as well as the phase behavior, on the molar ratio and total concentration of the components,

(4) rich phase behavior and structural diversity where the size of aggregates ranges from the nano to micrometer scale (mixed micelles, vesicles, tubules, liquid crystalline phases, etc.),

(5) spontaneous formation of stable vesicles, including, in some cases, equilibrium vesicles and

(6) precipitation of catanionic surfactant (CA) at/or near equimolar bulk composition:

$$CX + YA \rightarrow CA + X^- + Y^+ \qquad (1)$$

where CX and YA represent cationic and anionic surfactant, and X^- and Y^+ represent respective counterions.

Due to strong electrostatic attractions, addition of ionic surfactant to the solution of other, oppositely charged surfactants results in formation of tight ion pairs and removal of hydration water at the mixed aggregate/solution interface [8]. Thus, oppositely charged surfactant monomers form ion pairs, which can be described as pseudo double-tailed zwitterionic surfactants (**Figure 1**) [6].

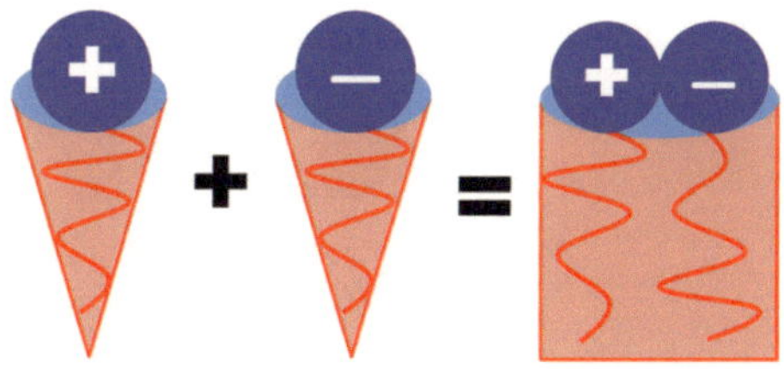

Figure 1. A schematic representation of ion pair formation in catanionic mixtures.

The formation of ion pairs has pronounced influence on the adsorption properties and self-assembly of catanionic mixtures. Unlike the solution of individual monomeric surfactants, in catanionic mixtures aggregates with minimal curvature, such as open and closed bilayers (vesicles), are spontaneously formed even at the low surfactant concentrations. For that reason, experimental investigations of dilute catanionic mixtures have made a key contribution to our understanding of the factors governing vesicle formation in surfactant systems [8, 9].

Spontaneous formation of stable vesicles in these systems can be explained by using packing parameter (P), which is defined with three nominal geometric parameters of surfactant molecules:

$$P = \frac{v_{hc}}{a_0 l_{hc}} \tag{2}$$

where a_0, v_{hc} and l_{hc} are the surface area per headgroup, the volume and fully extended length of the hydrophobic tail of the molecule, respectively [10]. The molecular shape and respective P will determine the type of preferred surfactants' aggregate: for $P = 0.33$, spherical micelles; for $P \approx 0.33$–0.5, cylindrical micelles; for $P \approx 0.5$–1, bilayer disks and vesicles; for cylinders (**Figure 1**) with $P \approx 1$ planar bilayers; and for $P > 1$, reverse structures. In other words, from the aspect of geometric constraints, the preferred structure of surfactant's aggregates, for a given hydrophobic tail size, strongly depends on the effective headgroup area. In general, the a_0 value depends on two opposite forces: (1) attractive hydrophobic interactions between hydrocarbon chains at the hydrocarbon-water interface and (2) repulsive electrostatic and/or steric interactions [8]. In the case of catanionic ion pairs, the effective headgroup area decreases, compared to the value of each of the surfactants individually, while the volume of the hydrophobic chain increases. As a result, the value of packing parameter approaches unity, which favors structures with low curvature like vesicles and flexible bilayers [10].

However, for better explanation of spontaneous vesicles' formation in catanionic mixtures, in addition to geometric parameters, the curvature free energy and its dependence on bilayer's composition should also be taken into account. In order for bilayer to have non-zero spontaneous curvature, its two individual leaflets should have equal and opposite spontaneous curvature (**Figure 2**). This is possible only when two leaflets have different compositions. Moreover, the composition should be such that average headgroup area in the outer leaflet is larger than the one in the inner. In catanionic mixtures, this is achieved by having higher molar ratio of the excess surfactant in the outer leaflet, resulting in larger headgroup spacing due to the electrostatic repulsions. In the inner leaflet, the higher fraction of paired surfactants reduces the headgroup area and results in positive curvature [5]. Therefore, spontaneously

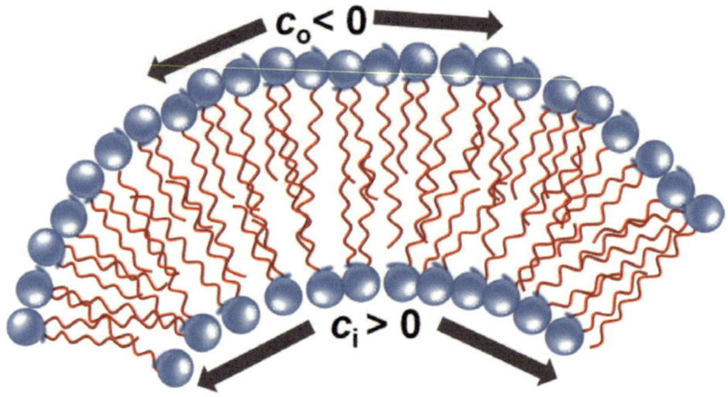

Figure 2. A schematic representation of catanionic bilayer with equal and opposite curvature of inner (c_i) and outer (c_o) leaflet (after [5]).

formed catanionic vesicles owe their stability to the non-ideal mixing of oppositely charged surfactants as well as to electrostatic effects.

The spontaneous formation of catanionic vesicles is of special interest due to their various applications in nanotechnology and pharmaceutical formulation, as will be discussed later in this chapter. Regarding the vesicle stability, different theoretical models have tried to rationalize why vesicles behave as true equilibrium aggregates rather than a dispersed form of a lamellar liquid crystal [11–13]. However, what is important for application purposes is that vesicles do readily form in catanionic mixtures and they appear to pose long-term stability. In addition, the low-cost and versatile physicochemical properties make them a good alternative to phospholipid vesicles, that is, liposomes [5].

A typical phase diagram for catanionic mixtures is schematically illustrated in **Figure 3**. However, there are numerous variations in the appearance of the catanionic phase diagram. The concentration regions in which vesicles form are represented by the lobes on both sides of the equimolar line. This indicates that vesicles are stabilized by the presence of excess surfactants. Catanionic vesicles usually have high degree of polydispersity, and their stability can be tailored by the choice of surfactant molecular structure, that is branched surfactants, and/or those containing a bulky group in alkyl tail usually form more stable vesicles [6]. Likewise, in asymmetric surfactant mixtures, in terms of different alkyl chain numbers or length as well as different chain morphology, the vesicle phase is often considerably enlarged and found in a broad concentration range [14–16]. The size, surface charge density and permeability of catanionic vesicles can be tailored by varying temperature, concentration and molar ratio, as well as chain length of surfactants [17].

As the molar mixing ratio of the two surfactants or the total surfactant concentration is varied, different phase transitions involving vesicles are found: micelle-to-vesicle, vesicle-to-lamellar and vesicle-to-solid phase transitions (**Figure 3**). At the highest excess of the mixture components, mixed micelles of various sizes and shapes can be found, including globular, elongated (worm-like) and branched ones [7]. The size and shape of mixed micelles depend on bulk composition and total surfactant concentration as well as geometry of the surfactants, temperature, salt content, etc. In surfactant mixtures, micelle-to-vesicle transition has been broadly found to occur through two pathways [18]. One path involves limited micellar growth and

http://dx.doi.org/10.5772/67998

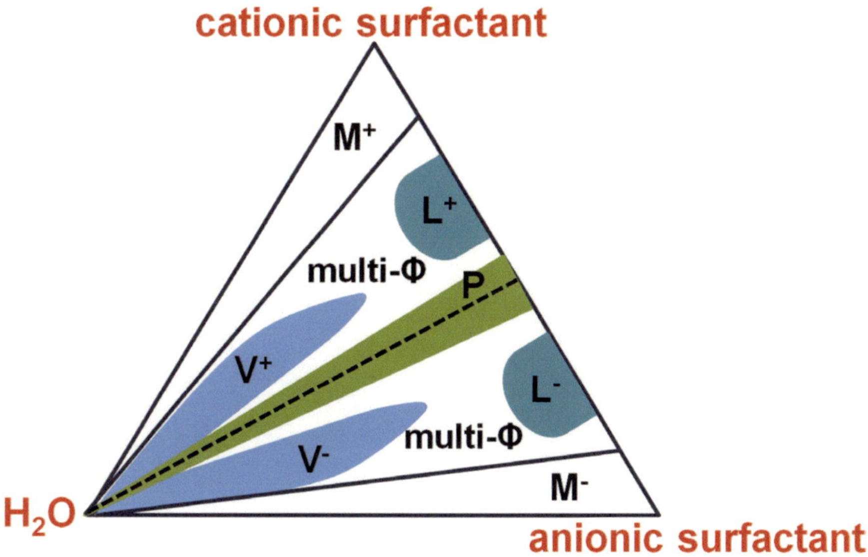

Figure 3. A schematic triangular phase diagram of symmetric catanionic mixture at constant temperature and pressure. The dashed line denotes the equimolar line dividing the diagram into the cationic-rich and the anionic-rich region. Close to the charge neutrality line, a solid precipitate (*P*) is usually formed, but excess charge in the system usually leads to vesicle stabilization (denoted as V^+ and V^-). Mixed micelles (denoted as M^+ and M^-) are usually formed at the highest excess of the mixture components. Multiphase regions (multi-Φ) often involve a lamellar phase occurring at higher concentrations (denoted as L^+ and L^-) (after [9]).

micelle/vesicle coexistence and is more common for systems with symmetric chain lengths [19]. Examples are dodecyltrimethylammonium bromide (C_{12}TAB)/sodium dodecyl sulfate (SDS) (**Figures 4** and **5**) [20], didodecyldimethylammonium bromide (DDAB, **Figure 6**)/SDS [21] mixtures and a few others involving amino acid-based surfactants [22, 23]. The second path involves strong micellar growth and is typical of highly asymmetric systems [19].

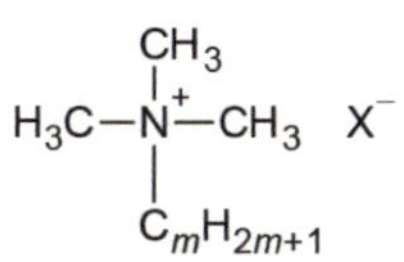

alkyltrimethylammonium halides (C_mTAX)

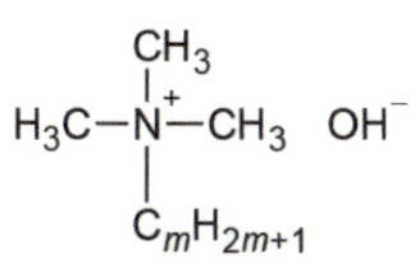

alkyltrimethylammonium hydroxides (C_mTAOH)

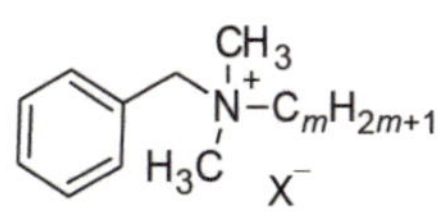

benzalkonium halides (C_mBzX)

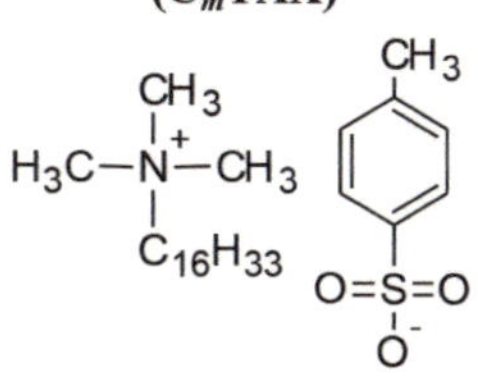

cetyltrimethylammonium tosylate (CTAT)

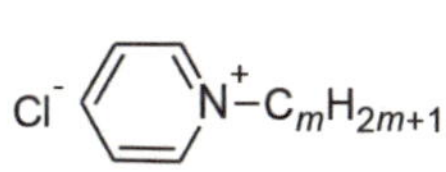

alkylpyridinium chlorides (C_mPC)

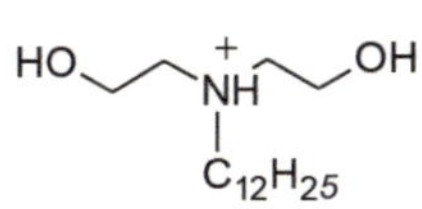

protonated *N*-dodecyldiethanolamine

Figure 4. Molecular structures of monomeric cationic surfactants—quaternary alkyl ammonium salts. m = number of C atoms in alkyl chains and s = number of C atoms in spacer.

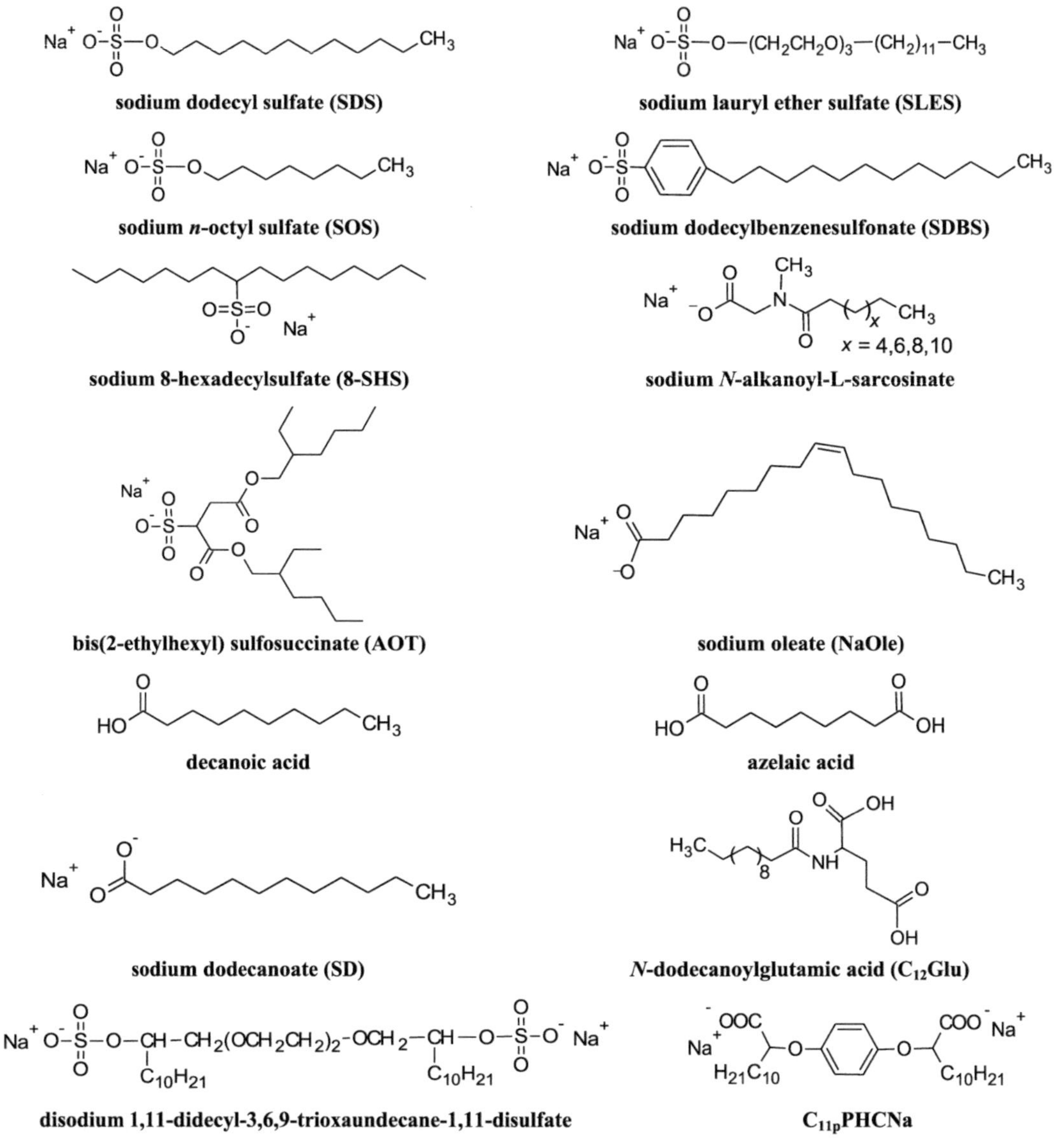

Figure 5. Molecular structures of various anionic amphiphiles.

In the majority of catanionic mixtures at equimolar concentrations precipitate, a new catanionic surfactant, which very often possesses lamellar structure, forms [2, 18]. It can form as only phase or in coexistence with (1) coacervates (small droplets in solution rich with surfactants) [6, 24] as well as (2) micelles or (3) lamellar phase, usually in asymmetric mixtures [14, 16, 22–28]. Formed precipitate can be redissolved by increasing concentration of one of the surfactants, leading to formation of micelles or vesicles. Although catanionic precipitate is generally found only near equimolar compositions or in samples below their Krafft temperature, it is

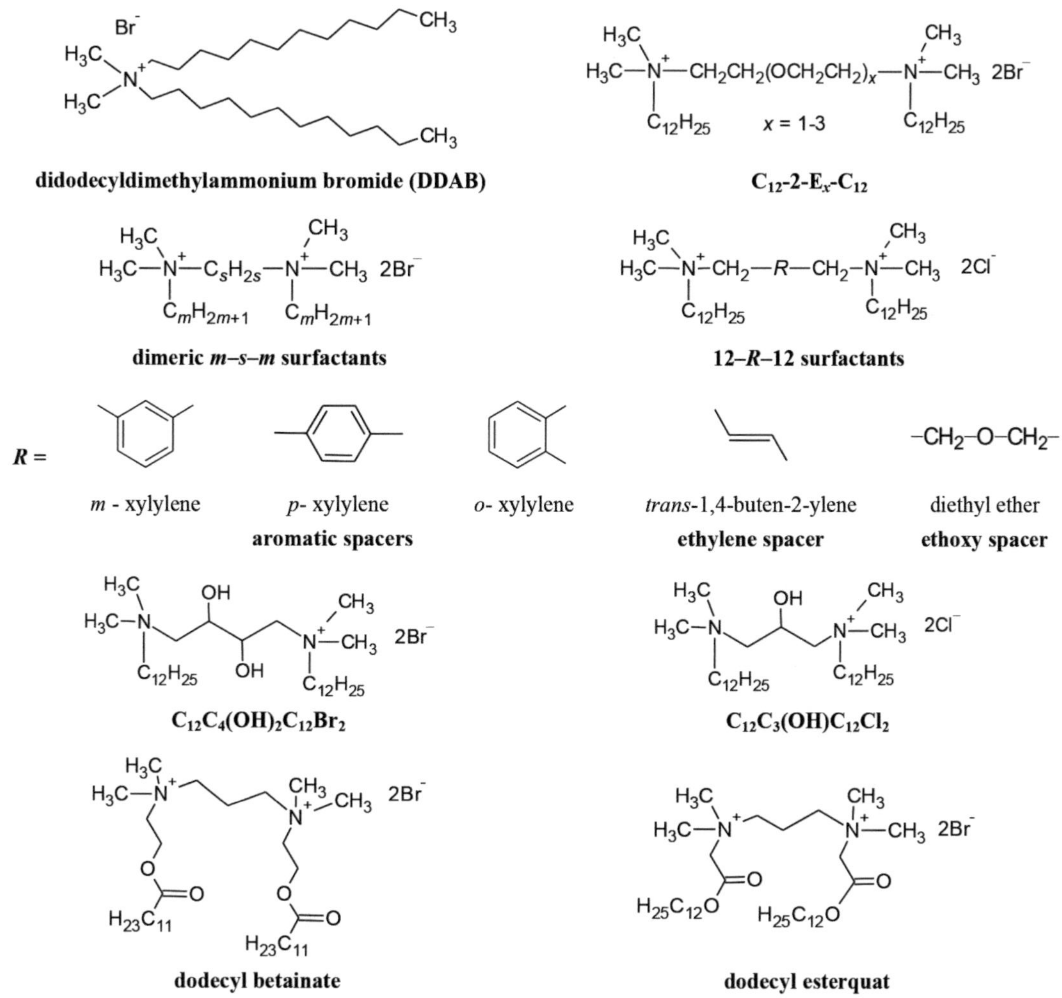

Figure 6. Molecular structures of double-tailed and dimeric cationic surfactants—quaternary alkyl ammonium salts. m = number of C atoms in alkyl chains and s = number of C atoms in spacer.

considered to be the main drawback for application of catanionic mixtures [3]. However, with the right selection of mixtures components, precipitation can be circumvented. Increased asymmetry in hydrophobic parts of surfactant molecules (different tail length or number, branched tails, rigid ring-based structures, etc.) weakens hydrophobic attractions among alkyl chains and prevents efficient packing into a crystalline lattice [14–16, 29]. As a result, precipitation does not occur and instead large micelles, vesicles or liquid crystalline phases can be formed.

Numerous studies showed that catanionic mixtures are characterized with solution behavior that considerably deviates from ideal mixing as well as pronounced synergism (high surface activity, enhanced adsorption, low critical micellization concentration (cmc), etc.) compared to other types of surfactant mixtures [1, 3–7]. For example, the cmc values in catanionic mixtures can be several orders of magnitude lower than the ones of single surfactants. It is not surprising considering that molecular interactions between cationic and anionic surfactants are generally dominated by the attractive electrostatic forces. Additionally, driving force for the mixed aggregates formation is large increase in entropy, which is a consequence of counterion release from both surfactants. On the contrary, in the case of single surfactant aggregate, the entropy decreases due to the condensation of counterions [2].

Synergism in catanionic mixtures can be quantitatively described using the regular solution theory (RST), which provides a thermodynamical approach to non-ideal mixing [1, 30–32]. Although frequently criticized on fundamental grounds, the RST still remains a helpful tool for description of the behavior of catanionic mixtures.

According to the RST and the standard state surface tension method, the mixed monolayer composition and the mixed monolayer interaction parameter can be calculated [1, 30–32]. The molar fraction of the cationic surfactant in the mixed monolayer (X_1) can be calculated according to the following equation:

$$\frac{X_1^2 \ln\left(\alpha_1 c_{1,2} / c_1 X_1\right)}{\left(1 - X_1\right)^2 \ln\left[\left(1 - \alpha_1\right) c_{1,2} / (1 - X_1)\, c_2\right]} = 1 \qquad (3)$$

where c_1, c_2 and $c_{1,2}$ are the molar concentrations in the solution phase of surfactant 1, surfactant 2 and their mixture, respectively, at the mole fraction α_1 of surfactant 1, required to produce a given surface tension (γ) value (see [31]).

The iterative solution of Eq. (3) gives X_1. The molecular interaction parameter in the mixed monolayer (β_{mon}) can be calculated according to the following equation:

$$\beta_{mon} = \frac{\ln\left(\alpha_1 c_{1,2} / c_1 X_1\right)}{\left(1 - X_1\right)^2} \qquad (4)$$

According to the RST, the deviation of experimentally obtained mixed micelle cmc value ($cmc_{1,2}$), from that calculated by assuming ideal mixing, can be represented by the molecular interaction parameter in the mixed micelle (β_{mic}). The molar fraction of the cationic surfactant in the mixed micelles (x_1) and $\beta_{mic,}$ can be calculated according the following equations:

$$\frac{x_1^2 \ln\left(\alpha_1 \mathrm{cmc}_{1,2} / \mathrm{cmc}_1 x_1\right)}{\left(1 - x_1\right)^2 \ln\left[\left(1 - \alpha_1\right) \mathrm{cmc}_{1,2} / (1 - x_1)\, \mathrm{cmc}_2\right]} = 1 \qquad (5)$$

$$\beta_{mic} = \frac{\ln\left(\alpha_1 \mathrm{cmc}_{1,2} / \mathrm{cmc}_1 x_1\right)}{\left(1 - x_1\right)^2} \qquad (6)$$

where cmc_1, cmc_2 and $cmc_{1,2}$ are the critical micelle concentrations of surfactant 1, surfactant 2 and their mixture, respectively, at the mole fraction α_1 [31].

The β parameters measure attractive net interaction between different surfactants relative to the self-interaction of the two surfactants under the same conditions before mixing. In other words, the β parameters describe the extent of non-ideal mixing. When parameter β is negative, the interaction is attractive; however, when it is positive, the interaction between two different surfactants is repulsive. To obtain valid β parameters, several conditions must be met as pointed out by Zhou and Rosen [31].

By applying the RST, additional parameters can be calculated for catanionic systems, such as activity coefficients and free energy of mixing, but for the sake of brevity, only the main principles of the theory are mentioned. A detailed discussion of the RST is far beyond the scope of this chapter.

3. Phase behavior and physicochemical properties of catanionic mixtures containing structurally different surfactants

3.1. Catanionic mixtures of oligomeric and monomeric surfactants

Surfactants, which have attracted considerable interest in last three decades, are oligomeric surfactants. These compounds are made up of two (dimeric or gemini surfactants) or more (higher oligomeric surfactants) amphiphilic moieties covalently linked at the level of the headgroups, or very close to them, by a spacer group (**Figure 7**) [33, 34]. Large interest for the investigation and synthesis of oligomeric surfactants is a consequence of their superior properties in comparison to the conventional ones [33, 34]:

(1) their cmcs are one or two orders of magnitude lower than for the corresponding monomeric surfactants,

(2) they are more efficient in lowering surface tension,

(3) their aqueous solution can have a very high viscosity or even show viscoelastic properties at relatively low surfactant concentrations, whereas the solutions of corresponding monomers remain low viscous as water,

(4) also, they have better solubilizing, wetting and foaming properties and

(5) the increase of the number of alkyl chains within oligomeric series, that is, the degree of oligomerization, enhances the above characteristics although the changes are becoming less significant with increase of degree of oligomerization above (2) [33, 34].

Due to their enhanced properties, catanionic mixtures containing various dimeric surfactants have been subject of numerous papers [25, 35–46]. The most investigated are catanionic mixtures containing bis-quaternary ammonium salts with the alkyl spacers. This type of surfactants is usually denoted as *m–s–m* where *m* represents the number of carbon atoms in the hydrophobic chain and *s* is the number of carbon atoms in the spacer (**Figure 6**). The great

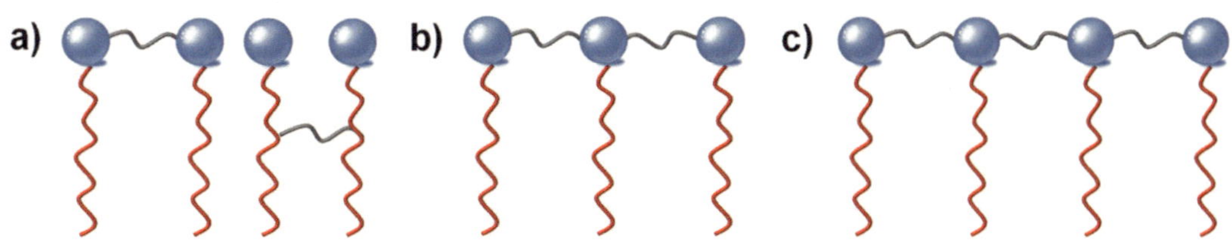

Figure 7. Schematic representations of (a) dimeric, (b) trimeric and (c) tetrameric surfactant molecule.

advantages of *m–s–m* surfactants are relative ease of their synthesis and possibility to tailor surfactant properties by changing spacer and chain length. Despite numerous papers dealing with catanionic mixtures containing dimeric surfactants, very few provide complete picture of the systems phase behavior.

Shang et al. determined the phase diagram for aqueous mixtures of 12–3–12 and SDS using freeze etching and negative staining on transmission electron microscope (TEM) [25]. Constructed phase diagram shows different phase regions in the majority of which coexistence of vesicles and micelles was found. As expected, the ratio of vesicles to micelles in diluted mixtures varies with bulk composition and total surfactant concentration. At higher surfactant concentrations, 12–3–12/SDS mixtures displayed very rich phase behavior, that is, regions of anisotropic phase, aqueous two-phase system (ATPS) as well as rod-like micelles and cylindrical clusters were detected. In order to corroborate experimental results, the authors used dissipative particle dynamics simulations. It was found that due to the finite size of the simulation box, results were somewhat different from that obtained by experiments [25].

Our group employed a variety of techniques: imaging by various microscopy techniques (light microscopy, confocal laser scanning microscopy (CLSM) and TEM) as well as dynamic (DLS) and electrophoretic light scattering (ELS) to determine phase diagram for 12–2–12/SDS system at the water-rich corner [35]. It was found that depending on bulk composition and total surfactant concentration in 12–2–12/SDS mixtures, various mixed nano and microaggregates form. The sequence of phases in the clear region in the SDS-rich side of the phase diagram is vesicles → the narrow coexistence region of vesicles and mixed micelles at the cmc_{SDS} → small mixed micelles. On the other hand, in the clear region in the 12–2–12-rich side, the sequence of phases is fragments of planar bilayers/lamellar sheets and vesicles → worm-like mixed micelles → transformation from worm-like to small mixed micelles above the $cmc_{12\text{–}2\text{–}12}$. In the precipitation region, two types of aggregates were detected, the tubules as prevailing aggregates on the 12–2–12-rich side and vesicles as prevailing aggregates on the SDS-rich side. The formation of tubules was ascribed to mutual influence of (1) specific molecular structure of 12–2–12 surfactant and (2) electrostatic interactions at the catanionic bilayer/solution interface. The microscopic observations indicated that tubular structures grow from rolled-up stacked catanionic bilayers [35].

Cheon et al. studied phase behavior in a very similar catanionic system, 12–2–12/sodium lauryl ether sulfate (SLES, **Figure 5**) by means of differential scanning calorimetry (DSC), UV-VIS spectroscopy, DLS, ELS and TEM [36]. These mixtures display less complex phase behavior compared to the 12–2–12/SDS system. In the phase diagram, isotropic molecular solution region, region of mixed micelles and vesicles formation as well as region of their coexistence were detected. Spontaneous vesicles formation has been attributed to electrostatic attractions and geometric packing constraints, that is formation of ion pair with "cuplike" structure that favors bilayer formation [36].

The phase behavior of 12–10–12/SDS system in diluted SDS-rich region using Langmuir trough, isothermal titration microcalorimetry (ITC), cryo-TEM and conductivity measurements has been investigated by Bai et al. [37]. The phase diagram shows three regions with a single type of aggregate (spherical and non-spherical micelles and vesicles), separated by two regions where two types of aggregates coexist (spherical/non-spherical micelles and non-spherical micelles/vesicles) and finally one multiphase region. Authors have concluded that observed phase transitions are consequences of asymmetric and uneven distributions of oppositely charged surfactants in vesicles' bilayers and non-spherical micelles, respectively [37].

Wang et al. determined phase diagram for 12–6–12/SDS system at the water-rich corner by employing turbidity measurements, ITC and TEM [38]. At constant total surfactant concentration, as the molar fraction of SDS increased, the morphology of mixed aggregates gradually changed from 12–6–12-rich micelles, through multiphase regions containing a precipitate (catanionic surfactant) and vesicle region, to SDS-rich micelles. Both TEM and ITC allowed identification of stable vesicles' region in the SDS-rich side of the phase diagram. Authors have argued that spontaneously formed vesicles in investigated mixture are consequences of (1) non-ideal mixing of cationic and anionic surfactant in bilayers as well as (2) a mechanism which involves an entropic stabilization in cases where the spontaneous curvature is not favorable but the bending penalty is not too high (soft bilayers) [38].

The same group of authors investigated monolayers formed in mixtures of *m*–2–*m* ($m = 12$, 14, 16 and 18) surfactants with SDS using the Langmuir trough technique [39], as well as micellization in mixtures of 12–*s*–12 ($s = 2$, 6 and 10) with several common anionic surfactants (SDS, sodium taurodeoxycholate (NaTDC, **Figure 8**) and sodium dodecanoate (SD, **Figure 5**)) by conductivity [40].

In *m*–2–*m*/SDS systems, it was found, from pressure-area, pressure-temperature and compression-expansion curves, that all the equivalent mixtures form highly stable monolayers with rich phase behavior and different desorption mechanisms [39]. Furthermore, it was established that if excess of cationic dimeric surfactants is present in 12–2–12/SDS and 14–2–14/SDS mixtures, the molecules in excess desorb from the monolayer so that the electroneutral composition of adsorbed film is maintained.

Results obtained by conductivity method revealed that all investigated systems containing 12–*s*–12 dimers and anionic surfactants show synergistic effects and have negative values of the molecular interaction parameter [40]. For the mixtures with 12–2–12, the strength of interaction increases in the order SD > SDS > NaTDC, while for 12–6–12, the order was SD ≈ SDS > NaTDC. Additionally, for the same anionic surfactant, the interaction with 12–2–12 is always stronger than that with 12–6–12. It is known that short spacers ($s < 10$) tend to lie flat at the water-hydrocarbon interface, which can lead to unfavorable packing constraints at the mixed micelles. Since 12–2–12 has a shorter spacer than 12–6–12, the packing constraints are slightly weaker and, together with a higher charge density of the headgroup region, lead to more favorable attractive interactions with anionic surfactants. Results reported for mixtures with 12–10–12 suggest that a catanionic solid is largely stabilized compared to mixed micelles when the alkyl spacer is long and flexible enough [40].

Aggregation behavior in mixtures of cationic dimeric surfactants derived from dodecyltrimethylammonium chloride ($C_{12}TACl$, **Figure 4**) and SDS by means of small-angle neutron

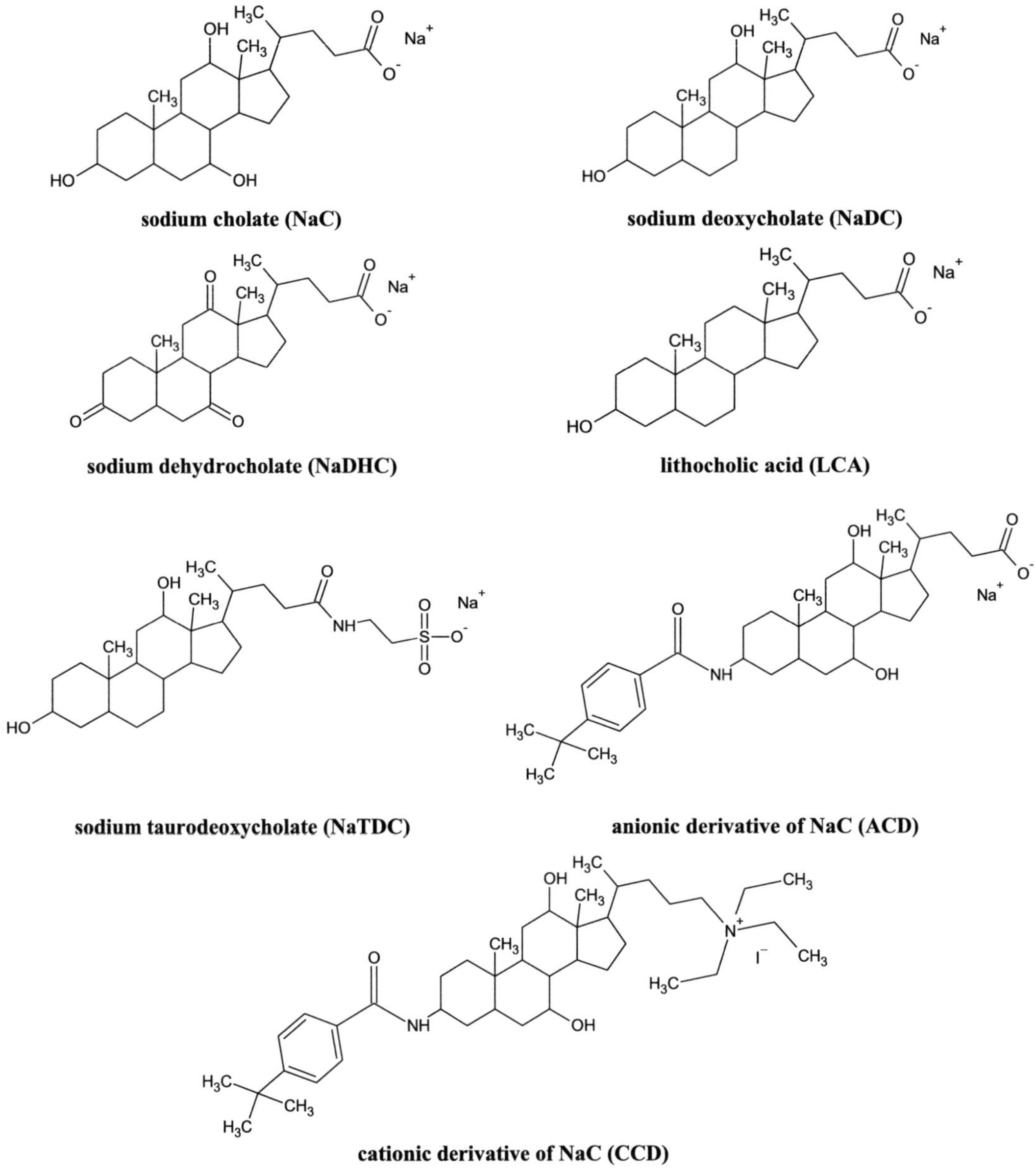

Figure 8. Molecular structures of bile salts.

scattering (SANS) and small-angle X-ray scattering (SAXS) was investigated by Prévost et al. [41]. Dimeric surfactants with spacers of different nature and geometry were used: *m*-, *p*- and *o*-xylylene (aromatic spacer), diethyl ether (ethoxy spacer) and trans-1,4-buten-2-ylene (ethylene spacer) (**Figure 6**). Authors concluded that among five spacers, due to their weak geometrical constraints and the ambivalent, hydrophilic and non-extensive lipophobic nature, ethoxy spacer is the most suitable for formation of vesicles in aqueous mixtures. On the contrary, the aromatic spacers with their low flexibility and higher apolarity, compared to ethoxy spacer, generally led to precipitation in investigated mixtures. Furthermore, it was established that all mentioned dimeric surfactants form colloidally more stable mixtures with SDS than their monomeric counterpart C_{12}TACl [41].

Ji et al. studied temperature induced phase transitions in aqueous mixtures of cationic dimeric surfactant, 1,4-bis(dodecyl-*N*,*N*-dimethylammonium bromide)-2,3-butane-diol ($C_{12}C_4(OH)_2C_{12}Br_2$, **Figure 6**), and anionic amino acid surfactant, *N*-dodecanoylglutamic acid (C_{12}Glu, **Figure 5**) at pH = 10.0 [42]. At 25 °C small spherical micelles, vesicles and entangled worm-like micelles were detected in the system. The main controlling factor for the aggregates transition at constant total surfactant concentration and varying molar ratio is strong electrostatic binding between oppositely charged surfactants which significantly reduces the headgroup area. Because both $C_{12}C_4(OH)_2C_{12}Br_2$ and C_{12}Glu carry two charges, strong electrostatic interactions in these mixtures are not surprising. At higher temperatures, mixed aggregates formed at 25 °C experience different transitions, that is, the following phase transitions occur: (1) small spherical micelles → large vesicles, (2) large vesicles → solid spherical aggregates → larger irregular aggregates and (3) entangled worm-like micelles → branched worm-like micelles. The larger irregular aggregates and branched micelles ultimately lead to precipitation and clouding phenomenon, respectively. All described transitions are thermally reversible and transition temperatures can be tuned by varying the molar ratio and/or the total surfactant concentration [42].

Aghdastinat et al. investigated self-assembly in cation-rich mixtures of ester-containing cationic dimeric surfactants, named dodecyl esterquat and dodecyl betainate (**Figure 6**), with SDS in the presence of salt, KCl [43]. Obtained results show that the position of ester bonds in surfactants' tail plays an important role in physicochemical properties and aggregation behavior in their mixtures with SDS. After mixing with SDS morphology of dodecyl esterquat, aggregates change from cubic nanoparticles (cubosomes) to cylindrical nanoparticles which coexist with cubosomes. On the contrary, upon mixing with SDS, no significant structural change can be observed in dodecyl betainate aggregates, that is, vesicles are formed in both cases. Authors explained observed changes in morphology of mixed aggregates using RST [43].

Investigation of higher oligomeric surfactants and their mixtures is hindered by the more complex synthesis and purification compared to the dimeric molecules [33, 34]. Very few reports can be found on mixtures containing trimeric or tetrameric quaternary ammonium surfactants [47–49].

Chen et al. studied self-assembly in mixtures of trimeric cationic surfactants, tri-(*N*-dodecyldimethylhydroxypropylammonium chloride) phosphate (PTA, **Figure 9**) and double-tailed anionic surfactant, bis(2-ethylhexyl) sulfosuccinate (AOT, **Figure 5**) by means of DLS and TEM [47]. Obtained results demonstrated that PTA/AOT vesicles are stable and can be found in a broad concentration range. The TEM micrographs revealed that at high surfactant concentrations, tubular microstructures, vesicle fusion and vesicle-tubular microstructure transition occurred. In addition, it was found that formation of tubular structures is more pronounced in aged samples. Authors have discussed the mechanism of vesicles and tubules formation from the viewpoint of molecular geometry and electrostatic interaction between oppositely charged surfactants [47].

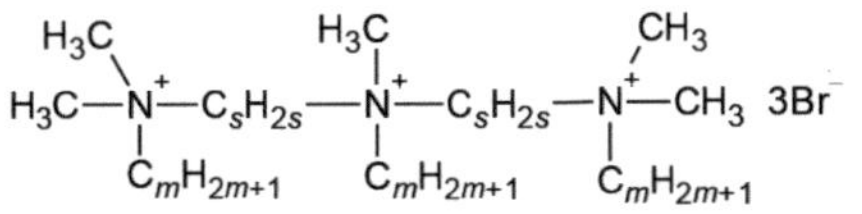

trimeric *m–s–m–s–m* surfactants

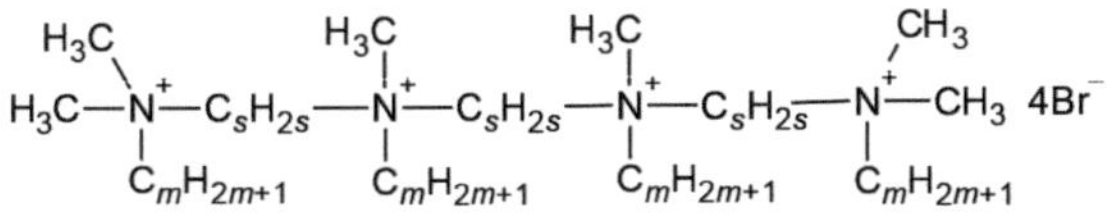

tetrameric *m–s–m–s–m–s–m* surfactants

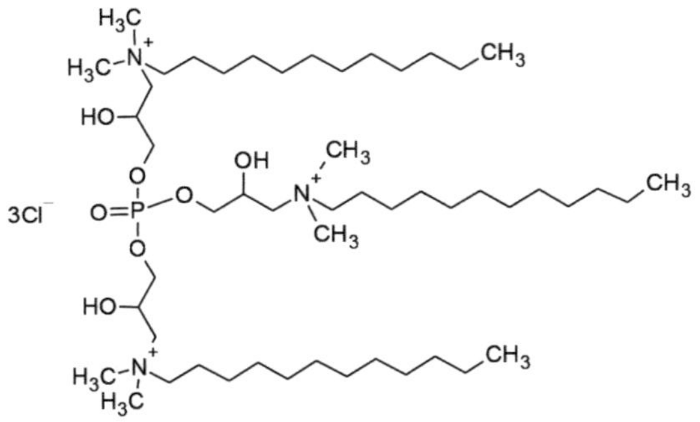

tri-(*N*-dodecyldimethylhydroxypropylammonium chloride) phosphate (PTA)

Figure 9. Molecular structures of higher oligomeric surfactants—quaternary alkyl ammonium salts. *m* = number of C atoms in alkyl chains and *s* = number of C atoms in spacer.

Yoshimura et al. investigated mixtures of trimeric cationic surfactants, *m*–2–*m*–2–*m* (*m* = 8, 10 and 12, **Figure 9**), and sodium *n*-octyl sulfate (SOS, **Figure 5**), employing several techniques such as static surface tension, fluorescence spectroscopy and DLS [48]. As expected, *m*–2–*m*–2–*m*/SOS mixtures show stronger micellization ability and lower cmc values compared with pure trimeric surfactants. In addition, the chain length of trimeric surfactants significantly influenced mixtures' properties at the air/water interface and in the solution. For example, *m*–2–*m*–2–*m*/SOS mixtures show a linear decrease in the cmc values with increasing alkyl chain length. Furthermore, the 8–2–8–2–8/SOS system exhibited a smaller surface area occupied by a surfactant molecule (a_{min}) compared to 10–2–10–2–10/SOS and 12–2–12–2–12/SOS mixtures [48].

Our group studied a series of quaternary ammonium bromide oligomers (from dimer to tetramer, **Figures 6** and **9**) with dodecyl chains connected at the level of headgroups by a short ethylene spacer and their mixtures with SDS [49]. In high excess of SDS (cationic surfactant: SDS = 1:9), negatively charged vesicles form in all mixtures regardless of the number of dodecyl chains in cationic surfactant. Contrary to vesicles, the mixed monolayer is enriched with cationic surfactant. Moreover, the increase in the number of dodecyl chains decreases the molar fraction of SDS in the mixed monolayer. Observed results can be explained by strong electrostatic headgroup interactions modulated by packing constraints imposed by the geometry of oligomeric surfactants [49].

Since anionic dimeric surfactants attract much less attention than cationic, there are only few reports describing catanionic mixtures containing dimeric surfactants as anionic components [44–46]. Back in 1996, Zana et al. investigated mixtures of disodium1,11-didecyl-3,6,9-trioxaundecane-1, 11-disulfate (**Figure 5**) and C_{16}TAB, in the presence of NaBr, employing conductivity, spectrofluorometry, time-resolved fluorescence quenching and cryo-TEM [44]. Obtained results proved that the aggregation numbers of the mixed micelles are larger than those of pure C_{16}TAB micelles even at very low molar ratio of dimeric surfactants. Apart from micelles, TEM micrographs revealed the presence of vesicles and very large aggregates which looked like distorted multi-bilayered vesicles with many defects [44].

Luo et al. and Zhao et al. studied interactions in catanionic mixtures containing anionic dimeric surfactant: O,O-bis(sodium 2-lauricate)-*p*-benzenediol (C_{11p}PHCNa, **Figure 5**) [45, 46]. It was found that large spherical aggregates form in C_{12}TAB/C_{11p}PHCNa mixtures and transform into branched and worm-like micelles with increasing NaBr concentration. In addition, authors established that due to the changes in morphology of mixed aggregates, viscosity of the C_{12}TAB/C_{11p}PHCNa mixtures gradually increases. Furthermore, it was reported that adsorption behavior in mixtures of two dimeric surfactants, C_{11p}PHCNa and (oligoona) alkanediyl-α,ω-bis(dimethyldodecylammonium bromide) (C_{12}-2-E_x-C_{12}, **Figure 6**), strongly depends on the molar ratio, that is, strong adsorption at the air/water interface is present in excess of cationic surfactant while in excess of anionic surfactant premicellization occurs.

In addition to properties which can be found in traditional catanionic systems, common features that can be drawn for catanionic mixtures of oligomeric and monomeric surfactants are:

(1) both physicochemical properties and phase behavior strongly depend on the length of the spacer (*s*) and alkyl chains (*m*) as well as the nature of the spacer (aromatic, hydrophilic, hydrophobic, etc.) in the oligomeric molecule,

(2) in general, all mixtures exhibit a strong synergistic effect which is manifested as high surface activity, enhanced adsorption and low cmc values but usually to a lesser extent than traditional mixtures of corresponding monomeric surfactants,

(3) due to high asymmetry between surfactant molecules in mixtures, in terms of different alkyl chains number or length, vesicle phase is often found in a very broad concentration range. Although this phenomenon was observed in mixtures of monomeric surfactants as well, it is more pronounced in oligomeric/monomeric surfactant mixtures due to increased asymmetry in molecular structures of components and

(4) in mixtures with dimeric surfactants, very often, elongated (cylindrical) and tubular aggregates can be found which were ascribed to a specific structure of dimeric molecules and electrostatic interactions in the catanionic micelles and bilayers. For example, it is well known that pure 12–2–12 in solutions forms structures with relatively less curvature, such as cylindrical and elongated worm-like micelles even at very low concentrations.

3.2. Catanionic mixtures of surface active ionic liquids (SAILs) and surfactants

In recent years, surface active ionic liquids (SAILs) have emerged as fascinating compounds due to their dual nature as well as unique and tunable physicochemical properties [50]. With combined properties of ionic liquids (ILs) and amphiphiles, SAILs represent a novel class of surfactants. The term ionic liquids refers to a class of substances formed by a poorly coordinated large organic cation with delocalized charge and either a small anion, such as Br^-, or relatively large one, such as $[(CF_3SO_2)N_2]^-$ [51]. Consequently, ILs possess melting points under 100 °C, often even lower than room temperature. Due to their unique characteristics such as high thermal stability, negligible vapor pressure, high conductivity and great ability to dissolve inorganic/organic compounds, ILs have attracted much interest for a variety of applications [52, 53].

Among different classes of SAILs, imidazolium-based compounds composed of the 1-alkyl-3-methylimidazolium cation ($[C_n mim]^+$, where n = number of carbon atoms in the hydrophobic chain, **Figure 10**) and their mixtures have been most extensively studied. Compared to the conventional alkyltrimethylammonium surfactants, imidazolium-based SAILs exhibit a stronger tendency to self-assemble and slightly better surface activity [50, 54]. Imidazol ring, which can be found, for example, in amino acid histidine, makes them also biologically interesting. Furthermore, it is known that SAILs exhibit low toxicity so their use as drug-delivery agents can represent a step forward in medicinal chemistry [50]. Recently, Sharma and Mahajan published a comprehensive review summarizing influence of various additives, including surfactants, on the physicochemical properties of imidazolium-based ILs [50]. Due to the large amount of published data, the aim of this section is to discuss only aqueous catanionic mixtures containing SAIL, although numerous reports of systems in which ILs acted as self-assembly media exist.

Zhao et al. reported the phase diagram of catanionic system composed by cationic SAILs, $[C_{16}mim]Cl$ and SDS [55]. Results from rheology and polarized optical microscopy observations demonstrated that a gel phase with quite high water content is formed in the $[C_{16}mim]Cl$-rich side of the phase diagram. On the contrary, in the SDS-rich side, lamellar phases were detected. The $[C_{16}mim]Cl$/SDS gel phase showed low ordering and similar rheological properties to vesicles usually formed in traditional catanionic systems [55].

Formation of gel phase was observed in a very similar system, $[C_{14}mim]Cl$/SDS mixtures, by Zhao et al. as well [56]. The SEM micrographs showed that gel phase is structured as a complex three-dimensional network. Authors argued that hydrophobic and electrostatic interactions present in the system are essential for gel-phase formation. In order to prove this thesis, mixtures in which $[C_{14}mim]Cl$ was replaced with $[C_4mim]Cl$ and SDS with SOS were also studied. In both of these systems, gel phase was not found. Performed control experiments demonstrated a key role of hydrophobic interactions in gel formation. In additional control experiments, 1-dodecanol was used instead of SDS to confirm the crucial role of electrostatic interactions in gel formation. In this case, the gel phase was also not found [56].

1-alkyl-3-methylimidazolium cation ($[C_n mim]^+$)

1-alkyl-1-methylpyrrolidinium cation ($[C_n MP]^+$)

Figure 10. Molecular structures of cations in surface active ionic liquids (SAILs). n = number of C atoms in alkyl chains.

The group of the same author also studied phase behavior in mixtures of *N*-dodecyl-*N*-methylpyrrolidinium bromide ([C_{12}MP]Br, **Figure 10**) and SDS by employing TEM, conductivity and rheological measurements [57]. It was found that at constant [C_{12}MP]Br concentration, as the molar fraction of SDS increases, the morphology of mixed aggregates changes as follows: mixed micelles → vesicles → coexistence of catanionic precipitate and vesicles → coexistence of catanionic precipitate and mixed micelles. Spontaneous vesicles formation was discussed in terms of packing parameter [57].

Micelle-to-vesicle transition induced by β-cyclodextrin (β-CD) in mixtures of [C_{16}mim]Cl and sodium oleate (NaOle, **Figure 5**) has been investigated by Dai et al. [58]. Cyclodextrins are structurally related cyclic oligosaccharides formed during bacterial digestion of cellulose [59]. It is known that the interior environment of β-CD is hydrophobic and its outer surface is hydrophilic. Authors established that in [C_{16}mim]Cl-rich mixtures, micelle-to-vesicle transition can be triggered by addition of sufficiently high β-CD concentration. The main factor governing this phase transition is formation of inclusion complexes in the system [58].

Chabba et al. employed various techniques such as tensiometry, steady-state fluorescence, DLS and SANS to study interactions of cationic SAILs, [$C_n$mim]Cl (n = 8, 10 and 12), with sodium dodecylbenzenesulfonate (SDBS, **Figure 5**) [60]. Results obtained by tensiometric and steady-state fluorescence measurements revealed strong synergism in the system. As well as in the classic catanionic mixtures, strong synergism between the cationic SAIL and anionic SDBS can be attributed to the strong electrostatic interaction between oppositely charged headgroups along with the hydrophobic interactions between the alkyl chains. However, authors argued that in addition to these forces, π–π and cation–π interactions between the imidazolium cation and the benzene ring of SDBS, as well as hydrogen bonding, between most acidic proton of imidazolium ring and sulfonate group of SDBS, also come into play. In general, ionic liquid cations frequently contain multiple donor sites able to participate in hydrogen bonding, resulting in H-bonds of varying strength and type [61]. Within an imidazolium cation, the H-bond donor is the C—H unit, the C2—H proton being the most acidic, followed by the other two hydrogens on the aromatic ring (C4—H and C5—H) and alkyl chain methyl hydrogens. The H atoms on imidazolium ring all participate in the formation of H-bonds with water molecules and in the ubiquitous H-bonds among the highly hydrated imidazolium cations even in very diluted IL solutions [62]. Similar increase in synergistic behavior due to additional non-covalent interaction was found in [C_{12}mim]Br/AOT, [C_{12}mim]Cl/ibuprofen (**Figure 11**) and [C_8mim]Br/SDBS mixtures [16, 63, 64]. Furthermore, as observed from DLS and SANS, the [$C_n$mim]Cl/SDBS mixtures exhibit micelle-to-vesicle transition dependent on the alkyl chain length and the molar ratio of the surfactants. It was found that vesicle region prevails in a broad range of concentrations and that mixtures show high stability towards precipitation [60].

Gehlot et al. studied mixtures of [C_8mim]Br and SDBS by employing a variety of techniques such as tensiometry, conductometry, UV-VIS spectroscopy, cryo-TEM, AFM, DLS, ELS, ITC, steady-state fluorescence and ^{1}H NMR measurements [64]. Based on the results obtained from these various physicochemical and imaging techniques, authors have concluded that: (1) spontaneously formed and differently shaped [C_8mim]Br/SDBS vesicles (sphere, tubes and ribbons) exist in a broad range of concentrations, (2) a negative value of interaction parameter and lower experimental cmc values, compared to the theoretically determined values,

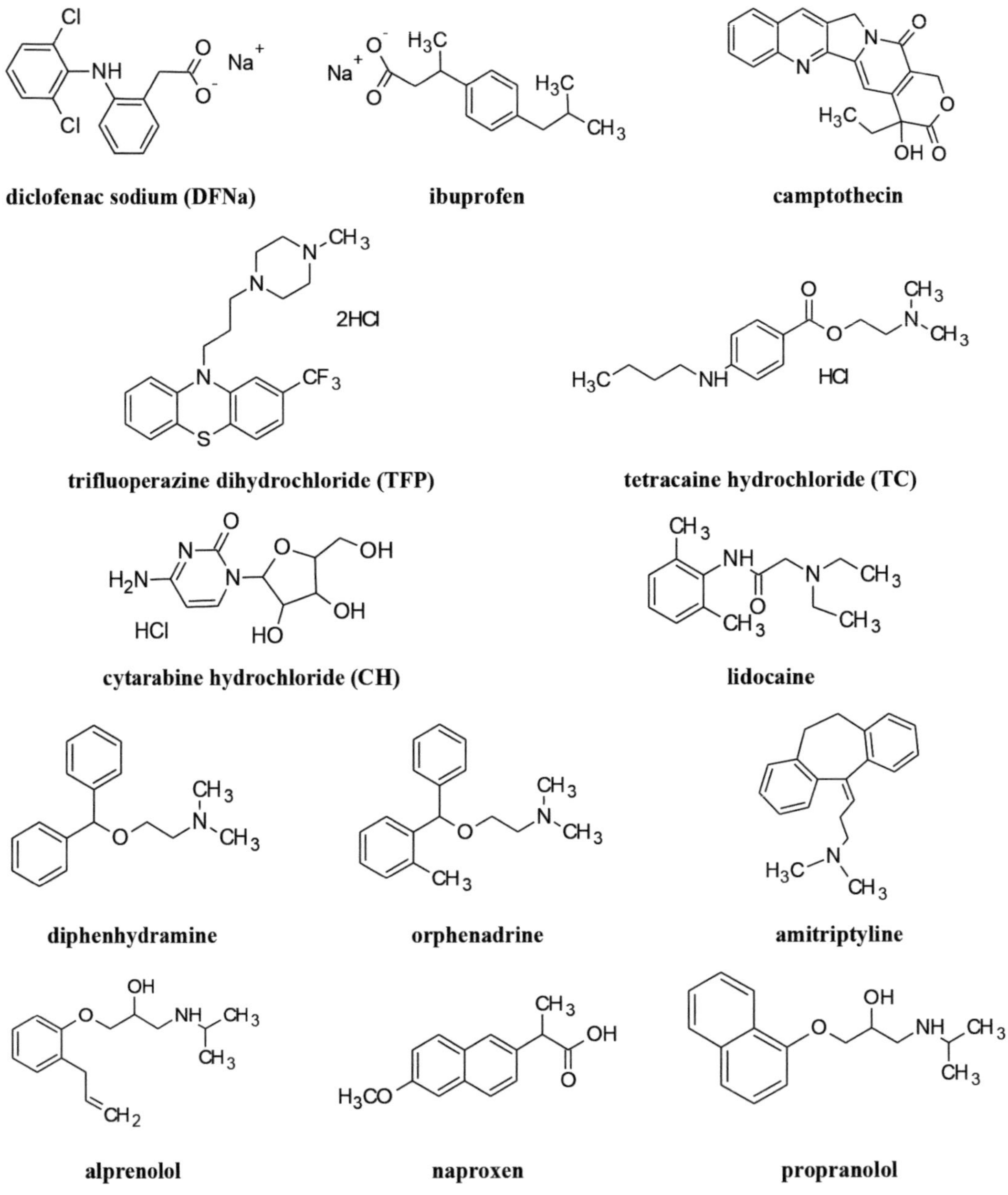

Figure 11. Molecular structures of amphiphilic drugs.

indicate high synergism in the system, (3) major forces responsible for synergism are electrostatic and hydrophobic interactions as well as π-π stacking of aromatic rings and (4) Br^- as a counterion in palisade layer assists in compact packing of ions, which leads to the formation of vesicles [64].

Our group studied phase transitions in mixtures of [C_{12}mim]Br and AOT using a multi-technique approach [16]. Depending on the bulk composition and total surfactant concentration, mixed micelles, coacervates, lamellar and inverse bicontinuous cubic liquid crystalline phase were observed. At stoichiometric conditions, coexistence of coacervates and vesicles was found at lower and bicontinuous cubic phase and vesicles at higher total surfactant concentrations. A mechanism was proposed in which phase transitions from a dispersed lamellar to inverse cubic bicontinuous phase occur as a consequence of charge shielding and closer packing of oppositely charged headgroups followed by a change in bilayers curvature. Additionally, along with electrostatic attractions and geometric packing constraints, additional non-covalent interactions in the system (hydrogen bonding, π-π stacking) enhanced attractive interactions and stabilized low curvature aggregates [16]. In the ternary diagram of the system similar to [C_{12}mim]/AOT, 1-butyl-3-methylimidazolium tetrafluoborate ([bmim]BF_4)/AOT, different phase behavior was observed. In the water-rich corner of the phase diagram, regions of isotropic fluid and lamellar phase were found [65]. In addition, substitution of BF_4 anion with Br^- causes the collapse of lamellar phase [66]. Murgia and co-workers employed a variety of techniques such as conductivity, optical microscopy, SAXS and NMR self-diffusion experiments to detect modifications on macro- and micro-scales within the systems upon the substitution of ILs counterion BF_4 with Br^-. Thus, the remarkable differences observed between the two systems appear to be mainly due to a specific counterion effect [65, 66].

Singh et al. reported structural changes induced by composition and dilution in aqueous catanionic mixtures containing SAILs ([C_{12}mim]Br and [C_{14}mim]Br) and a drug, diclofenac sodium (DFNa, **Figure 11**), as the anionic component [29]. The observed phase transitions were probed by SANS, DLS and ELS. The SAIL/DFNa systems display rich phase behavior and structural diversity of mixed aggregates, that is, depending on the bulk composition and total surfactant concentration, spherical and small micelles with prolate ellipsoidal shape as well as rod-shaped micelles and vesicles were detected. The ^{1}H NMR measurements revealed that DFNa intercalated into SAIL micelles via cation-π and π-π interaction in addition to hydrophobic interaction. It was found that increase in DFNa molar ratio increases aggregates curvature. Unlike conventional linear chain surfactants, a specific structure of DFNa does not allow effective packing of cationic-anionic pairs which prevented precipitation in equimolar mixtures [29].

Catanionic systems also containing a drug, ibuprofen and [C_{12}mim]Cl have been investigated by Sanan et al. [63]. Various techniques such as surface tension, steady-state fluorescence, UV-VIS spectroscopy, DLS and ^{1}H NMR measurements were used to provide a comprehensive knowledge about [C_{12}mim]Cl-ibuprofen interactions. The interactions between the SAIL and drug molecules are found to be highly synergistic both in the mixed micelles and in the mixed monolayer. The formation of highly surface active catanionic complexes of 1:1 stoichiometry ([C_{12}mim]Cl^+ibuprofen$^-$), stabilized largely by a combination of electrostatic, hydrophobic, cation-π and π-π interactions, was established through spectroscopic investigations. Depending on the bulk composition and total surfactant concentration of mixed micelles, unilamellar and multi-lamellar vesicles were detected in the system [63].

Vashishat et al. investigated mixtures of bile salts (sodium cholate (NaC) and sodium deoxycholate ((NaDC), **Figure 8**) with [C_{12}mim]Br [67]. From a biochemical point of view, bile salts play a vital role in many physiological processes, and more will be said about their catanionic mixtures in the next section. In order to obtain detailed information about interactions between [C_{12}mim]Br and bile salts in the mixed monolayer and in the mixed micelles, surface tension and steady-state fluorescence measurements were conducted. Various micellar and interfacial parameters, including cmc, β_{mic}, a_{min}, surface excess concentration (π_{max}) and surface pressure at cmc (π_{cmc}), were estimated. It was found that investigated mixtures exhibit pronounced synergism in mixed monolayer formation as well as micellization. Due to the more hydrophobic nature of NaDC, which allows its molecules to get deeply intercalated in the mixed micelles compared to NaC, mixture with NaDC showed stronger synergistic effect. In addition, this study aimed to determine the solubilization capacity of the poorly soluble drug, phenothiazine, in micellar media. It was found that solubility of phenothiazine is dependent on the hydrophobicity and the size of the micelles, with solubility increasing in the order: NaC < NaDC < [C_{12}mim]Br/NaC < [C_{12}mim]Br/NaDC [67].

Along with the properties of traditional catanionic mixtures, in systems containing SAIL, the following characteristics can be found:

(1) variation in the alkyl chain length of both SAIL (*n*) and surfactant (*m*) causes a significant change in the physicochemical properties and phase behavior of the systems,

(2) the magnitude of SAIL-surfactant interactions is larger for surfactants with aromatic moiety in their structure, for example, SDBS. The reason behind this is the involvement of cation-π and π-π interactions due to the π-electron cloud of the benzene ring in SDBS and the imidazolium ring in [$C_n$mim]$^+$ cation,

(3) along with electrostatic and hydrophobic interactions, additional non-covalent interactions (hydrogen bonding, π-π stacking) (1) enhance attractive interactions and (2) increase synergistic effect as well as (3) stabilize low curvature structures in SAIL-surfactant systems and

(4) apart from self-assembled aggregates commonly found in catanionic mixtures, formation of gel phase was observed in some SAIL-surfactant systems.

3.3. Catanionic mixtures of biologically active molecules and surfactants

Due to their important roles in various physiological processes and pharmaceutical applications amphiphilic biologically active molecules, as well as their catanionic mixtures, have been the subject of numerous papers. Historically, the most-studied catanionic mixtures of this type as biologically active molecules contain either (1) amphiphilic drug or (2) bile salt. Therefore, aqueous catanionic mixtures of such molecules are mostly summarized and discussed in this section.

In addition to [$C_n$mim]Br/DFNa and [C_{12}mim]Cl/ibuprofen systems [29, 63], mentioned in the previous section, Mahajan's group recently reported on: (1) interactions between the cationic drug, trifluoperazine dihydrochloride (TFP, **Figure 11**) and anionic surfactants, SDS and AOT [68] as well as (2) interactions prevailing in catanionic mixtures containing cationic drug, tetracaine hydrochloride (TC, **Figure 11**) and anionic surfactants, SDBS and sodium lauroyl sarcosinate (SLS, **Figure 5**) [69]. In all investigated mixtures, various micellar and interfacial parameters were determined from surface tension measurements and by applying the RST. Obtained results revealed strong synergism in systems' bulk and surface properties such as high surface activity and low cmcs. As expected, it was established that TC interacts more strongly with SDBS and that TC-SDBS complex possesses higher binding constant, as compared to TC/SLS mixture, due to the additional non-covalent interactions in the system (π-π stacking). Furthermore, Jiang et al. investigated aggregation behavior of vesicles formed by TC and AOT using conductivity, turbidity measurements, TEM, DLS and ELS [70]. The TC/AOT aggregates exhibited different morphology, charge properties, interaction enthalpies and drug release behaviors depending on the mixtures' bulk composition. Obtained drug release profiles indicated that investigated drug-containing vesicles have promising applications in drug delivery systems.

Zhao et al. determined various physicochemical parameters (cmc, Γ_{max}, a_{min}, surface tension at the cmc (γ_{cmc}), degree of counterion binding, etc.) from the surface tension and electrical conductivity measurements in mixtures of DFNa and DDAB [71]. The cmc and γ_{cmc} of mixed DDAB/DFNa systems were found to have values between that of pure DFNa and DDAB solutions. In addition, (1) the in vitro release results demonstrated that DDAB/DFNa vesicles exhibit good sustained drug release properties while (2) the hemolytic toxicity studies show that vesicles in mixtures with high DFNa molar ratio are safe for intravenous administration within the effective concentration [71].

Bile salts are well known and important biologically active surfactants, produced in the liver from cholesterol, which play an important role in emulsification of lipids, fats, fat soluble vitamins, etc. [67]. Due to their great importance in metabolism of insoluble molecules, such as phospholipids and monoglycerides, they have been extensively studied. They possess a unique molecular structure when compared with typical surfactant molecules, like the steroids, they have a nucleus composed of four fused rings, three cyclohexane rings and one cyclopentane ring as well as hydrophilic hydroxyl groups (**Figure 8**). Bile acids are favorable compounds for construction of supramolecular structures because of their biocompatibility, high structural rigidity, amphiphilicity and chirality [72].

Our group investigated 12–6–12/NaC and C_mTACl (m = 12, 14 and 16)/NaC mixtures employing a combination of techniques such as surface tension, conductometry, light microscopy, DLS and ELS [24, 73]. In all investigated systems, synergism in micellization and adsorption was observed. With increasing total surfactant concentration, in equivalent 12–6–12/NaC mixtures, morphology of mixed aggregates changes as follows: complexes → flexible cylindrical mixed micelles → coexistence of vesicles, coacervates and solid crystalline phase. In the high excess of cationic surfactant, the small 12–6–12 micelles are prevailing structures while with an increasing content of NaC, the long flexible mixed micelles are dominant. Obtained results demonstrated that interplay between (1) electrostatic effects, (2) geometry of molecules as well as (3) dissimilar separation of the hydrophobic and hydrophilic moieties in the surfactants dictates phase behavior of these systems [24].

The most interesting discovery in C_mTACl/NaC systems was that catanionic surfactants, precipitated in/or close to equimolar region, show a variety of morphologies including twisted ribbons and crystalline tubules, which are not commonly found in this kind of systems [73]. The three-dimensional structures that are yielded by the self-assembly of lipids and surfactants have recently drawn much research interest due to their applications in nanotechnology [74]. Formation of tubules in C_mTACl/NaC mixtures can be attributed to several factors: (1) chiral packing of NaC molecule in a bilayer, (2) strong attractive interactions between oppositely charged headgroups at the bilayer/solution interface and (3) hydrogen bonding at the bilayer surfaces, which enhance formation of multilayer sheets and their twisting and/or rolling up [73].

Long, fiber-like tubular structures instead of crystalline tubules were observed in the cationic-rich dilute region of DDAB/NaTDC system by Marques and Khan [75]. Authors suggested that formation of long tubular structures is a consequence of specific NaTDC's rigid ring-based structure with hydroxyl groups. Previously, Marques's group studied the phase behavior of the same catanionic pair but in the bile salt-rich area [26]. It was reported that the system displays coacervation instead of precipitation at equimolarity, consisting of a viscous isotropic solution in equilibrium with a very dilute solution. Formation of tubular structures was not detected in this part of the phase diagram.

Liu et al. reported that in lithocholic acid (LCA, **Figure 8**)/tetradecyltrimethylammonium hydroxide (C_{14}TAOH, **Figure 4**) system transition from vesicles to tubules was observed, while in mixtures of LCA with cetyltrimethylammonium hydroxide (C_{16}TAOH), transition from vesicles to helical ribbons occurred [72]. Thus, despite a difference of only two methylene groups in the alkyl chain of C_{14}TAOH and C_{16}TAOH, morphology of mixed aggregates in these two systems is largely different. In addition, it was found that time required for the phase transition depends on alkyl chain length as well. In the C_{14}TAOH/LCA systems, the transition from vesicles to tubules was completed within several hours, while in the C_{16}TAOH/LCA system, the vesicles were converted to helical ribbons after more than 4 days, depending on the concentration and temperature [72].

Motivated by their numerous potential applications in nanotechnology, Manghisi et al. prepared and characterized tubules in mixtures of anionic (ACD) and cationic (CCD) derivatives of NaC (**Figure 8**) [74]. It was found that charge of synthesized CCD/ACD tubules ranges from negative to positive values depending on the surfactant molar ratio in the mixtures. Analysis of the TEM micrographs revealed a correlation between the diameter and the composition of the tubules [74].

Bhattacharjee et al. investigated mixtures of cetylpyridinium chloride (C_{16}PC, **Figure 4**) and NaDC using DLS, SANS and SAXS [76]. It was shown that phase separation, i.e. coacervate phase, occurs near the equimolar composition at low surfactant concentrations and, contrary to expectations, disappears at higher concentrations. This associative phase separation has been explained on the basis of competition between electrostatic attraction and entropy of the components mixing. Additionally, based on the obtained results, authors suggested that structural features of bile salts are not favorable for formation of catanionic vesicles when combined with $C_{16}PC$. However, stable mixed micelles of widely differing morphologies were formed in a broad concentration range [76].

Fernández-Leyes et al. reported on physicochemical properties and phase behavior of DDAB/sodium dehydrocholate (NaDHC, **Figure 8**) and DDAB/NaDC mixtures using surface tension measurements, conductivity, DLS, ELS and TEM [77, 78]. The RST was applied for evaluating the non-ideal interactions between molecules in adsorbed monolayer and mixed micelles. All systems exhibited synergism in mixed monolayer formation as well as micellization. The obtained pC_{20} values, negative logarithms of the surfactant concentrations at which the surface tension of water is reduced by 20 mN m^{-1}, demonstrated that both mixed systems have analogous adsorption efficiencies, which are similar to the pure DDAB solutions and superior to that obtained for both bile salts. Nevertheless, difference in their adsorption effectiveness was observed: NaDC causes an increase of surface excess concentration, while NaDHC produces the opposite effect. The lower Γ_{max} values obtained for DDAB/NaDHC system are related to the deep penetration of the hydrophobic steroid backbone of NaDHC molecules that cause a great disturbance of DDAB hydrocarbon tails, that is larger a_{min} [77]. Furthermore, it was found that mixed aggregates in DDAB/NaDHC system are mainly composed of DDAB, regardless of the NaDHC solution molar fraction. Nevertheless, the gradual inclusion of NaDHC molecules leads to structural transformations in the system. The incorporation of NaDHC into DDAB bilayers had two effects: (1) the DHC^- and DDA^+ ions form ion pairs that are much less hydrated than separate ion headgroups, which consequently reduce the effective headgroup area and (2) the intercalation of the rigid ring-based structure of bile salts between DDAB chains causes an increase of chain repulsion due to steric effects [78].

Pereyra et al. analyzed $C_{16}TAB$/NaDHC system with two procedures: (1) the RST and (2) the EOMMM (Equation Oriented Mixed Micellization Modeling) [79]. Investigated system showed a non-ideal and asymmetric behavior with attractive interaction between the components, as reflected by the obtained interaction parameters. Moreover, it was established that the affinity of DHC^- ions for $C_{16}TAB$ micelles is stronger than that of $C_{16}TA^+$ ions for NaDHC ones [79].

Apart from the properties which can be found in traditional catanionic systems, common features of catanionic mixtures with biologically active molecules are:

(1) the rigid ring-based structure of amphiphilic drugs and especially, bile salts, as well as high asymmetry between surfactant molecules in mixtures, does not allow effective packing of cationic-anionic pairs and often prevents precipitation in equimolar mixtures. Instead, coacervates or vesicles can be found,

(2) for the same reason, synergism observed in these systems is less pronounced compared to traditional mixtures of monomeric surfactants,

(3) most catanionic mixtures containing bile salts revealed that apart from electrostatic interactions, the geometry and planar distribution of hydrophobic and hydrophilic properties of bile salts play a marked role in the construction of various mixed aggregates and

(4) frequent occurrence of tubules in catanionic mixtures containing bile salts can be attributed to several factors: (1) chiral packing of bile salt molecule in a bilayer, (2) strong attractive interactions between oppositely charged headgroups at the bilayer/solution interface and (3) hydrogen bonding at the bilayer surfaces, which enhance formation of multilayer sheets and their twisting and/or rolling up.

4. Applications of catanionic mixtures

In the past decades, a large number of systems for the controlled and targeted delivery of pharmaceutical compounds have been designed based on various self-assembled aggregates such as micelles, vesicles, liquid crystalline phases, tubules, etc. [80–82]. Catanionic systems, due to their rich phase behavior and numerous possibilities in mediating molecular self-assembly, by adjusting the mixing molar ratio and using appropriate geometry of surfactant molecules, offer considerable advantages in delivering biomolecules. For example, catanionic mixtures easily and spontaneously form vesicles at non-stoichiometric molar ratios. Vesicles are not only significant because they mimic biological membranes, but also due to their utility as drug carriers and targeted drug delivery systems.

As already mentioned, precipitation is considered to be the main drawback for application of catanionic mixtures [3]. However, as it can be seen from preceding sections, this drawback can be circumvented by using surfactants of largely different molecular structures. In addition, biologically active molecules, such as amphiphilic drugs, can be used as one of the mixtures' components, which provide a whole range of possibilities for designing novel drug delivery systems.

In addition to drug delivery applications, catanionic mixtures are drawing attention in the synthesis of novel materials, development of novel analytical methods and corrosion protection. As in drug delivery systems, vesicles play the most prominent role in these applications. However, other types of catanionic aggregates are becoming increasingly of more interest as structure-directing templates.

4.1. Pharmaceutical applications: drug delivery systems

Among different types of self-assembled drug delivery systems, vesicles remain one of the most common strategies for the delivery of drugs and genetic material in the human body [80, 83]. In general, vesicles can adsorb considerable amount of species needed to be transferred and efficiently bind to the cells. Also, it is possible to tune their physical state (gel, liquid, liquid crystalline) and in that way additionally control the release [83]. By far, most used vesicles are those composed of natural polar lipids—so-called liposomes. Liposomes possess excellent biocompatibility and biodegradability but often exhibit low stability, as they are susceptible to chemical degradation by hydrolysis and peroxidation. This is the key reason why catanionic vesicles, with their relative ease of preparation and long-term stability, attract attention as possible alternatives. In addition, catanionic vesicles can be made of biocompatible surfactants as well, such as amino acid-derived surfactants [19, 22, 23, 84].

However, despite the positive outlook for catanionic vesicles, the first studies have shown that they display number of problems such as (1) low encapsulation efficiency, both initial and long-term efficiency were not as high as for liposomes and (2) permeability, that is, occurrence of leakage due to the poor bilayers tightness [5, 7, 85]. Kaler et al. were the first to report spontaneous vesicles' formation from mixed cationic and anionic single-chain surfactants, that is, cetyltrimethylammonium tosylate (CTAT, **Figure 4**) and SDBS, as well as their potential to load glucose [86], while Caillet et al. investigated the encapsulation of anionic dye carboxyfluorescein (CF), riboflavine and glucose in C_{16}TAB/SOS vesicles [87]. These studies have shown that the permeability of vesicle membranes can be tailored by choosing appropriate surfactants' tail length. Surfactants with short alkyl chain enable higher permeability of amphiphilic films which in turn enable rapid and complete release, while longer tails increase vesicles' stability. These studies have also shown that expected specific interactions of ionic compounds with the surface of the vesicles can improve the entrapment efficiency [7]. Additionally, Wang et al. reported that CF can be encapsulated in the inner water pool as well as electrostatically adsorbed to the oppositely charged bilayers of CTAT-rich vesicles formed in CTAT/SDBS mixtures [88]. Moreover, achieved loading capacity was 10 times greater compared to phosphatidylcholine liposomes. However, no entrapment of CF was observed in SDBS-rich vesicles.

One way to overcome drawbacks of catanionic vesicles as drug delivery systems is preparation of mixtures in which one of the components is amphiphilic drug. For example, problem with permeability, that is integrity and tightness of the vesicles, is then reduced since drug molecule is incorporated into the catanionic bilayers. This approach also enables usage of the mixed micelles as self-assembled delivery systems [7].

As already discussed in the previous section, regarding the physicochemical properties and phase behavior of drug-surfactant mixtures, most of the recent research was done by Mahajan's group [29, 63, 68, 69]. Catanionic systems containing (1) anti-inflammatory drug for pain control and treatment of rheumatic diseases, diclofenac sodium [29, 71], (2) non-steroidal anti-inflammatory drug, ibuprofen [63], (3) antidepressant and antipsychotic drug, trifluoperazine dihydrochloride [68], as well as (4) tetracaine hydrochloride [69, 70], an anesthetic used in topical ophthalmic solutions, were investigated by a number of groups. In addition, Liu et al. established that amphiphilic anticancer drug, cytarabine hydrochloride (CH, **Figure 11**), and AOT can self-assemble into vesicles in the aqueous solution [89]. The parallel artificial membrane permeability assay (PAMPA) and hemolytic toxicity studies were carried out to evaluate the potential use of CH/AOT vesicles in drug delivery. The results indicate that catanionic vesicles can improve the permeability of CH about 160 times in PAMPA model and markedly decrease the hemolytic toxicity of both CH and AOT compared with their respective solutions. In addition, in vitro drug release behavior results for both CH/AOT vesicles and CH/AOT vesicles incorporated into the thermosensitive PLGA-PEG-PLGA hydrogel revealed them as good sustained drug release systems [89].

In most cases, two main strategies to improve release properties of catanionic vesicles are employed which are (1) incorporation of vesicles into the gels and (2) preparation of environment sensitive vesicles. Catanionic aggregates formed from drug and oppositely charged surfactant and then incorporated into the gel have been extensively studied by Edsman's group with the objective to utilize them for prolonged release [90–96]:

(1) Catanionic aggregates containing various drug compounds, diphenhydramine, lidocaine, ibuprofen, naproxen, alprenolol, propranolol or orphenadrine (**Figure 11**), and ionic surfactants, SDS, C_{14}TACl, C_{12}PC or benzalkonium chloride (C_mBzCl, **Figure 4**) incorporated in Carbopol® 940 or agar-agar gels, were studied. Obtained results demonstrated that both micelles and vesicles from the three systems examined in the release studies (lidocain/SDS, orphenadrine/SDS, ibuprofen/C_{14}TACl) helped to prolong the release between 10 and 100 times compared to the release of the pure drug from the gel [90].

(2) Constructed phase diagrams of the mixtures of three different cationic drug compounds, diphenhydramine, tetracaine and amitriptyline (**Figure 11**), with SDS, showed that although the diagrams may differ in some parts, vesicles and branched micelles are present in all three cases on the SDS-rich side. Drug release from Carbopol® 940 and agar gels revealed that sustained drug release may be accomplished by incorporation of investigated catanionic vesicles and micelles into the gels [91].

(3) Investigation of pH and ionic strength influence on the phase behavior of diphenhydramine/SDS and tetracaine/SDS mixtures, as well as study of drug release from drug/surfactant aggregates in Carbopol® gels, demonstrated that drug release in both systems was somewhat affected by changes in both pH and ionic strength but remained in all cases significantly prolonged compared to the release of the free drug [92].

(4) A study of controlled release of charged drugs from five different types of gels by adding surfactants (SDS, Brij 58, C_{12}BzBr) that can interact with the drug and polymer matrix demonstrated that interactions between the surfactant aggregates and the polymer can be used to further modify the drug release [93].

(5) When drug/SDS vesicles, drug substance being alprenolol or tetracaine, were mixed with polymers, one bearing hydrophobic modifications, one positively charged and one positively charged bearing hydrophobic modification, gels were form only in the case when negatively charge catanionic vesicles were mixed with positively charged polymer-bearing hydrophobic modification. In addition, the release of drug substance from these systems, where the vesicles are not trapped within the gel but constitute a founding part of it, could be significantly prolonged. The release rate was affected to a greater extent by variation of vesicles' concentrations than by variation in polymer concentration [94].

(6) Release profiles of (1) alprenolol/SDS aggregates incorporated into the SoftCAT and Carbopol® gels [95] and (2) tetracaine/SDS or capric acid aggregates incorporated into the SoftCAT and carbomer gels [96] have shown that prolonged drug release from this system enables prolonged skin penetration.

Regarding the preparation of environment-sensitive vesicles, Ghosh et al. [97] investigated pH-induced release of model drug (calcein, fluorescent dye) as well as hemocompatibility and cytotoxicity of catanionic vesicles containing anionic amino acid-based carboxylate surfactants, sodium *N*-alkanoyl-L-sarcosinate with varying chain length (**Figure 5**) and C_{12}TAOH or C_{16}TAOH. Obtained results demonstrated that with pH decrease ($pH \leq 5$), vesicles are transformed into small mixed micelles. It can be concluded that investigated vesicles are sensitive to pH change of the environment and interesting as drug delivery systems in which drug release is triggered by pH change. The hemocompatibility and cytotoxicity evaluation revealed that vesicles are hemocompatible and nontoxic.

Motivated with known antibacterial activity of anionic and cationic surfactants, Chaouat et al. prepared three component vesicles consisting of *N*-dodecyldiethanolamine, decanoic acid and azelaic acid (**Figures 4** and **5**) and evaluated their antimicrobial activity against different strains of bacteria [98]. Obtained results revealed that antimicrobial activity of catanionic vesicles displays synergistic effect compared with the activity of individual components.

Not only catanionic vesicles are considered of interest for designing drug delivery systems. The 1D structures, such as tubules, that are yielded by the self-assembly of lipids and surfactants are of particular interest for their applications in nanotechnology and pharmaceutical applications [74]. Lin et al. prepared multi-walled nanotubes using two anticancer drug amphiphiles in which drug camptothecin (CPT, **Figure 11**) was loaded [99]. Used amphiphiles contained one, two or four hydrophobic CPTs conjugated to a β-sheet-forming peptide sequence through a reducible disulfylbutyrate linker. The authors proposed that nanotubules were formed by combination of three occurrences: (1) 1D elongation, (2) formation of multilayers and (3) bilayer extension from helical ribbons due to mixing of oppositely charged drug amphiphiles.

The interaction between amphiphiles and DNA was studied over a long period of time in the area of gene therapy [7]. Likewise, due to their features, catanionic vesicles are of potential interest as non-viral gene carriers. Interactions of DNA and cationic vesicles result in complexes in which DNA molecule adopts more compact conformation and has reduced charge, facilitating its uptake through cell membranes. The fundamental framework for DNA/catanionic vesicles application has been established by Lindman's group [100–103]. In a number of studies, they have shown that:

(1) Positively charged C_{16}TAB/SOS vesicles can induce folding transition in large single linear DNA molecules, as well as adsorption of globular DNA. No such effects were observed in the presence of negatively charged vesicles. Most importantly, it was shown that the folding transition is reversible and that change in surfactants molar ratio results in DNA unfolding and release [100].

(2) Longer chain anionic surfactant (SDS vs. SOS) was more efficient in releasing DNA into the solution from catanionic vesicles [101], which was explained in terms of chain length dependence of surfactant self-assembly [102]. However, no influence of hydrophobicity of the cationic surfactant (C_{12}TAB, C_{14}TAB, C_{16}TAB) was observed.

(3) Interactions between DNA and positively charged vesicles are strong. Formed complexes withstand dilution or addition of excess surfactant or DNA and do not dissolve. Their structure resembled to other systems previously described, that is DNA molecules were packed between surfactant bilayers [103].

La Mesa's group [104] has shown that interaction with C_{16}TAB/SDS vesicles can protect a sensitive molecule, exogeneous RNA, from RNase, resulting in efficient delivery of RNA across the cell membrane. The efficiency of delivery increases when vesicles are formed in the presence of RNA. In a recent study [105] of DDAB/8-hexadecyl sulfate (8-SHS, **Figure 5**) vesicles interaction with calf thymus DNA, it was shown that strongly associating complexes are formed. Results revealed that their structure depended on DNA content. At low concentration, formed complexes resemble to bare vesicles, while at higher concentrations, multi-lamellar entities are formed in which adsorbed amount of DNA increases with its concentration. Further increasing DNA concentration leads first to formation of large clusters of vesicles and then to precipitation. DNA molecules undergo compaction process, which facilitates penetration into cell and at the same time protects it from nucelases action. The compaction process is reversible as addition of anionic surfactant induces DNA release [105].

4.2. Synthesis of advanced materials

Surfactants' role in the synthesis of nanomaterials renewed interest of research community for applying surfactants and self-assembled aggregates in the preparation of new materials. Surfactants have been used in the synthesis of inorganic materials, either as soft templates or in the surfactant-mediated synthesis [106]. Despite wide-ranging structural diversity of surfactants' aggregates, vesicles, and thus catanionic vesicles, are still frequently the template of choice. Due to the special structure of vesicles, inorganic material can be formed in different reaction environments: (1) the "bulk" solution outside the vesicles, (2) the inner chamber, (3) the outside surface or (4) the hydrophobic palisade layer of the vesicles [107, 108]. Different reaction environments enable formation of material of vastly different morphologies.

Recently, vesicles formed by SAIL, [C_{12}MP]Br (**Figure 10**), and a divalent metal surfactant, copper dodecyl sulfate ($Cu(DS)_2{\cdot}4H_2O$), were used for preparation of leaf-like CuO nanosheets [108].

Using vesicles composed of imidazolium-based SAIL, [C_{12}mim]Br (**Figure 10**) and SDS as structure-directing templates, Yuan et al. synthesized silica hollow spheres [107]. Silica hollow spheres of controlled size were previously synthesized in C_{12}TAB/SDBS mixtures by Kepczynski et al. [109]. Furthermore, catanionic vesicles formed in C_{16}TAOH/$Mg(DS)_2$ mixture were used for preparation of $Mg(OH)_2$ hollow nanospheres [110]. Interestingly, it was observed that encapsulation of $Mg(OH)_2$ particles, followed by crystal fusion, can induce the size and shape change of catanionic vesicles under non-equilibrium conditions. This phenomenon facilitated the direct observation of hydrophobic membrane fusion by means of TEM microscopy [110].

Not only inorganic hollowspheres were synthesized in the presence of catanionic vesicles. Morgan et al. developed the method for preparation of polymeric spheres by introducing polymerizable monomer into the vesicle's bilayer [111]. This method was later used for synthesis of polydisperse hollow polystyrene spheres in CTAT/SDBS and C_{16}TAB/SOS mixtures [112]. Additionally, a lot of research work employing the different preparation methods with polymerized ion pair amphiphile vesicles was done by Chung's group [113–117]. Recently, hollow microspheres of poly(3,4-ethylenedioxythiophene (PEDOT), ranging from 0.5 to 10 mm, were synthesized by oxidative polymerization in the presence of C_{16}TAB/SDBS vesicles. It was established that formation and size of microspheres were influenced by surfactant molar ratio. Moreover, it was shown that SDBS was incorporated in the polymer chain as dopant [118].

In addition to vesicles, catanionic micelles can be also effective templates for preparation of nanoparticles (NPs). For example, C_{16}TAB/SDS micelles were used in synthesis of mesoporous γ-Al_2O_3 NPs [119]. Authors demonstrated that the choice of the surfactant is important for the synthesis of organized mesoporous aluminas with a well-defined porosity, although it is unclear how the presence of micellar aggregates affects the final architecture in cationic-anionic double hydrolysis method [119].

Short-chain catanionic mixtures composed of C_{10}TAB and SOS were used in synthesis of highly ordered supermicroporous silica [120]. Pore size in the range 1–2 nm had hexagonal structure which was strongly dependent on the surfactants molar ratio. Previously, Ohkubo et al. reported synthesis of silica particles in which precise control of both, the pore size and the structure of pores, was achieved by changing C_{16}TAB/SOS mixing ratio [121]. Moreover, use of the cationic surfactant with longer alkyl chain, C_{18}TAB, shifted the point of phase transition from hexagonal phase to lamellar phase to lower concentration of SOS. Lind et al. reported on vesicle-like patterned, mesoscopically ordered silica synthesized in C_{16}TAB/decanoic acid mixtures with toluene used as the swelling agent [122]. Obtained results demonstrated that lower interfacial charge density of the mixed aggregates stabilizes structures of lower interfacial curvature and therefore facilitates a more controlled solubilization of toluene. In addition, it was shown that the pore size of the hexagonal phase could be controlled by changing the C_{16}TAB/decanoic acid and the C_{16}TAB/toluene molar ratios [122].

Using surfactants' aggregates as structure-directing templates in the synthesis of new materials is essentially a biomimetic approach [106]. Hard tissues in organisms, such as bones and teeth, are formed in the processes in which organic matrix (composed of surface active proteins, lipids, etc.) has a role of the template which determines morphology, size and orientation of inorganic phase. Therefore, it is not surprising that several attempts of biomineral synthesis in the presence of catanionic mixtures have been reported.

Prelote and Zemb used catanionic aggregates with hexagonal structures formed in mixtures of polyoxyethyleneoleyl ether phosphate ($POEPO_4$) and C_{14}TAB as structure-directing templates for synthesis of mesoporous hydroxapatite (HAP) with high surface area [123]. HAP is thermodynamically the most stable calcium phosphate phase which attracts attention due to its similarity to bone mineral. It is widely used as biomaterial for bone and dental tissue regeneration in the form of different ceramics formulations and as coating. In that sense, mesoporous HAP is of special interest as a 3D scaffold. Hexagonal network of cylindrical micelles formed in the C_{14}TAB/$POEPO_4$ mixture was preserved during the synthesis of HAP, which enabled formation of the precipitates with the structural characteristic of the hexagonal network. However, the repetition distance was low and obtained precipitates were not truly mesoporous material. In addition, the precipitates were not able to withstand calcification.

Tari et al. have shown that the morphology of HAP NPs in C_{16}TAB/SDS solution depends on surfactant molar ratio [124]. In the SDS-rich region, rod-like HAP NPs were obtained, while in C_{16}TAB-rich region, HAP nanosheets were formed.

Control of polymorphism and crystal morphology is not only important in the biomineral synthesis but also for fundamental understanding of biomineralization processes in vivo. Chen and Nang have shown that surfactants molar ratio in C_{16}TAB/SDS mixtures can be used to control both the morphology and polymorphism of $CaCO_3$ crystals [125]. Furthermore, Dong et al. obtained brick-like (dodecahedrons) and star-like (icositetrahedrons) calcium oxalate monohydrate ($CaC_2O_4{\cdot}H_2O$) crystals, not observed before, in mixtures of calcium dodecyl sulfate and C_{14}TAB with excess $CaBr_2$ [126].

4.3. Novel analytical methods

Several research groups investigated the use of catanionic aggregates in the development of new analytical and detection methods.

CTAT/SDBS vesicles, both positively and negatively charged, were used for highly efficient electrostatic sequestration of small molecules of similar weight but opposite charge, that is CF, lucifer yellow, sulforhodamine 101, doxorubicin and rhodamine 6G [127]. Authors have established that charge-dependent effect enables use of CTAT/SDBS vesicles for selective capture and separation of oppositely charged solute from a mixture of solutes.

Kahe et al. used C_{16}TAB/SDS mixtures in propanol water as a novel microextraction system for the preconcentration and determination of trace amounts of lead in (1) saline solutions and (2) food samples [128]. Since only small amount of propanol in water was used, both hydrophilic and hydrophobic sites in extraction solvent were available for interaction with analytes of various polarities enabling good efficacy. Obtained results confirm that the catanionic aggregate dispersive microextraction method can be used as a simple, safe, fast and low-cost technique for the microextraction of various organic and inorganic compounds from real samples [128].

Chen et al. employed coacervates formed by addition of hexafluoroisopropanol (HFIP) to C_{12}TAB/SDS mixtures for extraction of strongly polar sulphonamides (SAs) from environmental water samples [129]. Results demonstrated that even small amount of HFIP can induce coacervation and two-phase separation in a broad concentration range in C_{12}TAB/SDS system. In addition, analysis of real water samples confirmed that investigated method can be efficiently used for the preconcentration and determination of SAs traces.

With an aim to improve methods for Au(III) extraction, Wang et al. used $C_{12}C_3(OH)C_{12}Cl_2$ (**Figure 6**)/NaDC vesicles [130]. Through stepwise extraction and ligand-modified vesicles system, separation of Au (III), Cu (II) and Fe (III) from mixed solution was successfully achieved. The results collected in this study revealed great potential of catanionic aggregates in development of environmental friendly Au recovery method [130].

Gao et al. proposed a new method for determination of anionic surfactants based on in situ formation of catanionic aggregates in the presence of amphiphilic 2-(2-hydroxyphenyl) benzothiazolefluorogen probe [131]. Described approach enables quantitative determination of low anionic surfactant concentrations and can be extended to wash-free imaging of bacteria.

The role that carbohydrate-protein interactions have in biological processes and difficulties in their evaluation motivates development of novel analytical methods. Pond et al. applied CTAT/SDBS vesicles with incorporated glycans in the outer surface to form glycan array for investigating carbohydrate-lecitin interactions. The method proved to be facile and opens possibilities for characterizing unknown lecitins [132].

4.4. Corrosion protection

Catanionic mixtures also proved to be efficient in corrosion protection of mild steel. C_{16}TAB/SDS mixtures demonstrated better protective efficiency than the individual surfactants. This was explained by strong adsorption on the metal surface and formation of protective surfactant film. The strong adsorption was evidenced by more negative values of the adsorption free energy of C_{16}TAB/SDS mixtures compared to the individual surfactants [133].

Author details

Darija Domazet Jurašin*, Suzana Šegota, Vida Čadež, Atiđa Selmani and Maja Dutour Sikirć

*Address all correspondence to: djurasin@irb.hr

Laboratory for Biocolloids and Surface Chemistry, Division of Physical Chemistry, Ruđer Bošković Institute, Zagreb, Croatia

References

[1] Rosen MJ. Surfactants and interfacial phenomena (3th ed.). Hoboken, NJ: John Wiley & Sons; 2004. doi:10.1016/0166-6622(89)80030-7.

[2] Holmberg K, Jönsson B, Kronberg B, Lindman B. Surfactants and polymers in aqueous solution (2nd ed.). Chichester: John Wiley & Sons; 2002. doi:10.1002/0470856424.

[3] Kaler EW, Herrington KL, Iampietro DJ, Codren BA, Jung HT, Zasadzinski JA. Phase behavior and microstructure in aqueous mixtures of cationic and anionic surfactants. In: Abe M, Scamehorn JF, editors. Mixed surfactant systems (2nd ed.). Boca Raton: CRC Press and Taylor & Francis Group; 2004.

[4] Khan A, Marques EF. Synergism and polymorphism in mixed surfactant systems. Curr Opin Colloid Interf Sci 2000;4:402-410. doi:10.1016/S1359-0294(00)00017-0.

[5] Lioi SB, Wang X, Islam MR, Danoff EJ, English DS. Catanionic surfactant vesicles for electrostatic molecular sequestration and separation. Phys Chem Chem Phys 2009;11:9315-9325. doi:10.1039/b908523h.

[6] Kume G, Gallotti M, Nunes G. Review on anionic/cationic surfactant mixtures. J Surfactants Deterg 2008;11:1-11. doi:10.1007/s11743-007-1047-1.

[7] Bramer T, Dew N, Edsman K. Pharmaceutical applications for catanionic mixtures. J Pharm Pharmacol 2007;59:1319-1334. doi:10.1211/jpp.59.10.0001.

[8] Šegota S, Težak Đ. Spontaneous formation of vesicles. Adv Colloid Interf Sci 2006;121:51-75. doi:10.1016/j.cis.2006.01.002.

[9] Antunes FE, Marques EF, Miguel MG, Lindman B. Polymer-vesicle association. Adv Colloid Interf Sci 2009;147-148:18-35. doi:10.1016/j.cis.2008.10.001.

[10] Israelachvili JN. Intermolecular and surface forces (3th ed.). San Diego: Academic Press; 2011. doi:10.1016/B978-0-12-375182-9.10025-9.

[11] Safran SA, Pincus P, Andelman D. Theory of spontaneous vesicle formation in surfactant mixtures. Science (80-) 1990;248:354-356. doi:10.1126/science.248.4953.354.

[12] Lasic DD, Joannic R, Keller BC, Frederik PM, Auvray L. Spontaneous vesiculation. Adv Colloid Interf Sci 2001;89-90:337-349. doi:10.1016/S0001-8686(00)00067-1.

[13] Jung HT, Coldren B, Zasadzinski JA, Iampietro DJ, Kaler EW. The origins of stability of spontaneous vesicles. PNAS 2001;98:1353-1357. doi:10.1073/pnas.98.4.1353.

[14] Jurašin D, Vinceković M, Pustak A, Šmit I, Bujan M, Filipović-Vinceković N. Lamellar to hexagonal columnar liquid crystalline phase transition in a catanionic surfactant mixture: dodecylammonium chloride-sodium bis(2-ethylhexyl) sulfosuccinate. Soft Matter 2013;9:3349-3360. doi:10.1039/c3sm27665a.

[15] Yatcilla MT, Herrington KL, Brasher LL, Kaler EW, Chiruvolu S, Zasadzinski JA. Phase behavior of aqueous mixtures of cetyltrimethylammonium bromide (CTAB) and sodium octyl sulfate (SOS). J Phys Chem 1996;100:5874-5879. doi:10.1021/j100153a058.

[16] Novak S, Morasi Piperčić S, Makarić S, Primožič I, Ćurlin M, Štefanić Z, et al. Interplay of non-covalent interactions in ionic liquid/sodium bis(2-ethylhexyl) sulfosuccinate mixtures: from lamellar to bicontinuous cubic liquid crystalline phase. J Phys Chem B 2016;120:12557-12567. doi:10.1021/acs.jpcb.6b10515.

[17] Kuo A, Chang C. Recent strategies in the development of catanionic vesicles. J Oleo Sci 2016;65:377-384. doi:10.5650/jos.ess15249.

[18] Marques EF, Regev O, Khan A, Lindman B. Self-organization of double-chained and pseudodouble-chained surfactants: counterion and geometry effects. Adv Colloid Interf Sci 2003;100-102:83-104. doi:10.1016/S0001-8686(02)00068-4.

[19] Silva SG, Vale MLC do, Marques EF. Size, charge, and stability of fully serine-based catanionic vesicles: towards versatile biocompatible nanocarriers. Chem A Eur J 2015;21:4092-4101. doi:10.1002/chem.201406111.

[20] Herrington KL, Kaler EW, Miller DD, Zasadzinski JAN, Chiruvolu S. Phase behavior of aqueous mixtures of dodecyltrimethylammonium bromide (DTAB) and sodium dodecyl sulfate (SDS). J Phys Chem 1993;97:13792-13802. doi:10.1021/j100153a058.

[21] Marques EF, Regev O, Khan A, da Graça Miguel M, Lindman B. Vesicle formation and general phase behavior in the catanionic mixture SDS-DDAB-water. The anionic-rich side. J Phys Chem B 1998;102:6746-6758. doi:10.1021/jp980355t.

[22] Brito RO, Marques EF, Gomes P, Falcão S, Söderman O. Self-assembly in a catanionic mixture with an aminoacid-derived surfactant: from mixed micelles to spontaneous vesicles. J Phys Chem B 2006;110:18158-18165. doi:10.1021/jp061946j.

[23] Marques EF, Brito RO, Silva SG, Rodríguez-Borges EJ, do Vale ML, Gomes P, et al. Spontaneous vesicle formation in catanionic mixtures of amino acid-based surfactants: chain length symmetry effects. Langmuir 2008;24:11009-11017. doi:10.1021/la801518h.

[24] Jendrić M, Filipović-Vinceković N, Vinceković M, Bujan M, Primožič I. Phase behavior of bis(quaternary ammonium bromide)/sodium cholate/H_2O system. J Dispers Sci Technol 2005;26:39-51. doi:10.1081/DIS-200040213.

[25] Shang Y, Liu H, Hu Y, Prausnitz JM. Phase behavior and microstructures of the gemini (12-3-12,2Br-)-SDS-H_2O ternary. Colloids Surf A Physicochem Eng Asp 2007;294:203-211. doi:10.1016/j.colsurfa.2006.08.012.

[26] Marques EF, Regev O, Edlund H, Khan A. Micelles, dispersions, and liquid crystals in the catanionic mixture bile salt-double-chained surfactant. The bile salt-rich area. Langmuir 2000;16:8255-8262. doi:10.1021/la0002278.

[27] Šegota S, Težak Đ, Talmon Y. New catanionic mixtures of didodecyldimethylammonium bromide/sodium dodecylbenzene sulfonate/water with special reference to spontaneous formation of vesicles. II. Size and shape analysis by SAXS, light scattering, cryo-TEM, and light microscopy. Soft Mater 2006;3:51-69. doi:10.1080/15394450600683477.

[28] Šegota S, Heimer S, Težak Đ. New catanionic mixtures of dodecyldimethylammonium bromide/sodium dodecylbenzenesulphonate/water: I. Surface properties of dispersed particles. Colloids Surf A Physicochem Eng Asp 2006;274:91-99. doi:10.1016/j.colsurfa.2005.08.051.

[29] Singh O, Kaur R, Aswal VK, Mahajan RK. Composition and concentration gradient induced structural transition from micelles to vesicles in the mixed system of ionic liquid-diclofenac sodium. Langmuir 2016;32:6638-6647. doi:10.1021/acs.langmuir.6b01175.

[30] Rubingh DN. Mixed micelle solutions. In: Mittal K, editor. Solution chemistry of surfactants. New York: Plenum Press; 1979, pp. 337-354. doi:10.1007/978-1-4615-7880-2.

[31] Zhou Q, Rosen MJ. Molecular interactions of surfactants in mixed monolayers at the air/aqueous solution interface and in mixed micelles in aqueous media: the regular solution approach. Langmuir 2003;19:4555-4562. doi:10.1021/la020789m.

[32] Nguyen CM, Scamehorn JF. Thermodynamics of mixed monolayer formation at the air-water interface. J Colloid Interf Sci 1988;123:238-248.

[33] Zana R. Dimeric and oligomeric surfactants. Behavior at interfaces and in aqueous solution: a review. Adv Colloid Interf Sci 2002;97:205-253. doi:10.1016/S0001-8686(01)00069-0.

[34] Jurašin D, Dutour Sikirić M. Higher oligomeric surfactants-from fundamentals to applications. In: Lesieur C, editor. Oligomerization of chemistry and biology compound. Rijeka: InTech; 2014, pp. 133-172. doi:10.5772/57655.

[35] Jurašin D, Weber I, Filipović-Vinceković N. Phase behavior in mixtures of cationic dimeric and anionic monomeric surfactants. J Dispers Sci Technol 2009;30:622-633. doi:10.1080/01932690802598481.

[36] Cheon HY, Jeong NH, Kim HU. Spontaneous vesicle formation in aqueous mixtures of cationic Gemini surfactant and sodium lauryl ether sulfate. Bull Korean Chem Soc 2005;26:107-114. doi:10.5012/bkcs.2005.26.1.107.

[37] Bai G, Wang Y, Ding Y, Zhuo K, Wang J, Bastos M. Thermodynamics of self-assembling of mixture of a cationic Gemini surfactant and sodium dodecylsulfate in aqueous solution: calorimetry, conductivity and surface pressure measurements. J Chem Thermodyn 2016;94:221-229. doi:10.1016/j.jct.2015.11.017.

[38] Wang Y, Bai G, Marques EF, Yan H. Phase behavior and thermodynamics of a mixture of cationic Gemini and anionic surfactant. J Phys Chem B 2006;110:5294-5300. doi:10.1021/jp054323z.

[39] Wang Y, Marques EF, Pereira CM. Monolayers of Gemini surfactants and their catanionic mixtures with sodium dodecyl sulfate at the air-water interface: chain length and composition effects. Thin Solid Films 2008;516:7458-7466. doi:10.1016/j.tsf.2008.03.029.

[40] Wang Y, Marques EF. Non-ideal behavior of mixed micelles of cationic Gemini surfactants with varying spacer length and anionic surfactants: a conductimetric study. J Mol Liq 2008;142:136-142. doi:10.1016/j.molliq.2008.06.001.

[41] Prévost S, Wattebled L, Laschewsky A, Gradzielski M. Formation of monodisperse charged vesicles in mixtures of cationic Gemini surfactants and anionic SDS. Langmuir 2011;27:582-591. doi:10.1021/la103976p.

[42] Ji X, Tian M, Wang Y. Temperature-induced aggregate transitions in mixtures of cationic ammonium gemini surfactant with anionic glutamic acid surfactant in aqueous solution. Langmuir 2016;32:972-981. doi:10.1021/acs.langmuir.5b04211.

[43] Aghdastinat H, Javadian S, Tehrani-Bagha A, Gharibi H. Spontaneous formation of nanocubic particles and spherical vesicles in catanionic mixtures of ester-containing gemini surfactants and sodium dodecyl sulfate in the presence of electrolyte. J Phys Chem B 2014;118:3063-3073. doi:10.1021/jp4123908.

[44] Zana R, Lévy H, Danino D, Talmon Y, Kwetkat K. Mixed micellization of cetyltrimethylammonium bromide and an anionic dimeric (Gemini) surfactant in aqueous solution. Langmuir 1997;13:402-408. doi:10.1021/la9606963.

[45] Zhao J, Liu J, Jiang R. Interaction between anionic and cationic Gemini surfactants at air/water interface and in aqueous bulk solution. Colloids Surf A Physicochem Eng Asp 2009;350:141-146. doi:10.1016/j.colsurfa.2009.09.022.

[46] Luo HS, Wang N, Zhou LZ, Wang YL, Wang JB, Yan HK. Salt effect on the aggregation behaviors of an anionic carboxylate Gemini and a cationic surfactant. J Dispers Sci Technol 2008;29:787-791. doi:10.1080/01932690701756479.

[47] Chen WJ, Zhai LM, Li GZ, Li BQ, Xu J. Spontaneous vesicle formation and vesicle-tubular microstructure transition in aqueous solution of a poly-tailed cationic and anionic surfactants mixture. J Colloid Interf Sci 2004;278:447-452. doi:10.1016/j.jcis.2004.05.036.

[48] Yoshimura T, Ohno A, Esumi K. Mixed micellar properties of cationic trimeric-type quaternary ammonium salts and anionic sodium n-octyl sulfate surfactants. J Colloid Interf Sci 2004;272:191-196. doi:10.1016/j.jcis.2003.12.021.

[49] Jurašin D, Habuš I, Filipović-Vinceković N. Role of the alkyl chain number and head groups location on surfactants self-assembly in aqueous solutions. Colloids Surf A Physicochem Eng Asp 2010;368:119-128. doi:10.1016/j.colsurfa.2010.07.025.

[50] Sharma R, Mahajan RK. Influence of various additives on the physicochemical properties of imidazolium based ionic liquids: a comprehensive review. RSC Adv 2014;4:748-774. doi:10.1039/c3ra42228c.

[51] Goodchild I, Collier L, Millar SL, Prokeš I, Lord JCD, Butts CP, et al. Structural studies of the phase, aggregation and surface behaviour of 1-alkyl-3-methylimidazolium halide + water mixtures. J Colloid Interf Sci 2007;307:455-468. doi:10.1016/j.jcis.2006.11.034.

[52] Greaves TL, Drummond CJ. Ionic liquids as amphiphile self-assembly media. Chem Soc Rev 2008;37:1709-1726. doi:10.1039/b801395k.

[53] Patel DD, Lee JM. Applications of ionic liquids. Chem Rec 2012;12:329-355. doi:10.1002/tcr.201100036.

[54] Dong B, Li N, Zheng L, Yu L, Inoue T. Surface adsorption and micelle formation of surface active ionic liquids in aqueous solution. Langmuir 2007;23:4178-4182. doi:10.1021/la0633029.

[55] Zhao Y, Chen X, Jing B, Wang X, Ma F. Novel gel phase formed by mixing a cationic surfactive ionic liquid C16mimCl and an anionic surfactant SDS in aqueous solution. J Phys Chem B 2009;113:983-938. doi:10.1021/jp809048u.

[56] Zhao M, Gao M, Dai C, Wang S, Chen W, Liu Y, et al. A novel study on the gel phase formed in a catanionic surfactant system. J Surfactants Deterg 2016;19:519-525. doi:10.1007/s11743-016-1799-6.

[57] Zhao M, Yuan J, Zheng L. Spontaneous formation of vesicles by N-dodecyl-N-methylpyrrolidinium bromide (C12MPB) ionic liquid and sodium dodecyl sulfate (SDS) in aqueous solution. Colloids Surf A Physicochem Eng Asp 2012;407:116-120. doi:10.1016/j.colsurfa.2012.05.016.

[58] Dai C, Yang Z, Yang H, Liu Y, Fang J, Chen W, et al. Micelle-to-vesicle transition induced by β-cyclodextrin in mixed catanionic surfactant solutions. Colloids Surf A Physicochem Eng Asp 2016;498:1-6. doi:10.1016/j.colsurfa.2016.03.040.

[59] Jambhekar SS, Breen P. Cyclodextrins in pharmaceutical formulations I: structure and physicochemical properties, formation of complexes, and types of complex. Drug Discov Today 2016;21:356-362. doi:10.1016/j.drudis.2015.11.017.

[60] Chabba S, Kumar S, Aswal VK, Kang TS, Mahajan RK. Interfacial and aggregation behavior of aqueous mixtures of imidazolium based surface active ionic liquids and anionic surfactant sodium dodecylbenzenesulfonate. Colloids Surf A Physicochem Eng Asp 2015;472:9-20. doi:10.1016/j.colsurfa.2015.02.032.

[61] Hunt PA, Ashworth CR, Matthews RP. Hydrogen bonding in ionic liquids. Chem Soc Rev 2015;44:1257-1288. doi:10.1039/C4CS00278D.

[62] Zhang H, Li K, Liang H, Wang J. Spectroscopic studies of the aggregation of imidazolium-based ionic liquids. Colloids Surf A Physicochem Eng Asp 2008;329:75-81. doi:10.1016/j.colsurfa.2008.06.046.

[63] Sanan R, Kaur R, Mahajan RK. Micellar transitions in catanionic ionic liquid–ibuprofen aqueous mixtures; effects of composition and dilution. RSC Adv 2014;4:64877-64889. doi:10.1039/C4RA10840J.

[64] Gehlot PS, Rao KS, Bharmoria P, Damarla K, Gupta H, Drechsler M, et al. Spontaneous formation of multiarchitecture vesicles of [C8mim]Br + [Na]DBS in aqueous medium: synergic interplay of electrostatic, hydrophobic, and π-π stacking interactions. J Phys Chem B 2015;119:15300-15309. doi:10.1021/acs.jpcb.5b09850.

[65] Murgia S, Palazzo G, Mamusa M, Lampis S, Monduzzi M. Aerosol-OT forms oil-in-water spherical micelles in the presence of the ionic liquid bmimBF4. J Phys Chem B 2009;113:9216-9225. doi:10.1021/jp902970n.

[66] Murgia S, Palazzo G, Mamusa M, Lampis S, Monduzzi M. Aerosol-OT in water forms fully-branched cylindrical direct micelles in the presence of the ionic liquid 1-butyl-3-methylimidazolium bromide. Phys Chem Chem Phys 2011;13:9238-9245. doi:10.1039/c1cp20209j.

[67] Vashishat R, Sanan R, Mahajan RK. Bile salt-surface active ionic liquid mixtures: mixed micellization and solubilization of phenothiazine. RSC Adv 2015;5:72132-72141. doi:10.1039/C5RA09812B.

[68] Sharma R, Mahajan RK. An investigation of binding ability of ionic surfactants with trifluoperazine dihydrochloride: insights from surface tension, electronic absorption and fluorescence measurements. RSC Adv 2012;2:9571-9583. doi:10.1039/c2ra21020g.

[69] Vashishat R, Chabba S, Mahajan RK. Effect of surfactant head group on micellization and morphological transitions in drug-surfactant catanionic mixture: a multi-technique approach. Colloids Surf A Physicochem Eng Asp 2016;498:206-217. doi:10.1016/j.colsurfa.2016.03.058.

[70] Jiang Y, Li F, Luan Y, Cao W, Ji X, Zhao L, et al. Formation of drug/surfactant catanionic vesicles and their application in sustained drug release. Int J Pharm 2012;436:806-814. doi:10.1016/j.ijpharm.2012.07.053.

[71] Zhao L, Liu J, Zhang L, Gao Y, Zhang Z, Luan Y. Self-assembly properties, aggregation behavior and prospective application for sustained drug delivery of a drug-participating catanionic system. Int J Pharm 2013;452:108-115. doi:10.1016/j.ijpharm.2013.04.072.

[72] Liu C, Cui J, Song A, Hao J. A bile acid-induced aggregation transition and rheological properties in its mixtures with alkyltrimethylammonium hydroxide. Soft Matter 2011;7:8952-8960. doi:10.1039/c1sm05635b.

[73] Vinceković M, Jurašin D, Tomašić V, Bujan M, Filipović-Vinceković N. Interactions in aqueous mixtures of alkylammonium chlorides and sodium cholate. J Dispers Sci Technol 2006;27:1099-1111. doi:10.1080/01932690600858905.

[74] Manghisi N, Leggio C, Jover A, Meijide F, Pavel NV, Tellini VHS, et al. Catanionic tubules with tunable charge. Angew Chemie Int Ed 2010;49:6604-6607. doi:10.1002/anie.201000951.

[75] Marques EF, Khan A. Effect of a bile salt on the aggregation behavior of a double-chained cationic surfactant – the cationic-rich dilute region of the didodecyldimethylammonium bromide-sodium taurodeoxycholate-water system. In: Nylander T, Lindman B, editors. Lipid and polymer-lipid systems. Berlin and Heidelberg: Springer; 2002, pp. 83-91. doi:10.1007/3-540-45291-5_11.

[76] Bhattacharjee J, Aswal VK, Hassan PA, Pamu R, Narayanan J, Bellare J. Structural evolution in catanionic mixtures of cetylpyridinium chloride and sodium deoxycholate. Soft Matter 2012;8:10130-10140. doi:10.1039/c2sm25460c.

[77] Fernández-Leyes MD, Messina PV, Schulz PC. Bile salt structural effect on the thermodynamic properties of a catanionic mixed adsorbed monolayer. Colloid Polym Sci 2011;289:179-191. doi:10.1007/s00396-010-2336-1.

[78] Fernández-Leyes MD, Messina PV, Andreucetti N, Luengo C, Schulz PC. Aggregate structural transitions noticed for the didodecyldimetilammonium bromide-sodium dehydrocholate catanionic mixed system at low concentration. Soft Mater 2013;11:204-214. doi:10.1080/1539445X.2012.617642.

[79] Pereyra RB, Schulz EP, Durand GA, Ritacco H, Schulz PC. Thermodynamic analysis of an asymmetric system: aqueous sodium dehydrocholate-hexadecyltrimethylammonium bromide mixed micelles. Colloids Surf A Physicochem Eng Asp 2016;509:675-683. doi:10.1016/j.colsurfa.2016.08.092.

[80] Antonietti M, Förster S. Vesicles and liposomes: a self-assembly principle beyond lipids. Adv Mater 2003;15:1323-1333. doi:10.1002/adma.200300010.

[81] Malmsten M. Soft drug delivery systems. Soft Matter 2006;2:760-769. doi:10.1039/b608348j.

[82] Ainalem ML, Nylander T. DNA condensation using cationic dendrimers-morphology and supramolecular structure of formed aggregates. Soft Matter 2011;7:4577-4594. doi:10.1039/c0sm01171a.

[83] Barbetta A, La Mesa C, Muzi L, Pucci C, Risuleo G, Tardani F. Cat-anionic vesicle-based systems as potential carriers in nano-technologies. In: Phoenix DA, Ahmed W, editors. Nanobiotechnology, Manchester: One Central Press; 2014, pp. 152-179.

[84] Brito RO, Marques EF, Silva SG, do Vale ML, Gomes P, Araújo MJ, et al. Physicochemical and toxicological properties of novel amino acid-based amphiphiles and their spontaneously formed catanionic vesicles. Colloids Surf B Biointerf 2009;72:80-87. doi:10.1016/j.colsurfb.2009.03.017.

[85] Tondre C, Caillet C. Properties of the amphiphilic films in mixed cationic/anionic vesicles: a comprehensive view from a literature analysis. Adv Colloid Interf Sci 2001;93:115-134. doi:10.1016/S0001-8686(00)00081-6.

[86] Kaler EW, Murthy AK, Rodriguez BE, Zasadzinski JAN. Spontaneous vesicle formation in aqueous mixtures of single-tailed surfactants. Science (80-) 1989;245:1371-1374. doi:10.1126/science.2781283.

[87] Caillet C, Hebrant M, Tondre C. Sodium octyl sulfate/cetyltrimethylammonium bromide catanionic vesicles: aggregate composition and probe encapsulation. Langmuir 2000;16:9099-9102.

[88] Wang X, Danoff EJ, Sinkov NA, Lee JH, Raghavao SR, English DS. Highly efficient capture and long-term encapsulation of dye by catanionic surfactant vesicles. Langmuir 2006;22:6461-6464. doi:10.1021/la0605135.

[89] Liu J, Jiang Y, Cui Y, Xu C, Ji X, Luan Y. Cytarabine-AOT catanionic vesicle-loaded biodegradable thermosensitive hydrogel as an efficient cytarabine delivery system. Int J Pharm 2014;473:560-571. doi:10.1016/j.ijpharm.2014.07.032.

[90] Bramer T, Dew N, Edsman K. Catanionic mixtures involving a drug: a rather general concept that can be utilized for prolonged drug release from gels. J Pharm Sci 2006;95:769-780. doi:10.1002/jps.

[91] Bramer T, Paulsson M, Edwards K, Edsman K. Catanionic drug – surfactant mixtures: phase behavior and sustained release from gels. Pharm Res 2003;20:1661-1667.

[92] Bramer T, Karlsson G, Edwards K, Edsman K. Effects of pH and ionic strength on catanionic drug-surfactant mixtures used for prolonged release from gels. J Drug Deliv Sci Technol 2007;17:285-291. doi:10.1016/S1773-2247(07)50097-3.

[93] Paulsson M, Edsman K. Controlled drug release from gels vesicles formed from mixtures of charged surfactants. Pharm Res 2001;18:1586-1592.

[94] Dew N, Edwards K, Edsman K. Gel formation in systems composed of drug containing catanionic vesicles and oppositely charged hydrophobically modified polymer. Colloids Surfaces B Biointerf 2009;70:187-197. doi:10.1016/j.colsurfb.2008.12.021.

[95] Dew N, Edwards K, Eriksson J, Edsman K, Björk E. Gel formulations containing catanionic vesicles composed of alprenolol and SDS: effects of drug release and skin penetration on aggregate structure. Colloids Surf B Biointerf 2012;89:53-60. doi:10.1016/j.colsurfb.2011.08.022.

[96] Dew N, Edsman K, Björk E. Novel gel formulations with catanionic aggregates enable prolonged drug release and reduced skin permeation. J Pharm Pharmacol 2011;63:1265-1273. doi:10.1111/j.2042-7158.2011.01339.x.

[97] Ghosh S, Ray A, Pramanik N, Ambade B. Can a catanionic surfactant mixture act as a drug delivery vehicle? C R Chim 2016;19:951-954. doi:10.1016/j.crci.2016.03.020.

[98] Chaouat C, Balor S, Roques C, Franceschi-Messant S, Perez E, Rico-Lattes I. Antimicrobial catanionic vesicular self-assembly with improved spectrum of action. J Surfactants Deterg 2013;16:717-722. doi:10.1007/s11743-013-1451-7.

[99] Lin YA, Cheetham AG, Zhang P, Ou YC, Li Y, Liu G, et al. Multiwalled nanotubes formed by catanionic mixtures of drug amphiphiles. ACS Nano 2014;8:12690-12700. doi:10.1021/nn505688b.

[100] Mel'nikov SM, Dias R, Mel'nikova YS, Marques EF, Miguel MG, Lindman B. DNA conformational dynamics in the presence of catanionic mixtures. FEBS Lett 1999;453:113-118. doi:10.1016/S0014-5793(99)00699-7.

[101] Dias RS, Lindman B, Miguel MG. Compaction and decompaction of DNA in the presence of catanionic amphiphile mixtures. J Phys Chem B 2002;106:12608-12612. doi:10.1021/jp020392r.

[102] Miguel MG, Pais AACC, Dias RS, Leal C, Rosa M, Lindman B. DNA-cationic amphiphile interactions. Colloids Surf A Physicochem Eng Asp 2003;228:43-55. doi:10.1016/S0927-7757(03)00334-0.

[103] Dias RS, Lindman B, Miguel MG. DNA interaction with catanionic vesicles. J Phys Chem B 2002;106:12600-12607. doi:10.1021/jp020391z.

[104] Russo L, Berardi V, Tardani F, La Mesa C, Risuleo G. Delivery of RNA and its intracellular translation into protein mediated by SDS-CTAB vesicles: potential use in nanobiotechnology. Biomed Res Int 2013;2013:1-6. doi:10.1155/2013/734596.

[105] Pucci C, Scipioni A, Diociaiuti M, La Mesa C, Pérez L, Pons R. Catanionic vesicles and DNA complexes: a strategy towards novel gene delivery systems. RSC Adv 2015;5:81168-81175. doi:10.1039/C5RA15466A.

[106] Holmberg K. Surfactant-templated nanomaterials synthesis. J Colloid Interf Sci 2004;274:355-364. doi:10.1016/j.jcis.2004.04.006.

[107] Yuan J, Bai X, Zhao M, Zheng L. C12mimBr ionic liquid/SDS vesicle formation and use as template for the synthesis of hollow silica spheres. Langmuir 2010;26:11726-11731. doi:10.1021/la101221z.

[108] Zhao M, Yuan J, Zheng L. The formation of vesicles by N-dodecyl-N-methylpyrrolidinium bromide ionic liquid/copper dodecyl sulfate and application in the synthesis of leaflike CuO nanosheets. Colloid Polym Sci 2012;290:1361-1369. doi:10.1007/s00396-012-2664-4.

[109] Kepczynski M, Ganachaud F, Hémery P. Silicone nanocapsules from catanionic vesicle templates. Adv Mater 2004;16:1861-1863. doi:10.1002/adma.200400537.

[110] Liu F, Shen Q, Su Y, Han S, Xu G, Wang D. Transcriptive synthesis of $Mg(OH)_2$ hollow nanospheres and the non-equilibrium shell fusion assisted by catanionic vesicles. J Phys Chem B 2009;113:11362-11366.

[111] Morgan JD, Johnson CA, Kaler EW. Polymerization of equilibrium vesicles. Langmuir 1997;13:6447-6451. doi:10.1021/la970495e.

[112] McKelvey CA, Kaler EW, Zasadzinski JA, Coldren B, Jung HT. Templating hollow polymeric spheres from catanionic equilibrium vesicles: synthesis and characterization. Langmuir 2000;16:8285-8290. doi:10.1021/la000569d.

[113] Chung MH, Chung YC. Polymerized ion pair amphiphile that shows remarkable enhancement in encapsulation efficiency and very slow release of fluorescent markers. Colloids Surf B Biointerf 2002;24:111-121. doi:10.1016/S0927-7765(01)00223-5.

[114] Chung MH, Park C, Chun BC, Chung YC. Polymerized ion pair amphiphile vesicles with pH-sensitive transformation and controlled release property. Colloids Surf B Biointerf 2004;34:179-184. doi:10.1016/j.colsurfb.2003.12.014.

[115] Chung MH, Park JH, Chun BC, Chung YC. Polymerizable ion-pair amphiphile that has a polymerizable group at cationic ammonium chain. Colloids Surf B Biointerf 2004;39:165-170. doi:10.1016/j.colsurfb.2004.06.012.

[116] Chung M, Chung Y, Chul B. Highly pH-sensitive ion pair amphiphile vesicle. Colloids Surf B Biointerf 2003;29:75-80.

[117] Chung M, Park M, Chul B, Chung Y. Encapsulation and permeation properties of the polymerized ion pair amphiphile vesicle that has an additional carboxyl group on anionic chain. Colloids Surf B Biointerf 2003;28:83-93.

[118] Sui J, Zhang L, Travas-Sejdic J, Kilmartin PA. Synthesis of poly(3,4-ethylenedioxythiophene) hollow spheres in CTAB/DBS – mixed surfactant solutions. Macromol Symp 2010;290:107-114. doi:10.1002/masy.201050412.

[119] Jiao WQ, Yue MB, Wang YM, He MY. Catanionic-surfactant templated synthesis of organized mesoporous γ-alumina by double hydrolysis method. J Porous Mater 2012;19:61-70. doi:10.1007/s10934-010-9448-3.

[120] Wang R-L, Zhu Y-M, Zhang X-M, Ji H-W, Li L, Ge H-Y. An economic method for synthesis of highly ordered porous silica. J Colloid Interf Sci 2013;407:128-132. doi:10.1016/j.jcis.2013.06.033.

[121] Ohkubo T, Ogura T, Sakai H, Abe M. Synthesis of highly-ordered mesoporous silica particles using mixed cationic and anionic surfactants as templates. J Colloid Interf Sci 2007;312:42-46. doi:10.1016/j.jcis.2007.02.043.

[122] Lind A, Spliethoff B, Lindén M. Unusual, vesicle-like patterned, mesoscopically ordered silica. Chem Mater 2003;15:813-818. doi:10.1021/cm021243o.

[123] Prélot B, Zemb T. Calcium phosphate precipitation in catanionic templates. Mater Sci Eng C 2005;25:553-559. doi:10.1016/j.msec.2005.07.008.

[124] Tari NE, Motlagh MMK, Sohrabi B. Synthesis of hydroxyapatite particles in catanionic mixed surfactants template. Mater Chem Phys 2011;131:132-135. doi:10.1016/j.matchemphys.2011.07.078.

[125] Chen Z, Nan Z. Controlling the polymorph and morphology of $CaCO_3$ crystals using surfactant mixtures. J Colloid Interf Sci 2011;358:416-422. doi:10.1016/j.jcis.2011.02.062.

[126] Dong R, Weng R, Dou Y, Zhang L, Hao J. Preparation of calcium oxalate by vesicle modification in the catanionic surfactant system CDS/TTABr/H_2O. J Phys Chem B 2010;114:2131-2139. doi:10.1021/jp908736d.

[127] Danoff EJ, Wang X, Tung SH, Sinkov NA, Kemme AM, Raghavan SR, et al. Surfactant vesicles for high-efficiency capture and separation of charged organic solutes. Langmuir 2007;23:8965-8971. doi:10.1021/la070215n.

[128] Kahe H, Chamsaz M, Zavar MHA. A novel supramolecular aggregated liquid–solid microextraction method for the preconcentration and determination of trace amounts of lead in saline solutions and food samples using electrothermal atomic absorption spectrometry. RSC Adv 2016;6:49076-49082. doi:10.1039/C6RA03874C.

[129] Chen D, Zhang P, Li Y, Mei Z, Xiao Y. Hexafluoroisopropanol-induced coacervation in aqueous mixed systems of cationic and anionic surfactants for the extraction of sulfonamides in water samples. Anal Bioanal Chem 2014;406:6051-6060. doi:10.1007/s00216-014-8031-1.

[130] Wang S, Yin X, Yan Y, Xiang Z, Liu P, Chen Y, et al. Gold extraction through vesicles self-assembled by cationic Gemini surfactant and sodium deoxycholate. Ind Eng Chem Res 2016;55:8207-8214. doi:10.1021/acs.iecr.6b01712.

[131] Gao M, Wang L, Chen J, Li S, Lu G, Wang L, et al. Aggregation-induced emission active probe for light-up detection of anionic surfactants and wash-free bacterial imaging. Chem A Eur J 2016;22:5107-5112. doi:10.1002/chem.201505202.

[132] Pond MA, Zangmeister RA. Carbohydrate-functionalized surfactant vesicles for controlling the density of glycan arrays. Talanta 2012;91:134-139. doi:10.1016/j.talanta.2012.01.036.

[133] Javadian S, Yousefi A, Neshati J. Synergistic effect of mixed cationic and anionic surfactants on the corrosion inhibitor behavior of mild steel in 3.5% NaCl. Appl Surf Sci 2013;285:674-681. doi:10.1016/j.apsusc.2013.08.109.

Chapter 3

Hydrophobic Polymers Flooding

Abdelaziz N. El-hoshoudy, Saad M. Desouky,

Mohamed H. Betiha and Ahmed M. Alsabagh

Additional information is available at the end of the chapter

http://dx.doi.org/10.5772/intechopen.69645

Abstract

Crude oil and other petroleum products are crucial to the global economy today due to increasing energy demand approximately (~1.5%) per year and significant oil remaining after primary and secondary oil recovery (~45-55% of original oil in place, OOIP), which accelerates the development of enhanced oil recovery (EOR) technologies. Polymer flooding through hydrophobically associated polyacrylamides (HAPAM) is a widely implemented EOR-technique, so they attracted much attention on both academic and industrial scales. Hydrophobically associating polyacrylamide (HAPAM) prepared by free radical emulsion polymerization of acrylamide (AM) monomer, divinyl sulfone as hydrophobic crosslinked moiety and surfmers, to chemically anchor a surfmer and hydrophobic crosslinker moiety onto the back bone of acrylamide chain. After that, polymeric nanocomposite was prepared through copolymerization of prepared HAPAM with different molar ratios of silica nanoparticles through one shot synthesis. Rheological properties for the prepared composites were evaluated. Wettability evaluation carried through quantitative and qualitative techniques where the results indicate novel polymers ability to alter rock wettability from oil-wet to water- wet.

Keywords: hydrophobic polymers, wettability alteration, enhanced oil recovery, polymerization, surfmers

1. Introduction

Crude oil is the most critical energy source in the world, especially for transportation, provision of heat and light as there has not been a sufficient energy source to replace crude oil has broadly integrated (i.e., today's energy needs are met in large part by crude oil). Petroleum products are crucial to the global economy today due to increasing energy demand approximately 1.5% per year [1] associated with population growth and improving life styles, limited proven oil reserves (i.e., shortage of current oil resources), declining oil production since 1995, difficulties in finding a new oil field, nonproductive primary and secondary recovery, significant oil remaining after secondary recovery (~45–55% of original oil in place, OOIP), and forecasts for tightening oil supply which driving the need to maximize the extraction of the original oil in place for every reservoir, and accelerating the development of enhanced oil recovery (EOR) technologies. EOR can be defined as any processes that increase oil recovery by reduction of the residual oil saturation (S_{or}) after primary and secondary production. Indeed, EOR techniques refer to any process that involves the injection of a fluid not normally present in the reservoir (e.g., polymers, foams, and surfactants) where the injected fluids interact with a crude oil/brine/rock (COBR) system to create favorable conditions, which maximize oil recovery [2]. Tertiary or enhanced oil recovery techniques include chemical, thermal, and miscible flooding [3] for recovering up to 40% of the OOIP. Thermal EOR involves injection of steam or hot water to reduce heavy oils viscosity, thus improving its flow. Miscible methods employ hydrocarbon gases (natural gas and flue gas) nitrogen, supercritical CO_2 to displace oil from a depleted oil reservoir. Gas flooding improves oil recovery by dissolving in, swelling, and reducing the viscosity of oil. Chemical flooding was, up to 2000s, less common EOR method than thermal and gas flooding but now, huge projects are initiated or revisited. In chemical EOR methods, an agent that is not normally present in the reservoir is injected to enhance the oil displacement. The chemical flooding processes involve injection of three kinds of chemicals: alkaline, surfactant, and polymer (soluble and cross-linked polymers), in addition to other chemicals such as foaming agents, acids, and solvents [4] and/or combination of alkaline-surfactant-polymer (ASP) flooding, and surely the most important substance in these methods is polymer flooding [5]. In the polymer flooding method, water-soluble polymers aimed to reduce mobility of displacing fluid leading to a more efficient displacement of moderately viscous oils. Addition of surfactant to the polymer formulation may, under very specific circumstances, reduce oil-water interfacial tension (IFT) and hence remobilizing the trapped oil [6], changing the wettability of the surface, forming emulsions, so enhance the oil production. For all chemical flooding processes, inclusion of a viscosifier (usually a water-soluble polymer) is required to provide an efficient sweep of the expensive chemicals through the reservoir. To increase the oil recovery efficiency in oil-wet reservoirs (unswept regions), different techniques have to be pursued [7].

(1) Improving volumetric sweeping efficiency by adjusting the oil/water mobility ratio through polymer flooding agents, which increase displacing fluid viscosity in order to modify the viscous forces being applied to drive oil out of the pores [8], thus increasing the produced crude oil amount. A polymer solution has increased viscosity and decreased

relative permeability so it is an attractive option to decrease the mobility ratio and increase the volumetric sweep efficiency of the injection [9].

(2) Altering the wettability of porous reservoir rock surfaces to more water-wet (i.e., by letting the value of contact angle $\theta \leq 90°$) [10]. The success of wettability alteration is seen as the increment in percentage of recovered oil, depending on natural wettability [11].

(3) Increase the oil displacement effectiveness by overcoming the capillary barrier through viscous and gravitational forces, so water can invade the rock matrix (i.e., modifying permeability) and displace the oil. Reduction of capillary pressure forces can be achieved by surfactants flooding to lower the oil-water interfacial tension (IFT) to ultralow values ~10^{-3} dyne cm^{-1} [12] which allow spontaneous or nearly spontaneous emulsification and displacement of the oil [13, 14].

2. Surfactants nature and applications in EOR

Surfactants are surface-active components, low to moderate molecular weight polar compound, consisting of an amphiphilic molecule, with a water soluble hydrophilic part called "head" (anionic, cationic, amphoteric, or nonionic) and a water insoluble hydrophobic part called "tail" [15] as shown in **Figure 1**. Surfactants used in EOR applications in order to [16]:

(1) Reduce interfacial tension between the trapped oil and injected aqueous phase to ultralow values (0.001 mN M−1) thereby increasing the capillary number (N_c) by a factor of at least three order of magnitude and reduce capillary force, which decrease the oil contact angle so alters wettability [17] consequently, trapped oil solubilization and mobilization increase.

(2) Alteration of matrix rock wettability toward more water-wet conditions, which would increase the brine-imbibition rates.

(3) Modify the properties of polymeric systems for a variety of applications [18].

(4) Surfactants are widely used in oil recovery for particle dispersion, emulsion stabilization, foam generation, reservoir wetting, and many other applications [19, 20].

(5) Favorable surfactant should characterized by low adsorption on reservoir rocks and, long-term surfactant stability under reservoir conditions (temperature, brine salinity, and hardness) and appropriate compatibility with reservoir fluids, especially tolerance to divalent cations such as Ca^{2+} and Mg^{2+}.

Depending upon the nature of the hydrophilic head group, the surfactants are classified, as shown in **Figure 2**, into the following:

(1) **Anionic**: the surface-active portion of the molecule (hydrophilic group) bears a negative charge such as carboxyl ($RCOO^{-}M^{+}$), sulfonate ($RSO^{3-}M^{+}$), sulfate ($ROSO^{3-}M^{+}$), or phosphate ($ROPO^{3-}M^{+}$) (e.g., $RC_6H_4SO_3^{-}Na^{+}$, alkyl benzene sulfonates).

(2) **Cationic**: the surface-active portion bears a positive charge such as the quaternary ammonium halides ($R_4N^+X^-$) (e.g., $RNH^{3+}Cl^-$, salt of a long chain amine).

(3) **Amphoteric (zwitterionic)**: where the molecule contains both a negative and a positive charge on the principal chain (surface-active portion) such as the sulfobetaines, $RN^+(CH3)_2CH_2CH_2SO_3{}^-$ (e.g., $RN^+H_2CH_2COO^-$, long chain amino acid).

(4) **Nonionic**: where the surface-active portion (hydrophilic group) bears no apparent ionic charge (has no net charge) but gets its water solubility from highly polar groups such as polyoxyethylene (POE or $ROCH_2CH_2O$ $^-$), sugars, or similar groups (e.g., RCOOCH-$_2$CHOHCH$_2$OH, monoglyceride of long chain fatty acid).

Surfactant flooding in enhanced oil recovery processes is considered uneconomical and remains challenging, especially in a high-salinity, high-temperature environment due to the following drawbacks:

(1) Loss of chemicals by adsorption in porous media [21], which dictate, the economics of an oil recovery or remediation process.

(2) Surfactant aggregates exhibit relatively low tolerance to divalent ions, salinity, and high-temperature condition ≥90°C [22].

(3) As described by Austad and Taugbøl [23] and Berger et al. [24], high performance surfactants, greatly lower oil/water IFT, but do not favor capillary-driven imbibition during water flooding [25, 26].

These previously reported shortages made the one think in an alternative when speak about EOR project [27].

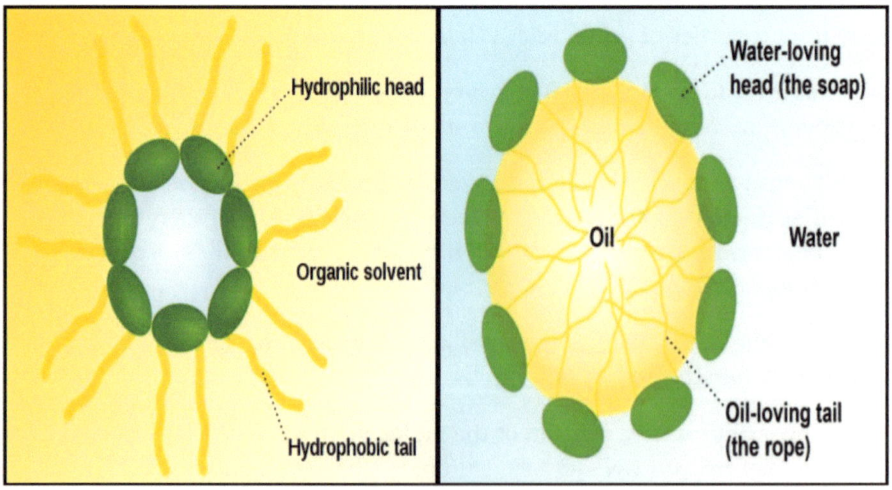

Figure 1. Surfactant skeletal structure (http://conf.sej.org/pollution-environmental-health/, 2011).

http://dx.doi.org/10.5772/intechopen.69645

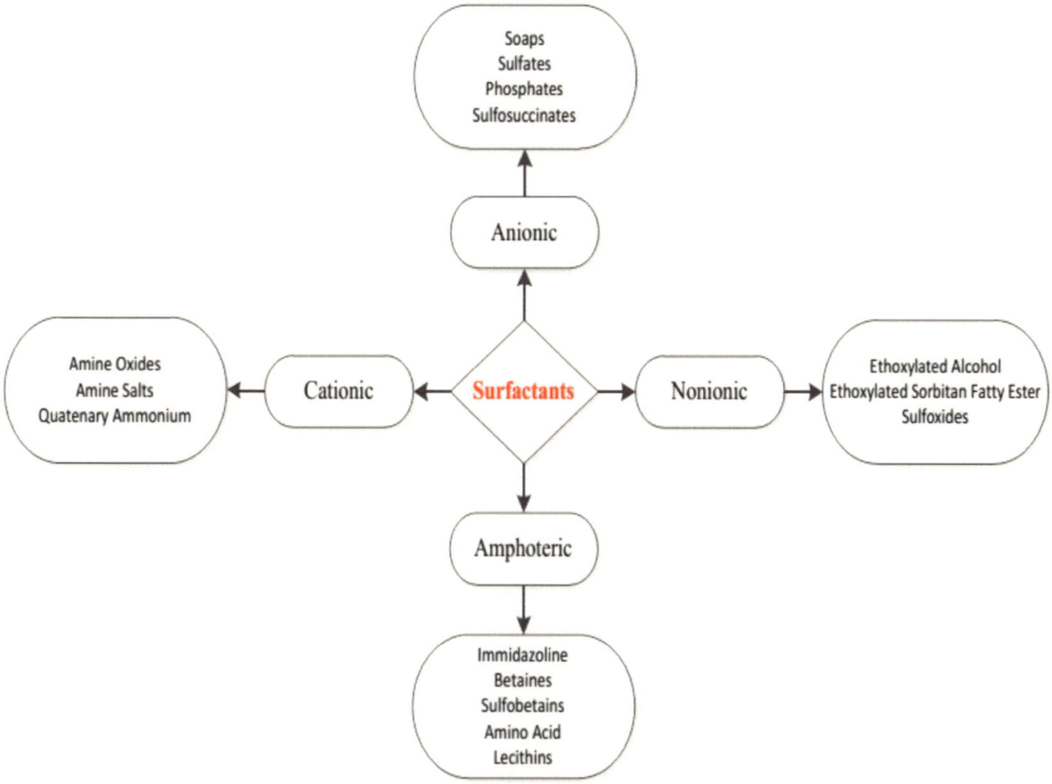

Figure 2. Classification of surfactants depending upon hydrophilic group nature.

3. Polymeric surfactants (surfmers)

Polymeric surfactants (surface-active monomers) are one kind of functional surfactants, which not only have amphiphilic structure composed of hydrophobic tail and hydrophilic head group [28], but also contain polymerizable vinyl double bonds [29] in their molecular architecture, resulting in novel physicochemical properties distinct from conventional surfactants [30] as follows:

(1) Analogous to common surfactants, they have surface activity and similar to vinyl monomers, they can be initiated and polymerized.

(2) Due to amphipathic property and polymerizability of surfmers, they can be used to synthesize inorganic/organic nanocomposite, can be applied to emulsion polymerization as polymerizable emulsifiers, to surface modification of solid substances, to synthesis of novel water-soluble hydrophobically associating polymers with strong thickening properties [31] so, they have great significance in enhanced oil recovery [32].

(3) Offer potential for developing hybrid-nanosized reaction and templating media with constrained geometries. Moreover, surfmer can be directly used as a hydrophobic monomer

to copolymerize with acrylamide (AM) forming hydrophobically associating polyacrylamide (HAPAM), which has been widely used in enhanced oil recovery, drilling fluids, coats, and paintings [33].

Freedman et al. [34] reported the first synthesis of vinyl monomers [35], which also functioned as emulsifying agents [36, 31]. Typical polymerizable groups that have been exploited are vinyl, allyl, acrylate, methacrylate, styryl, and acrylamide [37]. The position of the polymerizable group either "H-type" where the polymerizable group located in the hydrophilic head group, or "T-type" where the polymerizable group located in the hydrophobic tail have a profound effect on the surfactant self-assembly and properties [38, 39].

4. Wettability of porous media

Wettability defined as the preferential affinity of the solid matrix for either the aqueous phase or the oil phase "the tendency of one fluid (wetting fluid) to spread on or adhere to a solid surface in the presence of another immiscible fluid (nonwetting fluid)" [40]. Reservoir wettability is an important and elusive petrophysical parameter in all types of core analyses, which affect saturation and enhanced oil recovery processes [41]. There is a consensus in petroleum engineering that preferentially water-wet cores flood more efficiently than oil-wet cores; since, more oil is recovered from water-wet cores in the early flooding stages than from oil-wet cores [42]. This can be attained due to the strong wetting preference of the rock for water yields the most efficient oil displacement and due to imbibition phenomenon and other complex interactions occurring in the reservoir during production [43]. Modified polymers can affect markedly mineral wettability. The copolymer of acrylamide-methacrylamido propyl trimethyl-ammonium chloride can alter surface of oil-wet quartz with adsorbed dodecyl amine into water-wet one. Where the polymer masking the surfactant layer on the quartz particle accounts for the water wettability [44]. Consequently, in the present study, the authors try to prepare a copolymer and a nanocomposite modified by silica addition to alter wettability of sandstone rock from oil-wet to water-wet.

5. Principle and mechanism of polymer flooding for enhanced oil recovery

The polymer flooding process involves injection of polymer "slug" followed by continued long-term water flooding to drive the polymer slug and the oil bank in front of it toward the production wells, as shown in **Figure 3**. By water injection, it seeks layers of high permeability and since the oil is highly viscous as compared with injected water so, water will finger through oil resulting in decreased sweeping efficiency [5].

Mobility ratio (M) is defined as mobility of the displacing phase divided by the mobility of the displaced phase. Based on the principle of mobility ratio, water-soluble polymers reduce water mobility by two mechanisms: (1) increase the viscosity of the water phase and (2) reduce

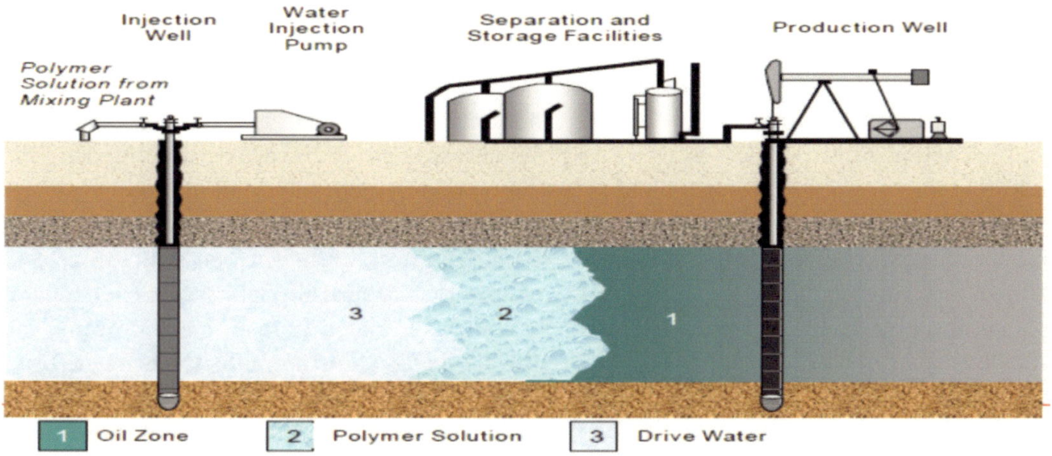

Figure 3. Polymer flooding mechanism.

the relative permeability of water to the porous rock by adsorption/retention of the polymer in the rock pore throats [45], and thereby creating a more efficient and uniform front to displace unswept oil from the reservoir (i.e., the mobility ratio (M) is inversely proportional to the water viscosity. With a reduced mobility ratio, the sweep efficiency is increased and, as a consequence, oil recovery is enhanced [46].

6. Hydrophobically associating polyacrylamide polymers (HAPAMs)

Hydrophobically associating polyacrylamide are prepared conveniently by micellar copolymerization of hydrophilic and hydrophobic monomers [47], or through grafting or incorporating hydrophobic chain cross-linking segments onto main chain of partially hydrolyzed polyacrylamide (HPAM). A lot of small molecule surfactants need to be added in order to enable the hydrophobic monomer to be solubilized into micelles, and the addition of small molecule surfactants brings many negative influences [30]. During HAPAM polymerization, hydrophilic surfmers dissolve in an aqueous phase resulting in homogeneous phase copolymerization of hydrophilic surfmers and acrylamide [48], which avoid drawbacks of surfactant addition. Moreover, above the critical micellar concentration (CMC) of surfmer, a microblock copolymerization mechanism carried out which means that a surfmer will be inserted into the backbone structure of acrylamide main chain, which gives rise to enhanced hydrophobic properties [49], stronger thickening property of HAPAM [50], and improved salinity resistance of HAPAM. As surfmer copolymerized with monomer and inserted in its main chain so surfmer separation from the polymer chain is prohibited [51]. These enhanced stability properties of polymers [52] have been reported for mechanical stability [53], electrolyte stability of the latex [54], a decrease of surfactant migration [51], and control of surface charge density [55]. Since surfactants are simply adsorbed onto the surface of particles in conventional emulsion polymerization, consequently increase emulsion stability by permanently fixing of the head groups. It has been long desired to obtain nanosized latexes containing higher polymer contents at lower surfactant concentrations [56]. Moreover, it was realized many years ago that polymerization of surfactants can lead to well-defined polymeric surfactants and potentially to polymerized micelles [57]. Recently, monomers composed of hydrophobic tail groups and hydrophilic head groups as well as a polymerizable group have been investigated [58]. Introduction of ionic groups contained in surfmer into polymer chains will improve the water solubility accompanying with perturbation of the hydrophobic association resulting in lowering the thickening effect [59]. The presence of phenyl group in the surfmer structure is well known to induce stronger van der Waals interactions than typical aliphatic groups due to their planar and polarizable structure, so the incorporation of one or more aromatic group(s) can stabilize hydrophobic associations involving the alkyl chain. Furthermore, the benzene rings can act as spacers, increasing the rigidity of polymer chains [60]. Consequently, incorporation of phenyl rings into the polyacrylamide (PAM) backbone through surfmer will improve its flooding characteristics in EOR applications. In addition, introduction of cationic groups into the PAM structure increases water solubility [61, 62] and decreases the water phase permeability (K_w) as it flows through porous media, which improve oil recovery in oil-displacing applications [63]. Hydrophobic polymers have attracted much attention on both academic and industrial laboratories for polymer flooding in enhanced oil recovery [64, 65] owing to their unique characteristics [66] which can be summarized as follow;

(1) In aqueous solutions, above a critical association concentration (C^*), their hydrophobic groups develop intermolecular hydrophobic associations in nanodomains, leading to building up of a 3D-transient network structure [67] in high ionic strength medium, so providing excellent viscosity building capacity [68, 69], remarkable rheological properties, and better stability with respect to salts than the unmodified HPAM precursors [70].

(2) Reduce interfacial tension at the solid/liquid interface, since hydrophobic moiety associates forming aggregates or micelles.

(3) Shows an unusual adsorption isotherm [71] so can be considered as a wettability modifier.

(4) Does not undergo mechanical degradation under high shear stress such as those encountered in pumps and near the well bore area, since the physical links between chains are disrupted before any irreversible degradation occurs, also they reform and retain their viscosity upon shear decreasing [72].

(5) High resistance to physicochemical conditions (temperature, pH, and ion content) prevailing around the wells, so considered a prospective EOR candidate as thickeners or rheology modifiers in high-temperature, high-pressure reservoirs [73–75], reservoir stimulation [76], and tertiary oil recovery [77].

In the present chapter, the authors try to overcome the shortage in chemical EOR candidates through synthesis of a novel surfmers (H-type) by the reaction of a 1-vinyl imidazole as a polymeric moiety containing double bond and 4-dodecyl benzene sulfonic acid surfactant, then hydrophobically associating polyacrylamide (HAPAM) prepared by free radical emulsion polymerization of acrylamide (AM) monomer, divinyl sulfone as a hydrophobic cross-linked moiety and surfmers, to chemically anchor a surfmer and hydrophobic cross-linker moiety onto the hydrophilic backbone of acrylamide chain. After that a hydrophobically associating polyacrylamides-SiO_2 (HAPAM-SiO_2) nanocomposite was prepared through copolymerization of acrylamide monomer with silica nanoparticles through one-shot synthesis. The rheological properties of copolymer solutions were investigated with respect to the polymer concentration, shear rate, shear stress, temperature, and salinity. Moreover, evaluation of behavioral characteristics and performance of these copolymers solution on wettability alteration, mobility ratio reduction, interfacial tension (IFT) reduction, and recovered oil amount under harsh reservoir are also reported [78, 79].

7. Experimental design and procedure

7.1. PROTOCOL 1: synthesis of polymeric surfactant (surfmer)

The addition reaction was carried out in a 250 ml three-necked Erlenmeyer flask equipped with a reflux condenser, mechanical stirrer, and nitrogen inlet/outlet. Note that 0.106 mol of 1-vinylimidazole was added dropwise to 0.106 mol of 4-dodecyl benzene sulfonic acid in ethyl acetate (150 ml) in an ice bath under a N_2 atmosphere. The reaction mixture maintained at 0°C for 2 h and then stirred for 12 h at 45°C. The white product was precipitated and recrystallized in 50 ml ethyl acetate upon cooling [78]. The yield was about 73%. The proposed structure is shown in **Scheme 1**.

7.2. PROTOCOL 2: synthesis of HAPAM copolymer

An aqueous solution of acrylamide in distilled water was gently bubbled with nitrogen gas for 30 min. The emulsion polymerization was carried out as previously reported in our literature [78] where designated reactants as listed in **Table 1** were added in a jacketed autoclave under an inert nitrogen environment for 12 h at 60°C. After reaction completion, viscous polymer gel was precipitated by acetone, redissolved in water, and reprecipitated in acetone then subjected to Soxhlet extraction with methanol for 24 h until a white solid obtained. The proposed structure is shown in **Scheme 2**.

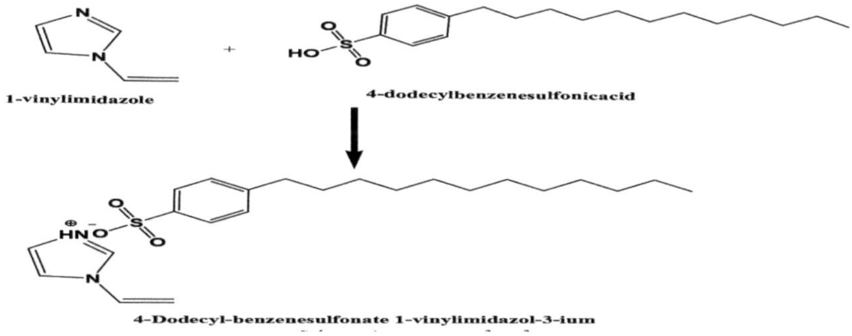

Scheme 1. Structure of surfmer.

Run #	A	×B	C	D	E	F	G	H
1	4.22×10^{-1}	8.44×10^{-3}	1.52×10^{-3}	2.84×10^{-3}	40	5.4	12	260
2	8.44×10^{-1}	1.69×10^{-2}	3.03×10^{-3}	5.68×10^{-3}	50			
3	1.69	3.38×10^{-2}	6.07×10^{-3}	1.14×10^{-2}	60			
4	3.38	6.75×10^{-2}	1.21×10^{-2}	2.27×10^{-2}	65			
5	6.75	1.35×10^{-1}	2.43×10^{-2}	4.54×10^{-2}	70			

Notes: × Surfmer concentration is (3–45) times its CMC value. A; monomer (acrylamide) concentration, mol L^{-1}. B; surfmer concentration, mol L^{-1}. C; initiator (KPS) concentration, mol L^{-1}. D; cross-linker (DVS) concentration, mol L^{-1}. E; temperature, °C. F; pH-value. G; reaction time, h. H; deionized water, g.

Table 1. Reactants concentration and reaction conditions in the case of HAPAM.

7.3. PROTOCOL 3: preparation of HAPAM-SiO_2 nanocomposite

After determination of optimum polymerization conditions and optimum reactants concentration, (3-aminopropyl) triethoxy silane was added in different molar ratios, as shown in **Table 2.** The polymerization procedure was carried out typically as shown previously in **PROTOCOL 2** with regarding addition of 3-aminopropyl triethoxy silane and KPS-initiator at the same time individually at reaction temperature of 60°C. The proposed structure is shown in **Scheme 3.**

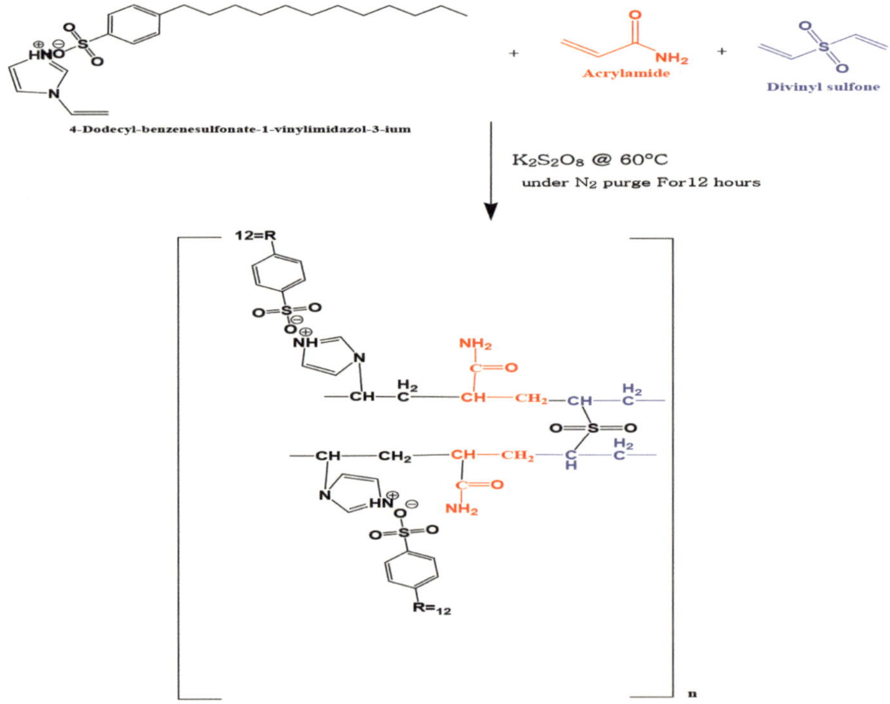

Scheme 2. Structure of HAPAM copolymer.

http://dx.doi.org/10.5772/intechopen.69645

Run #	A	B	C	D	E	F	G	H	I
1	1.69	3.38×10^{-2}	6.07×10^{-3}	1.14×10^{-2}	60	5.4	12	260	7.53×10^{-4}
2									1.51×10^{-3}
3									3.01×10^{-3}
4									6.02×10^{-3}
5									1.20×10^{-2}

Note: **I**; silica concentration, mol L^{-1}.

Table 2. Reactants concentration and reaction conditions in the case of HAPAM-SiO_2.

Further discussions of chemical synthesis and spectroscopic characterization of surfmer, HAPAM copolymer, and nanocomposite by means of FTIR, ^{1}H-NMR, ^{13}C-NMR, transmission electron microscope (TEM), scanning electron microscope (SEM), X-ray diffraction (XRD), and differential scanning calorimetry (DSC) as well as optimum polymerization conditions are reported in our previous literature [9, 78, 80]. Moreover, critical micelle concentration, surface excess concentration, and surface area of prepared surfmer and original surfactant indicate higher surface activity of prepared surfmer, which increases latex stability [78].

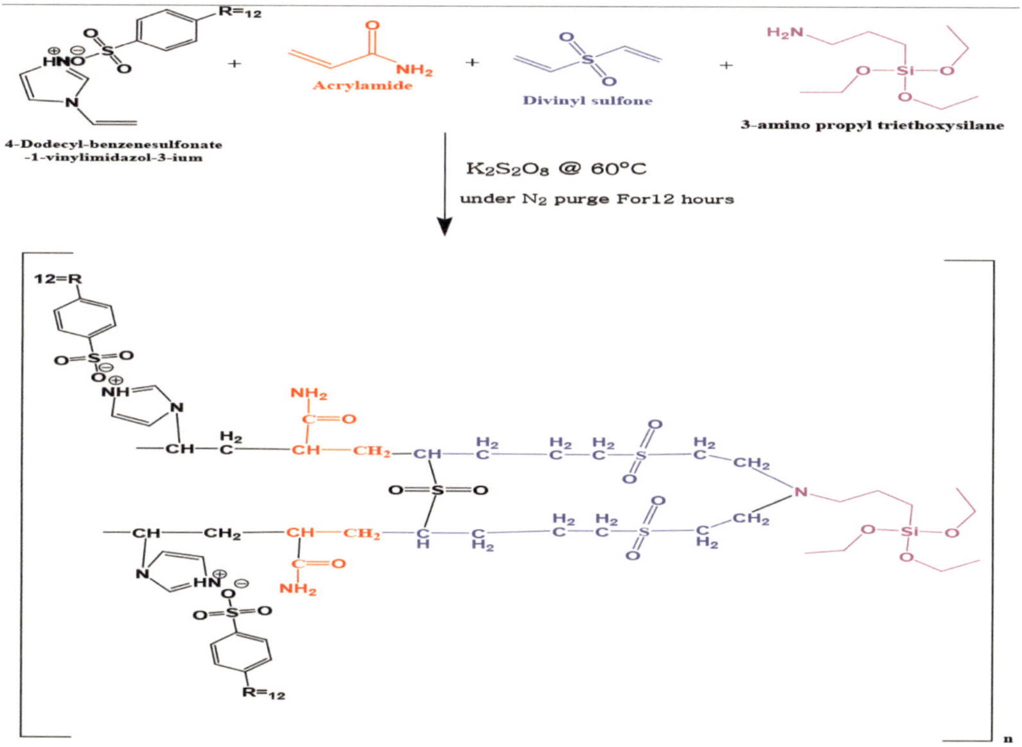

Scheme 3. Structure of HAPAM-SiO_2 nanocomposite.

Rheological and solution properties were evaluated under simulated reservoir conditions as a function of polymer concentration and reservoir salinity, temperature, and shear rate. The results show good salt and temperature resistance, interfacial tension reduction and enhanced viscosity characteristics. The capability of polymer and nanocomposite to increase oil recovery was assessed through a linear packed sandstone model, as previously reported [78, 80]. Core flood tests were carried out under simulated reservoir conditions where a sand cleaning procedure, packing, flooding experiments, and recovered oil amount were discussed elsewhere [78, 80].

8. Rock wettability evaluation

8.1. Quantitative assessment

Wettability was evaluated by measuring the contact angle between oil droplet and rock surface at temperature = 90°C and salinity = 40,000 ppm) and polymer solution concentration of 2 g L^{-1}. The contact angles measured through a static sessile drop method for a spherical core plate for a period of 2 days as reported in **Figure 4**. After aging with crude oil for a day at elevated temperature, the plate was found to be oil-wet. The plate is then immersed in polymer/nanocomposite-brine solution under reservoir conditions, where oil droplet hanged on the plate lower surface and photographed for 48 h. Images are analyzed mathematically to calculate the contact angle. It is observed that advancing contact angle decreases with time and stabilizes at a value of about 74 and 68° in the case of HAPAM and HAPAM-SiO_2, respectively [78]. Wettability alteration by can be explained on basis of (1) by polymer/nanocomposite adsorption on the rock surface, physicochemical properties altered, where thin wetting water film becomes unstable at the interface [81] and ruptured so, creating a continuous oil path for oil displacement which in turn increases oil recovery, (2) positively charged nitrogen bases adsorb on negatively charged sandstone rock surface at pH = 6, so wettability

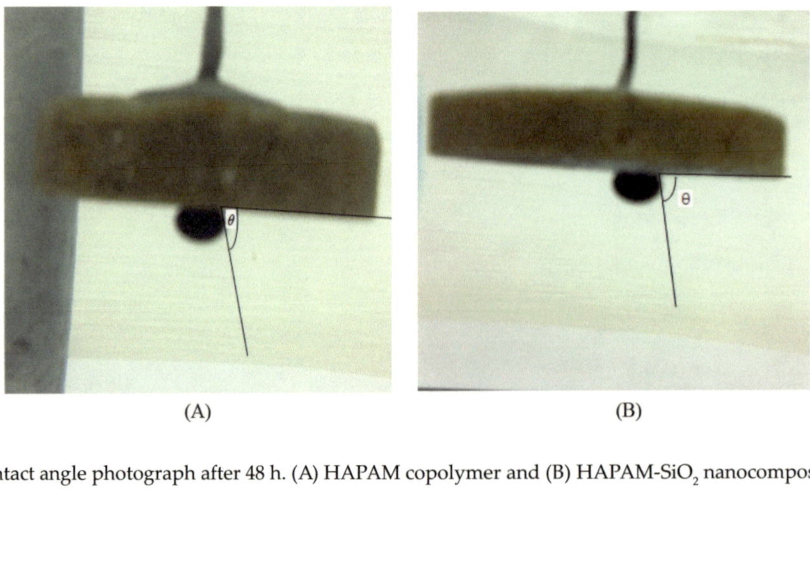

(A) (B)

Figure 4. Contact angle photograph after 48 h. (A) HAPAM copolymer and (B) HAPAM-SiO_2 nanocomposite.

changed from oil-wet to water-wet, (3) flood detergency improved by SiO_2-nanoparticles addition leading to higher recoveries of residual oil, since associating polymer chain with nanoparticles enables a nanofluid to act as wetting agents, demulsifiers, and surface tension reducers at the very smallest of contact angles, which greatly enhances the removal of "foreign" materials such as oil, paraffin, and polymer residues, leaving the substrate water-wet. This is confirmed by reducing contact angle to nearly 74 and 68° in the case of HAPAM and HAPAM-SiO_2, respectively [78].

8.2. Qualitative assessment

Figure 5 shows qualitative evaluation of wettability through a two-phase separation test, it is shown that grinded sandstone grains are oil-wet as it is dispersed in the oil phase in the case of oil and brine solution, as shown in **Figure 5A**. While the sandstone grains sink into the aqueous phase of polymer solution in the case of polymer/nanocomposite, as shown in **Figure 5B** and **C**. This means that sandstone grains become water-wet. So, we can conclude that the polymer/nanocomposite able to alter wettability of the rock from oil-wet to water-wet so, improve recovered oil amount [78].

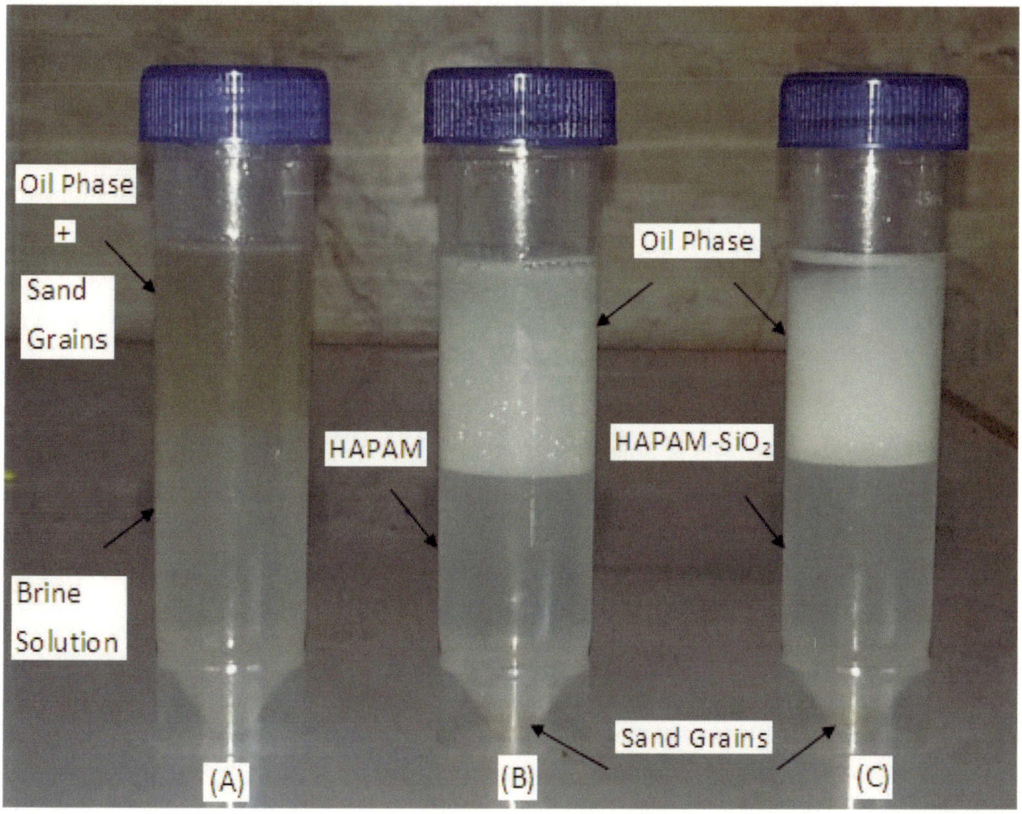

Figure 5. Two-phase separation test.

9. Conclusion

Improved oil recovery by polymer flooding involves injection of a mobility control agent (e.g., polyacrylamide and its hydrophobically associated derivatives) in order to displace the mobilized oil to the producing well, and improve seeping efficiency. In this chapter, the authors reported about synthesis of hydrophobically associating polyacrylamide (HAPAM) prepared by free radical emulsion polymerization and its modified nanocomposite derivative. Chemical structure of the prepared latexes was proven through different techniques such as FTIR, ^{1}H-NMR, ^{13}C-NMR, scanning electron microscope (SEM), transmission electron microscope (TEM), and X-ray diffraction, while particle size and particle size distribution were characterized by dynamic light scattering (DLS) and thermal properties characterized by thermal gravimetric analysis (TGA), and differential scanning calorimetry (DSC) as reported in our previous literature [78, 80]. Rheological properties were assessed in accordance with salinity and temperature tolerance, polymer concentration, and shear rates. Core flooding carried out via a linear pressurized packed model [9, 78–80]. Based on the experimental results, the following conclusions can be drawn:

(1) HAPAM-SiO_2 nanocomposite prepared by introducing silica nanoparticles through one-shot synthesis via Aza-Michael addition reaction, so we can overcome shortages arising from agglomeration and coagulation of modified silica particles during emulsion polymerization reactions.

(2) The prepared HAPAM copolymer and HAPAM-SiO_2 nanocomposite had the perfect property of retaining the viscosity and strong non-Newtonian behaviors (i.e., exhibit shear thinning behavior); so they can be considered as a promised EOR candidates for polymer flooding projects.

(3) They respond to *in situ* reservoir stimuli (temperature, ionic strength, pH, and shear stress) also, show good thermal, rheological, and salt resistant properties even under reservoir conditions, and consequently improve sweeping efficiency.

(4) They effectively reduce interfacial tension to ultralow values, so increase mobilization of residual crude oil, which resemble the behavior of interfacial tension agents.

(5) Wettability assessment by a static sessile drop method indicates that the HAPAM copolymer and HAPAM-SiO_2 nanocomposite can alter rock wettability from oil-wet to water-wet, which in turn will increase a recovery factor as there is a consensus in petroleum engineering that water-wet reservoirs recover more oil than oil-wet ones.

In addition to the aforementioned aspects, and to the best of our knowledge, no polymers had previously reported to alter sandstone rock wettability, consequently the novel copolymer and nanocomposite considered as a promising candidates for EOR applications as a wettability-modifying agent in high-temperature and high-mineralization oil fields as compared to currently applied commercial polyacrylamides. On an industrial scale, we hope that a novel polymer applied as an EOR candidate to solve some of energy shortages as recovered oil amount reach to 26% from original oil in place (OOIP).

Author details

Abdelaziz N. El-hoshoudy*, Saad M. Desouky, Mohamed H. Betiha and Ahmed M. Alsabagh

*Address all correspondence to: azizchemist@yahoo.com

Egyptian Petroleum Research Institute, Cairo, Egypt

References

[1] Tabary R, Bazin B. Advances in chemical flooding IFP-OAPEC Joint Seminar; Improved Oil recovery (IOR) Techniques and Their Role in Boosting the Recovery Factor, France; 2007

[2] Green DW, Willhite GP. Enhanced Oil Recovery: Henry L. Doherty Memorial Fund of AIME. TX: Society of Petroleum Engineers Richardson; 1998

[3] Marle CM. Oil entrapment and mobilization. Critical Reports on Applied Chemistry, Basic Concepts in Enhanced Oil Recovery Processes. New York: Elsevier Applied Science. 1991;**33**:39

[4] Schramm LL. Fundamentals and Applications in the Petroleum Industry; Foams. Cambridge: Cambridge University Press; 1994. p. 355

[5] Abidin A, Puspasari T, Nugroho W. Polymers for enhanced oil recovery technology. Procedia Chemistry. 2012;**4**:11-16

[6] Chen H, Lucas L, Nogaret L, Yang H, Kenyon D, editors. Laboratory monitoring of surfactant imbibition using computerized tomography. In: SPE International Petroleum Conference and Exhibition in Mexico; 1-3 February, Villahermosa, Mexico. Society of Petroleum Engineers; 2000

[7] Aladasani A. Updated EOR screening criteria and modeling the impacts of water salinity changes on oil recovery. 2012

[8] Nguyen P-T, Do B-PH, Pham D-K, Nguyen Q-T, Dao D-QP, Nguyen H-A, editors. Evaluation on the EOR potential capacity of the synthesized composite silica-core/polymer-shell nanoparticles blended with surfactant systems for the HPHT offshore reservoir conditions. In: SPE International Oilfield Nanotechnology Conference and Exhibition; 12-14 June, Noordwijk, the Netherlands. Society of Petroleum Engineers; 2012

[9] El-Hoshoudy A, Desouky S, Elkady M, Alsabagh A, Betiha M, Mahmoud S. Investigation of optimum polymerization conditions for synthesis of cross-linked polyacrylamide-amphoteric surfmer nanocomposites for polymer flooding in sandstone reservoirs. International Journal of Polymer Science. 2015;**2015**

[10] Al-Hadhrami HS, Blunt MJ, editors. Thermally induced wettability alteration to improve oil recovery in fractured reservoirs. In: SPE/DOE Improved Oil Recovery Symposium; 3-5 April, Tulsa, Oklahoma. Society of Petroleum Engineers; 2000

[11] Ayirala SC, Vijapurapu CS, Rao DN. Beneficial effects of wettability altering surfactants in oil-wet fractured reservoirs. Journal of Petroleum Science and Engineering. 2006;**52**(1):261-274

[12] Goddard III WA, Wu Y, Shuler PJ, Tang Y. An experimental study of wetting behavior and surfactant EOR in carbonates with model compounds. SPE Journal. **2008**;13:26

[13] Poettmann FH. In Improved Oil Recovery. Oklahoma City: Interstate Compact Commission; 1983. pp. 173-250

[14] Lake LW. Enhanced oil recovery. 1989

[15] Malmsten M. Surfactants and Polymers in Drug Delivery. CRC Press; 2002

[16] Hirasaki GJ, Miller CA, Puerto M, editors. Recent advances in surfactant EOR. In: SPE Annual Technical Conference and Exhibition; 21-24 September; Denver, Colorado, USA. Society of Petroleum Engineers; 2008

[17] Wu Y, Iglauer S, Shuler P, Tang Y, Goddard III W. Branched alkyl alcohol propoxylated sulfate surfactants for improved oil recovery. Tenside Surfactants Detergents. 2010;**47**(3):152-161

[18] Holmberg K. Novel Surfactants: Preparation Applications and Biodegradability, Revised and Expanded. CRC Press; 2003

[19] Maia Filho DC, Ramalho JB, Spinelli LS, Lucas EF. Aging of water-in-crude oil emulsions: Effect on water content, droplet size distribution, dynamic viscosity and stability. Colloids and Surfaces A: Physicochemical and Engineering Aspects. 2012;**396**:208-212

[20] Schramm LL. Surfactants: Fundamentals and Applications in the Petroleum Industry. Cambridge University Press; 2000

[21] Somasundaran P, Zhang L. Adsorption of surfactants on minerals for wettability control in improved oil recovery processes. Journal of Petroleum Science and Engineering. 2006;**52**(1):198-212

[22] Zhang DL, Liu S, Puerto M, Miller CA, Hirasaki GJ. Wettability alteration and spontaneous imbibition in oil-wet carbonate formations. Journal of Petroleum Science and Engineering. 2006;**52**(1):213-226

[23] Austad T, Taugbøl K. Chemical flooding of oil reservoirs. 2. Dissociative surfactant-polymer interaction with a negative effect on oil recovery. Colloids and Surfaces A: Physicochemical and Engineering Aspects. 1995;**103**(1-2):73-81

[24] Berger PD, Berger CH, Hsu IK. Anionic surfactants based on alkene sulfonic acid. Google Patents. 2000

[25] Cao Y, Li H. Interfacial activity of a novel family of polymeric surfactants. European Polymer Journal. 2002;**38**(7):1457-1463

[26] Seethepalli A, Adibhatla B, Mohanty K, editors. Wettability alteration during surfactant flooding of carbonate reservoirs. In: SPE/DOE Symposium on Improved Oil Recovery; 17-21 April; Tulsa, Oklahoma. Society of Petroleum Engineers; 2004

[27] El-hoshoudy A, Desouky S, Elkady M, Alsabagh A, Betiha M, Mahmoud S. Hydrophobically associated polymers for wettability alteration and enhanced oil recovery—Article review. Egyptian Journal of Petroleum. 2016

[28] Gao B, Guo H, Wang J, Zhang Y. Preparation of hydrophobic association polyacrylamide in a new micellar copolymerization system and its hydrophobically associative property. Macromolecules. 2008;**41**(8):2890-2897

[29] Reb P, Margarit-Puri K, Klapper M, Müllen K. Polymerizable and nonpolymerizable isophthalic acid derivatives as surfactants in emulsion polymerization. Macromolecules. 2000;**33**(21):7718-7723

[30] Guyot A. Advances in reactive surfactants. Advances in Colloid and Interface Science. 2004;**108**:3-22

[31] Summers M, Eastoe J. Applications of polymerizable surfactants. Advances in Colloid and Interface Science. 2003;**100**:137-152

[32] Wang Y, Wu F. Amphiphilic acrylamide-based copolymer with porphyrin pendants for the highly selective detection of Hg^{2+} in aqueous solutions. Polymer. 2015;**56**:223-228

[33] Xue W, Hamley IW, Castelletto V, Olmsted PD. Synthesis and characterization of hydrophobically modified polyacrylamides and some observations on rheological properties. European Polymer Journal. 2004;**40**(1):47-56

[34] Freedman HH, Mason JP, Medalia A. Polysoaps. II. The preparation of vinyl soaps. 1. The Journal of Organic Chemistry. 1958;**23**(1):76-82

[35] Suresh KI, Bartsch E. Effect of sulfonated 3-pentadecyl phenyl acrylate as surfmer in the emulsion polymerization of styrene: Synthesis and polymer properties. Colloid and Polymer Science. 2013;**291**(8):1843-1853

[36] Tsvetkov N, Lebedeva E, Lezov A, Podseval'nikova A, Akhmadeeva L, Zorin I, et al. Macromolecules of poly-(12-acryloylaminododecanoic acid) in organic solvent: Synthesis and molecular characteristics. Polymer. 2014;**55**(7):1716-1723

[37] Benbayer C, Saidi-Besbes S, de Givenchy ET, Amigoni S, Guittard F, Derdour A. Copolymerization of novel reactive fluorinated acrylic monomers with styrene: Reactivity ratio determination. Colloid and Polymer Science. 2014;**292**(7):1711-1717

[38] Samakande A, Hartmann P, Sanderson R. Synthesis and characterization of new cationic quaternary ammonium polymerizable surfactants. Journal of Colloid and Interface Science. 2006;**296**(1):316-323

[39] Samakande A, Hartmann PC, Cloete V, Sanderson RD. Use of acrylic based surfmers for the preparation of exfoliated polystyrene–clay nanocomposites. Polymer. 2007;**48**(6):1490-1499

[40] Khusainova A, Nielsen SM, Pedersen HH, Woodley JM, Shapiro A. Study of wettability of calcite surfaces using oil–brine–enzyme systems for enhanced oil recovery applications. Journal of Petroleum Science and Engineering. 2015;**127**:53-64

[41] Das S, Thundat T, Mitra SK. Asphaltene migration and separation in presence of aggregation in electroosmotic–electrophoretic microchannel transport. Colloids and Surfaces A: Physicochemical and Engineering Aspects. 2014;**446**:23-32

[42] Yanjun RE, Jiang G, Zheng D, Shilin SU, Yuxiu AN, Chunlei WA. Characteristics of ethoxylated fatty amine emulsion: Effects on the wettability and permeability of silicate formation under various pH conditions. Petroleum Exploration and Development. 2015;**42**(1):138-141

[43] Zhao Y, Song K, Bai M, Sun N. A new approach to obtaining relative permeability curves during chemical flooding process in a low-permeable reservoir. Environmental Earth Sciences. 2015;**73**(11):7001-7009

[44] Mandal B, Moulik SP, Ghosh S. Influence of aquo-organic solvent media on the self-aggregation of sodium dodecyl sulfate (SDS) and its interaction with polyvinylpyrrolidone (PVP). Colloid and Polymer Science. 2014;**292**(10):2485-2495

[45] Sorbie KS. Polymer-Improved Oil Recovery. Glasgow and London: Blackie and Son Ltd; 1991

[46] Zheng C, Gall B, Gao H, Miller A, Bryant R, editors. Effects of polymer adsorption and flow behavior on two-phase flow in porous. In: SPE/DOE Improved Oil Recovery Symposium; 19-22 April; Tulsa, Oklahoma. Society of Petroleum Engineers; 1998

[47] Camail M, Margaillan A, Martin I, Papailhou A, Vernet J. Synthesis of N-alkyl-and N-arylalkylacrylamides and micellar copolymerization with acrylamide. European Polymer Journal. 2000;**36**(9):1853-1863

[48] Gao B, Yu Y, Jiang L. Studies on micellar behavior of anionic and surface-active monomers with acrylamide type in aqueous solutions. Colloids and Surfaces A: Physicochemical and Engineering Aspects. 2007;**293**(1):210-216

[49] Yasin A, Zhou W, Yang H, Li H, Chen Y, Zhang X. Shape memory hydrogel based on a Hydrophobically-Modified polyacrylamide (HMPAM)/α-CD mixture via a host-guest approach. Macromolecular Rapid Communications. 2015;**36**(9):845-851

[50] Gao B, Wu N, Li Y. Study on tercopolymer of acrylamide containing strong anions and hydrophobic association blocks. Acta Polymerica Sinica. 2004;**1**(5):736-742

[51] Xu XJ, Chen F. Modified emulsifier-free emulsion polymerization of butyl methacrylate with ionic or/and nonionic comonomers. Journal of Applied Polymer Science. 2004;**92**(5):3080-3087

[52] Texter J, editor. Reactions and Synthesis in Surfactant Systems. CRC Press; 2001

[53] Borzenkov M, Mitina N, Lobaz V, Hevus O. Synthesis and properties of novel surface active monomers based on derivatives of 4-hydroxybutyric acid and 6-hydroxyhexanoic acid. Journal of Surfactants and Detergents. 2015;**18**(1):133-144

[54] Atta AM, Dyab AK, Al-Lohedan HA, AlJenady KA. Novel reactive polymerizable nonyl phenol ethoxylate surfactants as emulsifier in non-aqueous emulsion polymerization. Polymer Science Series B. 2014;**56**(6):770-787

[55] Borzenkov M, Hevus O. Surface Active Monomers: Synthesis, Properties, and Application. Springer; 2014

[56] Xu XJ, Goh HL, Siow KS, Gan LM. Synthesis of polymerizable anionic surfactants and their emulsion copolymerization with styrene. Langmuir. 2001;**17**(20):6077-6085

[57] Yuan J, Antonietti M. Poly (ionic liquid) latexes prepared by dispersion polymerization of ionic liquid monomers. Macromolecules. 2011;**44**(4):744-750

[58] Liu YX, Du ZJ, Li Y, Zhang C, Li HQ. Covalent functionalization of multiwalled carbon nanotubes with poly (acrylic acid). Chinese Journal of Chemistry. 2006;**24**(4):563-568

[59] Li J, Wu F-p, Wang E-j. Hydrophobically associating polyacrylamides modified by a novel self-associative cationic monomer. Chinese Journal of Polymer Science. 2010;**28**(2):137-145

[60] Zhang Q, Crosland E, Fabris D. Nested Arg-specific bifunctional crosslinkers for MS-based structural analysis of proteins and protein assemblies. Analytica Chimica Acta. 2008;**627**(1):117-128

[61] Zhang P, Wang Y, Chen W, Yu H, Qi Z, Li K. Preparation and solution characteristics of a novel hydrophobically associating terpolymer for enhanced oil recovery. Journal of Solution Chemistry. 2011;**40**(3):447-457

[62] Wever D, Picchioni F, Broekhuis A. Polymers for enhanced oil recovery: A paradigm for structure–property relationship in aqueous solution. Progress in Polymer Science. 2011;**36**(11):1558-1628

[63] Li LW, Bin NY, Hua SG, Jun SL. Oil-displacing experiments and mechanisms of amphoteric tercopolymers of acrylamide AP-81 and AP-125. Oilfield Chemistry. 2000;**1**:015

[64] Bastiat G, Grassl B, Khoukh A, François J. Study of sodium dodecyl sulfate–poly (propylene oxide) methacrylate mixed micelles. Langmuir. 2004;**20**(14):5759-5769

[65] Feng Y, Billon L, Grassl B, Bastiat G, Borisov O, François J. Hydrophobically associating polyacrylamides and their partially hydrolyzed derivatives prepared by post-modification. 2. Properties of non-hydrolyzed polymers in pure water and brine. Polymer. 2005;**46**(22):9283-9295

[66] Niu Y, Jian O, Zhu Z, Wang G, Sun G, editors. Research on Hydrophobically Associating water-soluble polymer used for EOR. In: SPE International Symposium on Oilfield Chemistry; 13-16 February; Houston, Texas. Society of Petroleum Engineers; 2001

[67] Candau F, Selb J. Hydrophobically-modified polyacrylamides prepared by micellar polymerization. Advances in Colloid and Interface Science. 1999;**79**(2):149-172

[68] Shashkina YA, Zaroslov YD, Smirnov V, Philippova O, Khokhlov A, Pryakhina T, et al. Hydrophobic aggregation in aqueous solutions of hydrophobically modified polyacrylamide in the vicinity of overlap concentration. Polymer. 2003;**44**(8):2289-2293

[69] Gao B, Wu N, Li Y. Interaction between the strong anionic character of strong anions and the hydrophobic association property of hydrophobic blocks in macromolecular chains of a water-soluble copolymer. Journal of Applied Polymer Science. 2005;**96**(3):714-722

[70] Maia AM, Borsali R, Balaban RC. Comparison between a polyacrylamide and a hydrophobically modified polyacrylamide flood in a sandstone core. Materials Science and Engineering: C. 2009;**29**(2):505-509

[71] Hu S-S, Zhang L, Cao X-L, Guo L-L, Zhu Y-W, Zhang L, et al. Influence of crude oil components on interfacial dilational properties of hydrophobically modified polyacrylamide. Energy & Fuels. 2015;**29**(3):1564-1573

[72] Evani S, Rose G. Water soluble hydrophobe association polymers. Polymer Materials Science & Engineering. 1987;**57**:477-481

[73] Rodrigues JDA, Lachter ER, de Sá CH, de Mello M, Nascimento RSV, editors. New multifunctional polymeric additives for water-based muds. In: SPE Annual Technical Conference and Exhibition; 24-27 September, San Antonio, Texas, USA. Society of Petroleum Engineers; 2006

[74] Zhong C, Huang R, Xu J. Characterization, solution behavior, and microstructure of a hydrophobically associating nonionic copolymer. Journal of Solution Chemistry. 2008;**37**(9):1227-1243

[75] Gouveia LM, Paillet S, Khoukh A, Grassl B, Müller AJ. The effect of the ionic strength on the rheological behavior of hydrophobically modified polyacrylamide aqueous solutions mixed with sodium dodecyl sulfate (SDS) or cetyltrimethylammonium p-toluenesulfonate (CTAT). Colloids and Surfaces A: Physicochemical and Engineering Aspects. 2008;**322**(1):211-218

[76] Dalrymple ED, Eoff LS, Everett DM, editors. Conformance while fracturing tight gas formations. In: SPE Tight Gas Completions Conference; 9-11 June; San Antonio, Texas, USA. Society of Petroleum Engineers; 2008

[77] Ranjbar M, Schaffie M. Improved treatment of acrylamide co-and terpolymers for water control in gas-producing and storage wells. Journal of Petroleum Science and Engineering. 2000;**26**(1):133-141

[78] El-hoshoudy A, Desouky S, Betiha M, Alsabagh A. Use of 1-vinyl imidazole based surfmers for preparation of polyacrylamide–SiO_2 nanocomposite through aza-Michael addition copolymerization reaction for rock wettability alteration. Fuel. 2016;**170**:161-175

[79] El-hoshoudy AN, Desouky SE, Alsabagh AM, Betiha MA, MY EK, Mahmoud S. Evaluation of solution and rheological properties for hydrophobically associated polyacrylamide copolymer as a promised enhanced oil recovery candidate. Egyptian Journal of Petroleum. 2016

[80] El Hoshoudy A, Desouky S, Al-sabagh A, El-kady M, Betiha M, Mahmoud S. Synthesis and characterization of polyacrylamide crosslinked copolymer for enhanced oil recovery and rock wettability alteration. International Journal of Oil, Gas and Coal Engineering. 2015;**3**(4):43-55. DOI: 10.11648/j.ogce.20150304.11

[81] Xu W, Ayirala SC, Rao DN. Wettability alterations due to crude oil composition and an anionic surfactant in petroleum reservoirs. Journal of Adhesion Science and Technology. 2006;**20**(7):693-704

Chapter 4

Multifunctional Gemini Surfactants: Structure, Synthesis, Properties and Applications

Bogumil E. Brycki, Iwona H. Kowalczyk,
Adrianna Szulc, Olga Kaczerewska and Marta Pakiet

Additional information is available at the end of the chapter

http://dx.doi.org/10.5772/intechopen.68755

Abstract

Gemini cationic surfactants are compounds which are composed of two hydrophilic head groups and two hydrophobic tails linked by a spacer at the head groups or closed to them. The spacer can be either hydrophobic or hydrophilic. It can be rigid or flexible. The neutral charge of the molecule is retained by the presence of organic or inorganic counterions. Critical micelle concentrations (CMCs), surface tension (γ) and minimal inhibitory concentration (MIC) are dozen times lower than corresponding parameters of monomeric surfactants. The unique properties of gemini surfactants with a wide range of hydrophilic-lipophilic balance (HLB) make them a very useful, innovative material in detergents, cosmetics, personal care products, additives for paints and coatings, biocides, material science, organic synthesis, pharmacy, textiles, enhanced oil recovery, nanotechnology, petroleum and many other branches of life. A large number of papers concerning gemini surfactants have been published so far. This review presents a synthetic look at current work devoted to structure, synthesis and applications of gemini surfactants.

Keywords: gemini surfactant, surface activity, antimicrobial activity, corrosion inhibitors, smart materials

1. Introduction

Everything in the world involves chemistry and chemicals. Chemistry is essential for our life and our existence in the material world. Without chemistry there would literally be nothing. The diversity of life and material forms is based on versatility of chemical compounds and interactions between them. Therefore, the better we know chemistry, the better we know our world. From among of over 120 millions of currently known organic and inorganic compounds [1], some of them have a very special meaning to facilitate our life. One of the very important and it seems to be irreplaceable groups of chemicals are surfactants.

The surfactant molecules contain at least two moieties, hydrophobic and hydrophilic one. Hydrophobic moiety is usually a straight or branched hydrocarbon or fluorocarbon chain with 8–18 carbon atoms, whereas hydrophilic moiety is a polar or ionic group. The balance between hydrophobic and hydrophilic parts, hydrophilic-lipophilic balance (HLB), is responsible for special properties of these amphiphilic compounds in solutions such as adsorption on the surfaces and interfaces and formation of self-assembly aggregates. The driving force for amphiphiles' adsorption is the lowering of the free energy of the phase boundary that provides to lowering the surface and interface tension.

This fundamental feature of amphiphiles is a base of their very wide practical applications. Surfactants are used in almost every field of our activities. They find application in detergents [2], in personal care products [3], as additives for paints and coatings [4], as dyestuffs [5, 6], as biocides [7–10], in material science [2], in organic synthesis [11, 12], in pharmacy [5, 13], in textiles and leather [2, 14, 15], in agrochemicals [16], in fibres [17–19], in plastics [20], in food processing [21, 22], in petroleum industry for enhanced and tertiary oil recovery [23–25], in environmental protection (*oil slick dispersant*) [26, 27] and in explosives [28]. Surfactants are also used to replace traditional solvents, giving lower risk and reduced environmental impacts [29]. Surfactants can also play a key role in the development of technologies such as nano- and smart materials [30].

Currently, the global surfactant market has been segmented into anionic, cationic, non-ionic and amphoteric [31]. Anionic surfactants, like alkylbenzene sulfonates, α-olefin sulfonates, sulphates and ether sulphates, carboxylates, isethionates, taurates and phosphate surfactants held around 50% share of the global surfactant market. These surfactants are largely used in industrial and institutional cleaners and detergents. Cationic surfactants, like quaternary alkylammonium salts, exhibit mainly softening, antistatic, soil-repellent, antibacterial and corrosion inhibitory effects, whereas non-ionic surfactants, that is, alcohol ethoxylates, are suitable for cleaning purposes, as they are not sensitive to water hardness. Amphoteric (zwitterionic) surfactants, mainly derivatives of trimethyl glycine, are pH sensitive and have excellent dermatological properties. Besides, of these four main groups, there are also special surfactants, that is, fluorocarbon and silicone surfactants, sugar-based surfactants derived from mono- and polysaccharides, biosurfactants and polymeric surfactants. A very special group of surfactants are naturally occurring in living organisms' amphipathic molecules, *phospholipids*, like phosphatidylcholine (lecithin) (**Figure 1**), phosphatidylserine, phosphatidylethanolamine (cephalin), phosphatidylinositol and sphingomyelin, with the main applications in drug delivery systems [32].

The global surfactant market has been exceeded 15 million tons [31] and is expected to reach a valuation of US$28.8 billion by 2023, increasing at a 4.20% compound annual growth rate (CAGR) upon its 2014 value of close to US$20.3 billion [33].

An increasing use of surfactants is mainly driven by higher demand for personal care products, detergents, cleaners and industrial—anticorrosion and biocidal—products. This, in turn, is expected to lead to the introduction of innovative, more effective, surfactant-based products in

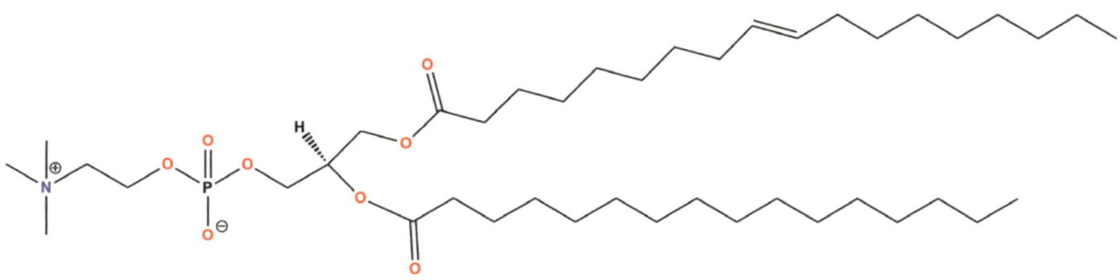

Figure 1. Structure of lecithin.

the near future. The higher efficacy of surfactants is directly related to lower CMC and surface tension as well as the efficient emulgation and solubilization effects. Such profile of innovative surfactants is accomplished to a high extent by gemini surfactants [34–36]. These compounds contain two hydrophilic head groups and two hydrophobic tails linked by a spacer at the head groups or closed to them. The structure of linker and its affinity to solvents can vary in a wide range. The gemini alkylammonium salts show unique surface and interfacial properties in aqueous solution. Critical micelle concentrations (CMCs) of gemini surfactants are usually much lower, up to hundred times, than CMCs of corresponding monomeric surfactants. The effectiveness of dimeric surfactants in lowering the surface tension is also much better than their monomeric analogues. The values of C_{20}, that is, surfactant concentration at which the surface tension (γ) is lowered by 20 mN/m, are dozen times smaller for gemini surfactants than monomeric surfactants. Moreover, gemini surfactants can form in solution many morphological structures, like spherical, ellipsoidal, rod shape and worm-like micelles as well as vesicles and helical or tubular forms. These unusual properties of gemini surfactants are ground of their applications as emulsifiers, dispersants, coating agents and corrosion inhibitors. Dimeric quaternary ammonium salts are also the excellent microbiocides. The antimicrobial activity (minimal inhibitory concentration - MIC) of quaternary ammonium salts strongly depends on their hydrophilic-lipophilic balance (HLB) and the length of the spacer. The longer the spacer, the better the antimicrobial activity. It is because gemini surfactants with longer spacers are more flexible and easily connect with the negative-charged surface of bacteria or fungi.

To better understand the fascinating physicochemical and biological properties of gemini surfactants and their wide potent applications, we present a review of synthesis, structure, properties and applications of these compounds.

2. Structure

Gemini surfactants contain two hydrophilic head groups and two hydrophobic tails linked by a spacer at the head groups or closed to them. When both hydrophobic parts are the same and hydrophilic groups are identical, then gemini surfactant forms symmetric structure (**Figures 2–4**) [37].

In contrast to symmetric dimeric surfactants are *heterogeminis* with two different, or the same, polar head groups and two different, or the same, hydrophobic groups (**Figures 5** and **6**) [38].

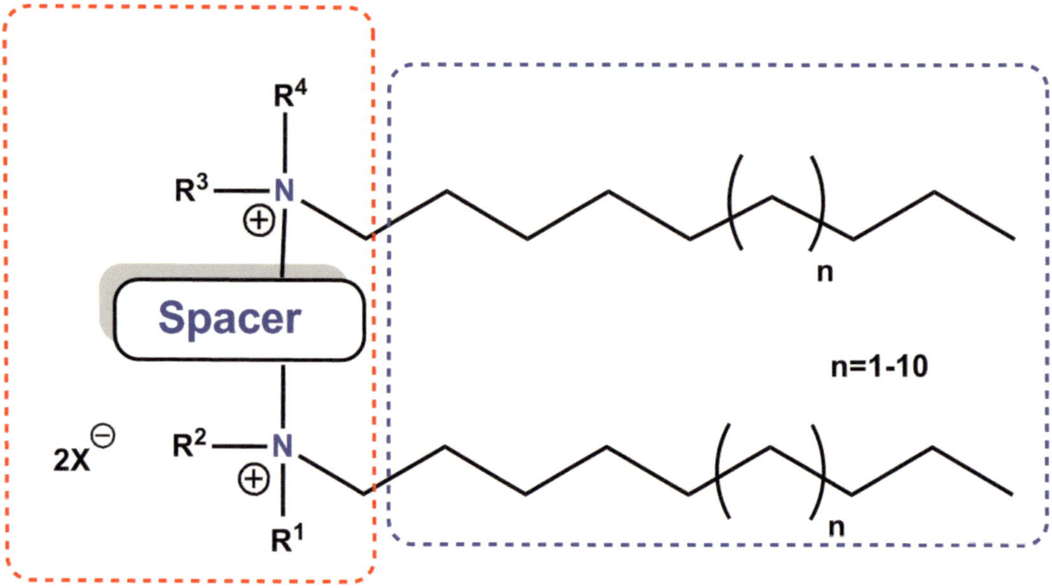

Figure 2. Structure of gemini surfactant with spacer at head groups.

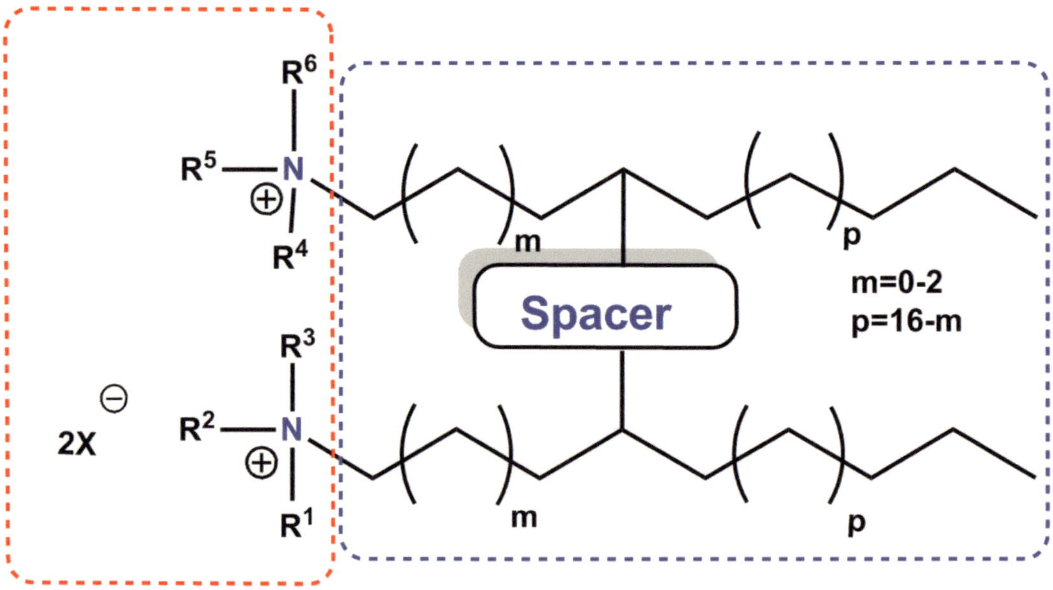

Figure 3. Structure of gemini surfactant with spacer in hydrophobic part.

The substituents in gemini surfactants are responsible to high extent for behaviour of these compounds in solution and their possible applications. Some examples of a large group of substituents, both hydrophobic and hydrophilic, are shown in **Figure 7**.

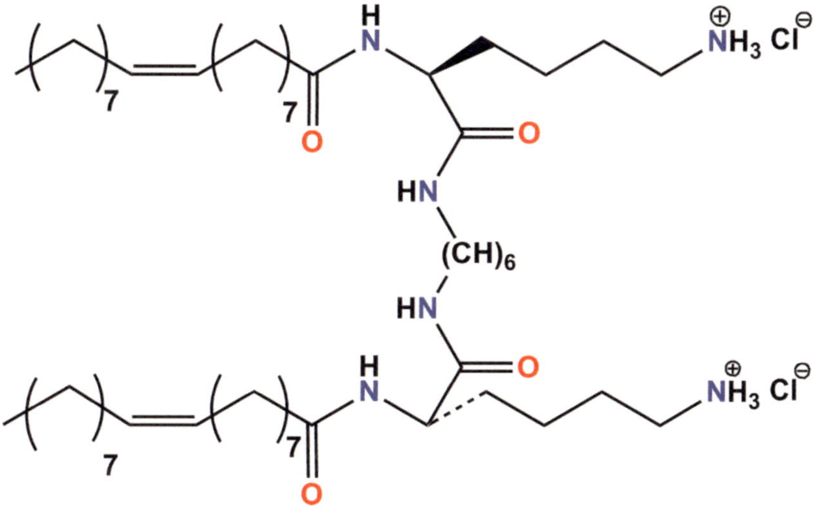

Figure 4. Bolaform of gemini surfactant and lysine-based gemini surfactant.

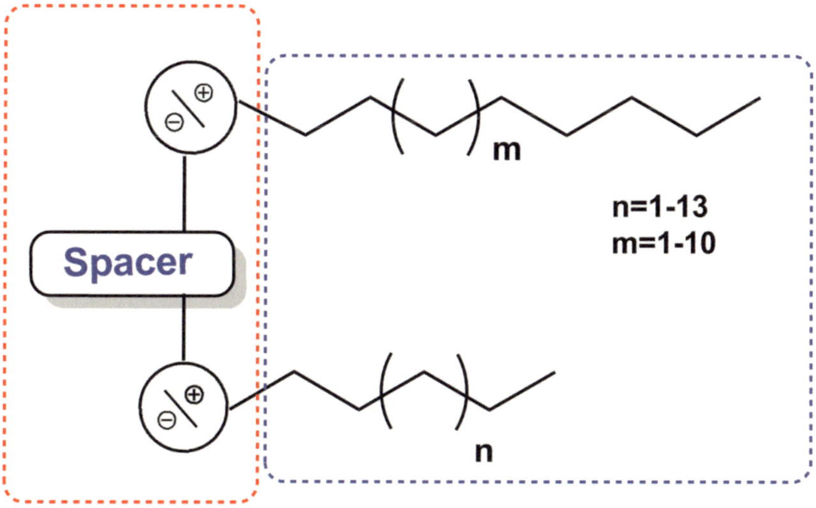

Figure 5. Structure of heterogemini surfactant.

Quaternary nitrogen atom usually exists in acyclic forms; however, there are many geminis with nitrogen involved in saturated and unsaturated rings (**Figure 8**).

Compounds with nitrogen involved in annulene unsaturated ring have a very special character because a ring plays to some extent a role of spacer (**Figure 9**) [39].

The spacer can be either rigid or flexible with tendency to hydrophobicity or hydrophilicity (**Figure 10**). It is a very important part of gemini surfactant which regulates the adsorption on the surfaces and interfaces and formation of self-assembly aggregates.

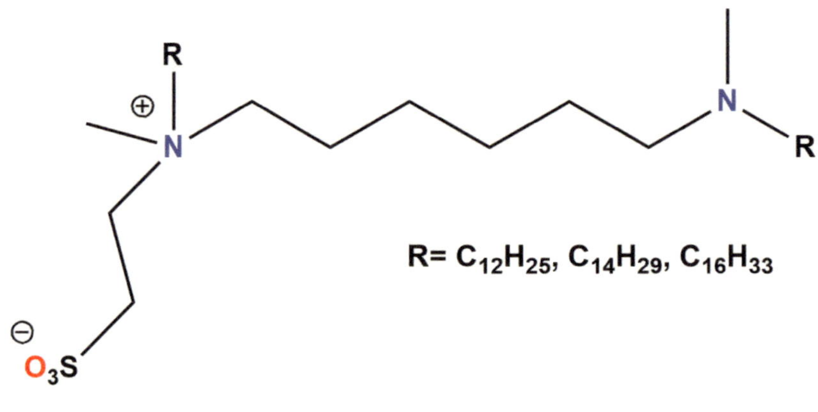

Figure 6. Example of dissymmetric surfactant.

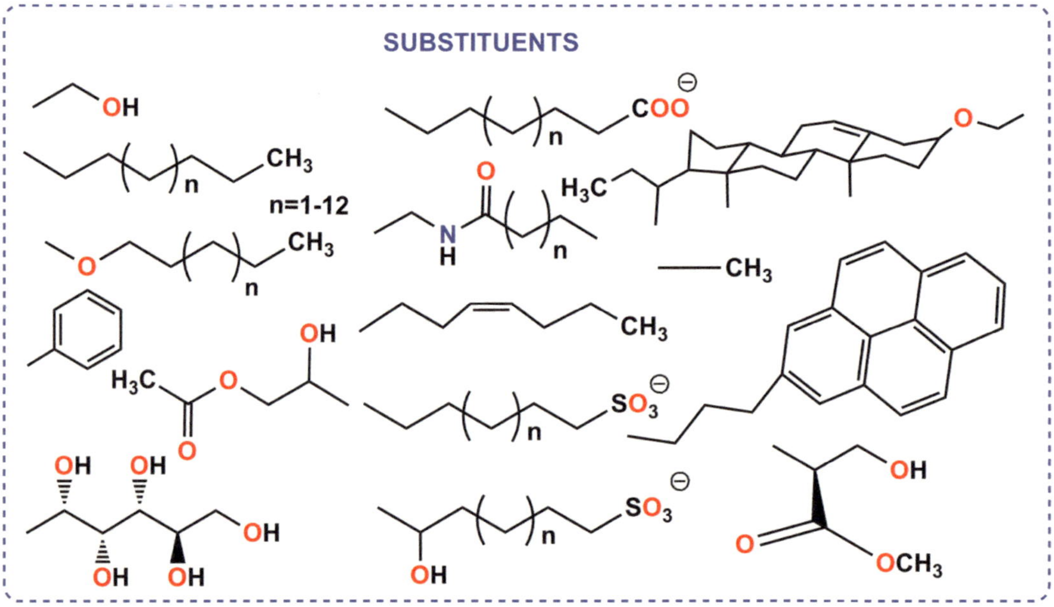

Figure 7. Examples of substituents in gemini surfactants.

The neutral charge of the molecule is retained by the presence of counterions, which can be organic or inorganic ones (**Figure 11**).

To get the anticipated properties of gemini surfactants, the structure has to be optimized by modification of HLB. It can be done by introduction of balanced polar or hydrophobic groups both to substituents and spacers. Polarity can be increased by ester, ether, amide, sulphide and cyclodextrin group [39–41]. To increase hydrophobicity some fluorine groups [42] or a dehydroabietylamine derivative [43] can be introduced to substituent. To increase biodegradability some amide or ester groups, which facilitate the biodegradation should be also introduced [44].

http://dx.doi.org/10.5772/intechopen.68755

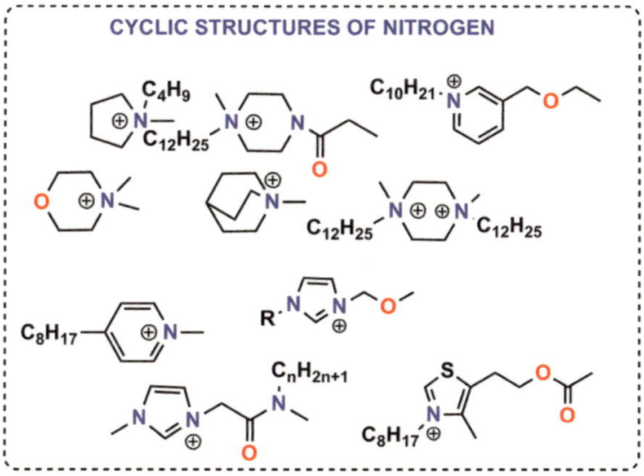

Figure 8. Nitrogen involved in saturated and unsaturated rings.

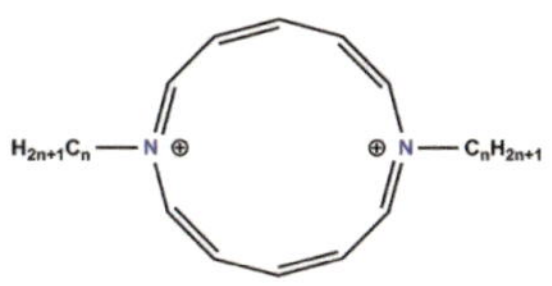

Figure 9. Annulene gemini surfactants.

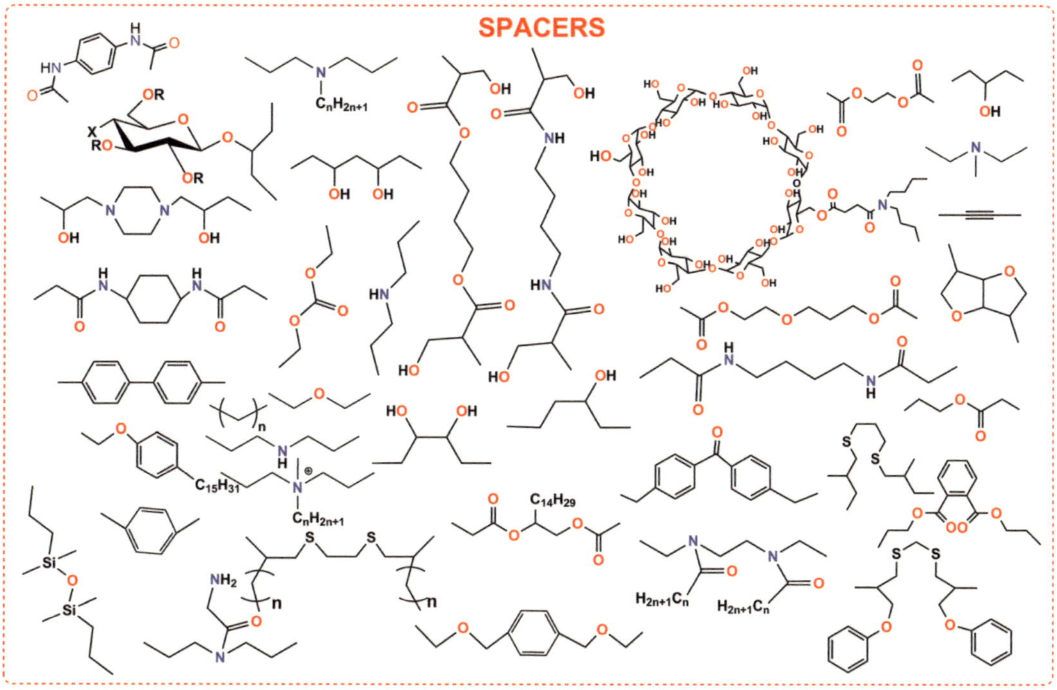

Figure 10. Examples of spacer in gemini surfactants.

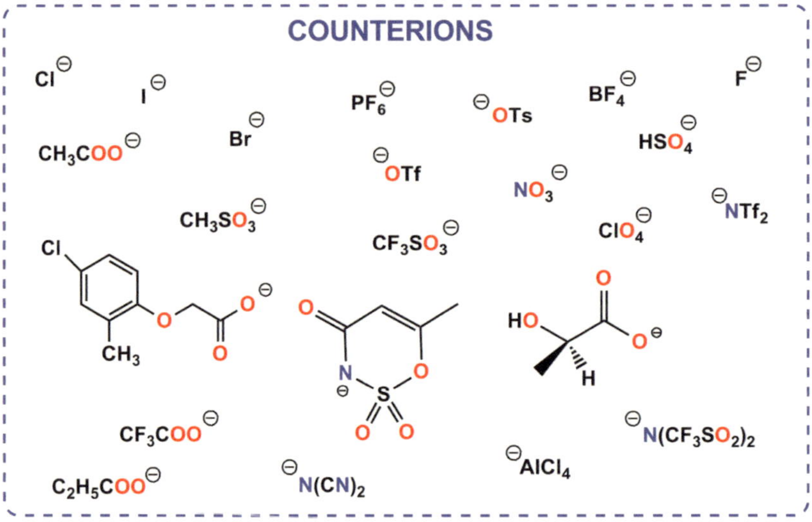

Figure 11. Organic and inorganic counterions in gemini surfactants.

3. Synthesis

The study of bisquaternary ammonium surfactants—gemini surfactants—has been commenced by Bunton and collaborators in 1974 [45]. They described the synthetic approach and kinetic of these nucleophilic reactions. Some years later Devinsky et al. synthesized a great variety of bisquaternary ammonium surfactants and investigated their surface activity and micellization [46]. A unique self-assembly behaviour of gemini surfactants in comparison to their monomeric analogues has been perceived by Zana [47] and Esumi et al. [48]. The first anionic dimeric salts with two sulphate groups and two alkyl chains have been synthesized by Okahra in 1990 [49]. Currently there are three main routes to obtain symmetric gemini surfactants, that is, (1) reaction of long-chain tertiary amines with dihalogenated substrates such as organic dibromides or dichlorides, (2) reaction of *N,N,N′,N′*-tetramethylpolymethylene diamines with alkyl halides and (3) reaction of long-chain tertiary amines with a haloalkylene oxide substrate, commonly epichlorohydrin (**Figure 12**).

The yield of the synthesis of the symmetrical gemini surfactants mainly depends on reactivity of dihalogenoalkanes and polarity and protic character of solvent [50–52]. The best results are achieved in aprotic and polar solvents. Some of these reactions can also be carried out without solvent in mild conditions with very high yields [53].

Cationic gemini surfactants with ester bond as a spacer can be synthesized by the method given by Liao [54] and Gao (**Figure 13**) [55].

The gemini ester quats (ethylene-bis-alanine-n- alkylesterquats bromides) TMEAL-n (Br) and (1,3-propylene-bis-alanine-n-alkylesterquats bromides) TMPAL-n (Br) were synthesized in two steps. In the first step, intermediates—alkyl 2-bromopropionates with 6, 8, 10, 12 and 14 carbon atoms in their alkyl chain—were obtained by acylation of the appropriate alkanols with 2-bromo-

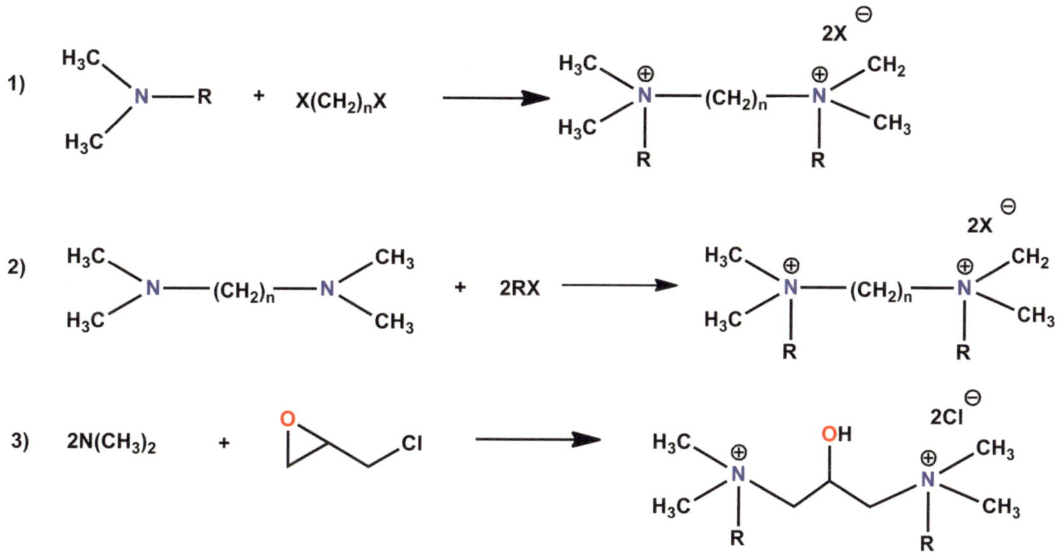

Figure 12. The general routes to prepare symmetric gemini surfactants.

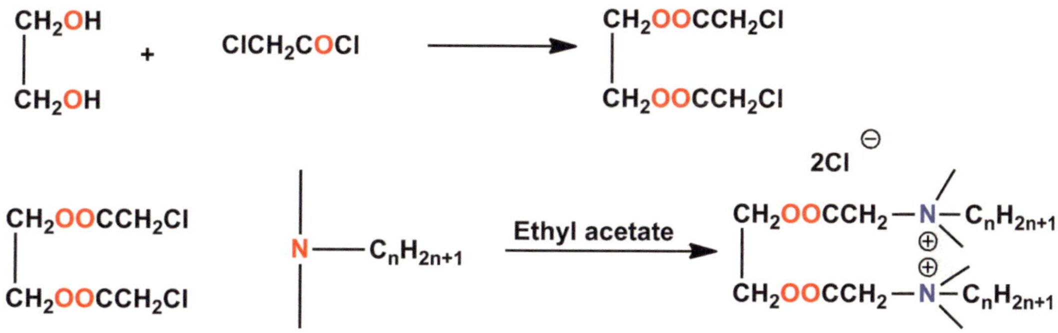

Figure 13. The synthesis method for preparing ester derivatives of gemini surfactants.

propanoyl bromide. In the second step, the alkyl 2-bromopropionates were reacted with *N,N,N′,N′*-tetramethyl-ethylene diamine (for TMEAL-n synthesis) or *N,N,N′,N′*-tetramethyl-propylene diamine (for TMPAL-n synthesis). Bis-quaternization was performed in the acetonitrile solution (**Figure 14**) [56].

Most of the amino acid-based gemini surfactants synthesized so far are *N*-alkylamides and ester derivatives of the amino acids (*N*-alkanoyl derivatives, *N*-alkylamides and O-alkyl esters). These compounds are prepared by condensation reactions at either the amino or the carboxyl group of the amino acid [57].

Cationic serine-based gemini surfactants were obtained by the reductive amination of glutaraldehyde with the O-protected amino acid. To avoid the cyclization reaction, the dialdehyde substrates must contain very short or very long alkyl substituent. Another possibility is to prepare *N*-alkyl derivatives before introduction of the linker [58].

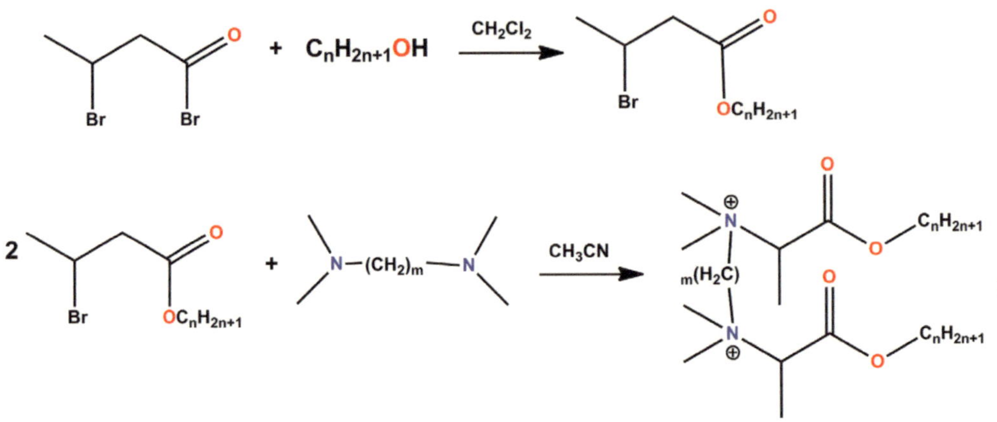

Figure 14. Bis-quaternization ester derivatives of gemini surfactant.

Most studies on the synthesis and biological evaluation of the amino acid gemini surfactants address arginine derivatives [59]. A few reports on lysine-, glycine- and cystine-based gemini surfactants have also been published [60–63].

Sugar-based gemini surfactant (polymethylene-α,ω-bis(*N*,*N*-dialkyl-*N*-deoxy-D-glucitolammonium iodides)) was synthesized in multistep reactions, by a condensation of D-glucose with diamine, followed by reduction of D-glucopyranosyle ring with sodium borohydride and a reductive alkylation with aliphatic aldehydes, containing from 6 to 12 carbon atoms, in the presence of sodium cyanoborohydride as a selective reducing agent. Quaternization of nitrogen atoms by aliphatic *n*-iodides was the last step of the reaction procedure [64].

Zwitterionic gemini surfactants contain positive and negative atoms inside one molecule [64]. The synthesis of zwitterionic geminis is quite complicated; therefore only a few reports appeared so far. The work of Peresypkin and Menger [65] concerns a preparation of zwitterionic gemini surfactant with phosphodiester as a negatively charged group and a positively charged quaternary ammonium salt separated by two pairs of methylene groups [C_x–PO_4^-–$(CH_2)_2$–$N^+(CH_3)_2$–C_y, where x + y = 22]. Yoshimura et al. [66, 67] synthesized sulfobetaine-type zwitterionic gemini surfactants and heterogemini zwitterionic surfactants containing ammonium and carboxylate head groups. Xie et al. [68] offered a simple method for synthesizing alkylbetaine zwitterionic gemini surfactants based on 1,2-bis[*N*-methyl-*N*-carboxymethyl-alkylammonium] ethane (CnAb, where n represents a hydrocarbon chain with a length of 8, 10, 12 or 14) that were synthesized by alkylation of *N*,*N*-dimethylethylenediamine with an alkyl bromide, followed by reaction with sodium 2-bromoacetate.

Quagliotto et al. [69] synthesized a series of pyridinium cationic gemini surfactants by quaternization of the 2,2′-(R,ω-alkanediyl)bispyridines with *N*-alkylating agents. Limei Zhou et al. [70] synthetized novel gemini pyridinium surfactants by using 1,4-dibromobutane and R-alkyl pyridine.

Gemini surfactants with a non-Hückel diaza[12]annulene core were synthesized by treating *N*-(2,4-dinitrophenyl) pyridinium chloride with long-chain amines (**Figure 15**) [71, 72].

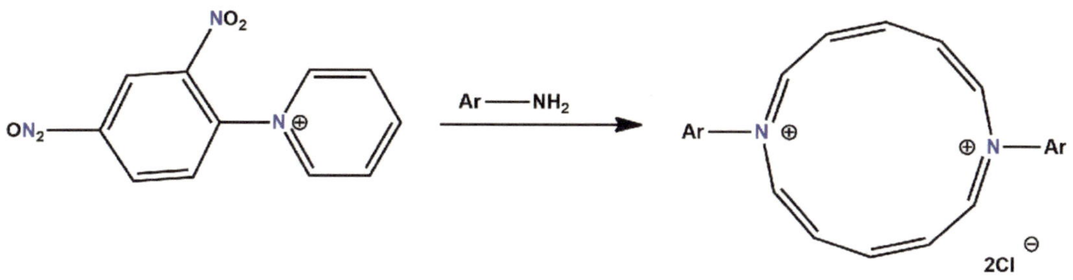

Figure 15. Synthesis of annulene derivatives.

Dissymmetric gemini surfactants contain two nonidentical polar head groups and two different (or identical) lengths of alkyl tails according to Alami and Holmberg [73]. The first heterogemini surfactants containing quaternary ammonium and carboxylate head groups and two dodecyl or tetradecyl tails have been obtained by Jaeger et al. [74]. Some other dissymmetric gemini cationic surfactants with hydroxyl group in the spacer and different long carboxylic acid dimethylethylamine esters as cationic parts [75].

Surface activity of heterogeminis strongly depends on degree of the asymmetry. For pyrene-based dissymmetric gemini surfactants synthesized in five-step reactions (**Figure 16**) [76], the Krafft temperatures increase with the increase of the alkyl chain length. Similarly, CMC values are much lower than those of their symmetrical counterparts.

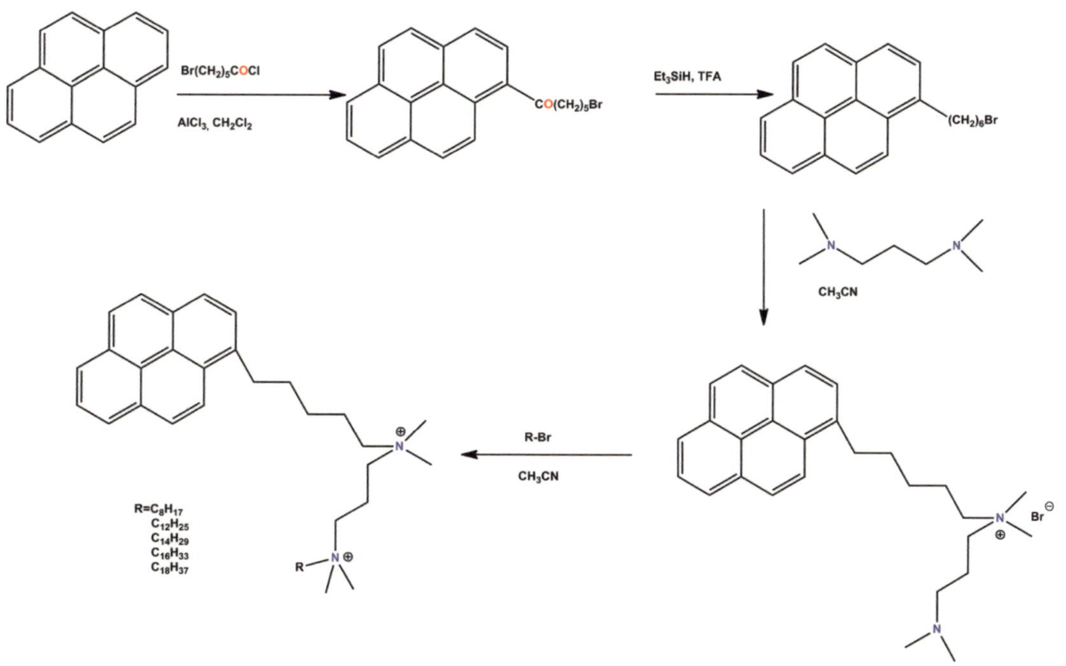

Figure 16. Preparation of dissymmetric gemini surfactant.

Gemini surfactants can be also prepared in microwave-assisted organic syntheses. The usefulness of 5.8-GHz microwaves is demonstrated by the solvent-free synthesis of 2-allylphenol through a Claisen rearrangement process and by the synthesis of the C_{12}–C_2–C_{12} [77].

4. Analytical methods

A large number of analytical methods can be applied to study gemini surfactants and their structure, surface behaviour and interaction with polymers and other materials.

To determine structures of gemini surfactants, the standard spectroscopic methods, nuclear magnetic resonance (NMR) spectroscopy (^{1}H, ^{13}C and ^{31}P), mass spectrometry (fast atom bombardment) and Fourier-transformed infrared spectroscopy (FT-IR) are mostly used.

For cationic gemini surfactants, ^{1}H NMR chemical shift values (δ/ppm) are generally higher for head group protons because of the proximity of the positive charge on the nitrogen atom. In turn, protons of hydrocarbon chain at highly hydrophobic section of the surfactant residue in the core portion of micelle are highly shielded; hence ^{1}H NMR peaks are observed in lower ppm regions [78].

The variation of the chemical shifts due to the hydrophobicity can be used as a method to study of the aggregation process. For example, the correlation between ^{1}H NMR shifts and ([gemini]/CMC)$^{-1}$ for the terminal methyl group of the chains suggests the presence of transient proximity between the methyl group and the annulene ring system. Diffusion coefficients from pulsed gradient spin-echo (PGSE) NMR experiments reveal that the annulene gemini micelles are similar in size and shape to those of simple monomeric surfactants [71].

The very interesting results have also been observed for ^{1}H NMR study of two series of quaternary ammonium gemini surfactants 12-*s*-12 and 14-*s*-14 at concentrations below their CMC in aqueous solutions. The analysis of self-diffusion coefficients, changes in chemical shift, line width and line shapes indicates the premicellization of these two series of geminis below their CMC values [79]. NMR technique has been also applied to study of the binding isotherms of the surfactants to the polymers [79–82]. McLachlan and Marangoni have investigated the interactions between poly(styrenesulfonate) (PSS), dodecyltrimethylammonium bromide (DTAB) and cationic gemini surfactants (12-s-12) [80]. For the gemini cationic surfactants, the NMR chemical shifts indicate that the manner in which the gemini surfactants self-assemble with the polymer is dependent on the spacer length of the surfactant. The ^{1}H chemical differences indicate that the manner in which the DTAB and the long-spacer gemini surfactants interact with the PSS is fundamentally similar, whereas the short-spacer gemini surfactants have a different ^{1}H chemical shift difference pattern for the spacer and chain protons; this may indicate subtle differences in the nature of the binding of these cationic surfactants to the polyanions.

The unusual self-assembly behaviour of gemini surfactants possesses challenging puzzles to theoretical investigations [83]. In view of the above, the cationic gemini surfactant designated as 16-E2-16 (ethane-1,2-diyl bis(*N*,*N*-dimethyl-*N*-hexadecylammoniumacetoxy)-dichloride) was obtained and investigated as a corrosion inhibitor for mild steel (MS) in 1 M HCl solution by refined analytical methods and weight loss measurements. Moreover, the inhibition effect of the investigated compound was analysed by DFT method [84].

Cationic gemini surfactants of the m-2-m type have been investigated with luminescence probing and neutron scattering [85]. Dynamic light scattering (DLS) shows that the surfactant interacts with the polymer at low concentrations and 12-2-12 mixed systems grow to large aggregates with surfactant concentration. It is also confirmed that the longer the hydrocarbon chain length of surfactant, the stronger the interactions.

The molecular composition of each G12-s and G18:1-s gemini surfactant was determined by quadrupole time-of-flight mass spectrometry analysis (QqToF-MS). The fragmentation pattern of the investigated compounds was done by QqToF-MS=MS and showed that the geminis share fragmentation patterns that are specific to their respective gemini surfactant families. At present, a study of some gemini surfactant families are directed to identify for each gemini surfactant two or three product ions with unique m/z values which can be utilized in multiple-reaction monitoring and analysis of biological samples [86].

The combined MS and DFT methods can be very useful for studying competitive S_N2 and E2 reactions in the gas phase. The M2+X- pairs formed from hexadecyldiyl-R,ω-bis(dimethylalkylammonium) surfactants are stable in the ion trap of spectrometer, which is consistent with DFT computations of the bolaform analogues. It shows that M2+X- pairs are extremely stable in the gas phase [87].

The self-aggregation behaviour of gemini surfactant 12-2-12 (ethanediyl-1,2-bis(dimethyldodecylammonium bromide)) in water was investigated by dielectric relaxation spectroscopy (DRS) over a frequency range from 40 Hz to 110 MHz [88]. A defined, widely distributed dielectric relaxation was observed in the 107–108 Hz frequency range for all micelle suspensions; the relaxation mechanism was recognized as the interfacial polarization between the micelles and solution medium.

Currently, the most common method for quantitative determination of surfactants is high-performance liquid chromatography (HPLC) [89, 90] and GC-MS method [91].

Many HPLC methods for the determination of quaternary alkylammonium compounds have been reported. Wee and Kennedy reported a normal phase method for the determination of cationic surfactants without separation of the homologous series in environment samples [92]. One of the methods which can differentiate and quantitate the homolog mixture is high-performance capillary electrophoresis (HPCE) that separates compounds in an electric field according to their charge and size [93, 94].

One of the simplest methods of determining the amount of surfactants in the sample is titration. The first method is turbidimetric titration. In this method cationic surfactants are titrated with anionic surfactants. The next method is a two-phase colorimetric titration. Two-phase titration was first described by Epton in 1947 [95]. Soon it became a commonly used method. This method was developed as a standard method and published as ASTM, BSI and DIN standards [96–99].

Another method is potentiometric titration in the aqueous phase. The potential of a solution containing surfactants is measured as a function of added titrant. The potential of the sample is measured by means of electrode-sensitive surfactants [100].

5. Surface properties

One of the fundamental properties of surfactants is their tendency to adsorb at interfaces. It affects surface tension reduction because of their dual chemical nature [35, 101, 102]. The surface tension (γ) of pure water is 72 mN/m [103]. The ability to reduce it by surfactants depends on the replacement of molecules of solvent at the interface by molecules of surfactants. Mechanism of action of cationic gemini surfactants based on the adsorption of hydrophilic groups (positively charged nitrogen atoms) onto a polar phase and hydrophobic groups in a nonpolar phase. These phenomena are characterized by an efficiency factor pC_{20} which is a concentration of surfactants when the tension is reduced by 20 mN/m [101]. Gemini surfactants are better at lowering the surface tension than their monomeric analogues. The value of pC_{20} for DTAB is 2.3 (C_{20} = 5.25 mM) whilst for 12-2-12 is 3.78 (0.16 mM) [53, 104]. Higher pC_{20} means lowering surface tension by 20 mN/m at lower concentrations. Usually the value of surface tension is given at CMC (γ_{CMC}) which is a critical micelle concentration [105]. It is the concentration when monomeric molecules of surfactants abruptly assemble into aggregates called micelles [106] and they are in the balance. Critical micelle concentration can be estimated using different methods: conductivity measurements, surface tension measurements, UV-absorption spectroscopy, fluorescence spectroscopy, dynamic light scattering or dye solubilization [106]. Conductometry and tensiometry are the most popular and the easiest methods [35]. There are several factors affecting the value of CMC such as structure of surfactant (hydrophilic group, hydrophobic group and spacer) and temperature.

Gemini surfactants exhibit lower CMC values than conventional QAC which is connected with the number of hydrophilic groups. Monomeric quaternary ammonium salts have higher critical micelle concentrations than dimeric one [107, 108]. The relation between number of positively charged nitrogen atoms and the CMC is presented in **Figure 17** [109].

Lower CMC values for dimeric and oligomeric surfactants are connected with their packing into a micelle. Monomeric salts need more molecules to form micelles than dimeric. It is showed in **Figure 18**.

The structure of the hydrophobic groups has also a big impact on the CMC. Its value decreases as the carbon atom numbers increase [110]. The value is halved after addition of one methylene group to a straight hydrophobic chain [101]. The relationship between CMC and the length of the hydrocarbon tail, for compounds of m-s-m type, is shown in **Figure 19** [111].

Elongating the hydrocarbon chain makes the molecule more hydrophobic. The greater the hydrophobicity of the surfactant, the greater tendency to form micelles [112].

CMC can be also controlled by the type and the length of the spacer. At first, CMC gradually increases as the spacer becomes longer, up to four carbon atoms, and then for longer spacers, CMC again decreases [113]. The relationship between CMC and the length of the spacer for

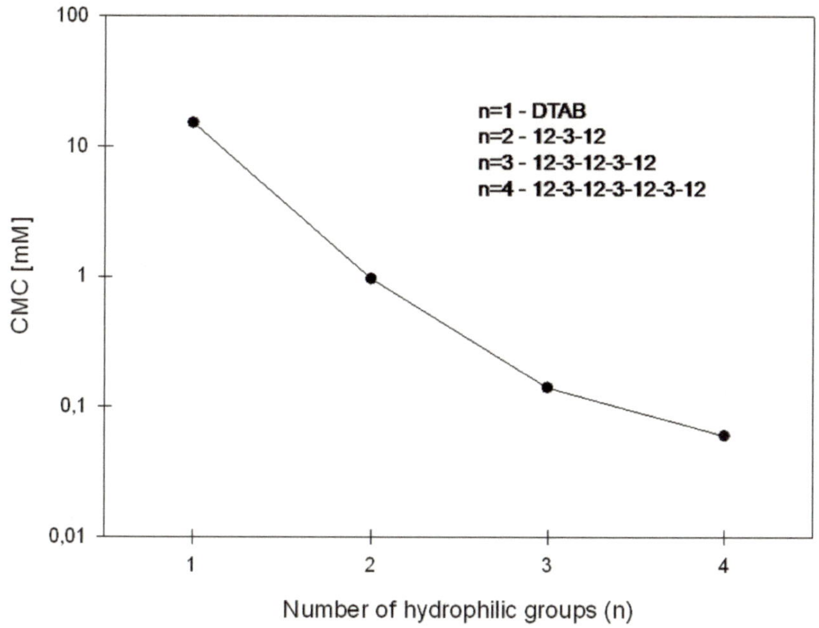

Figure 17. The correlation between number of hydrophilic groups and critical micelle concentration.

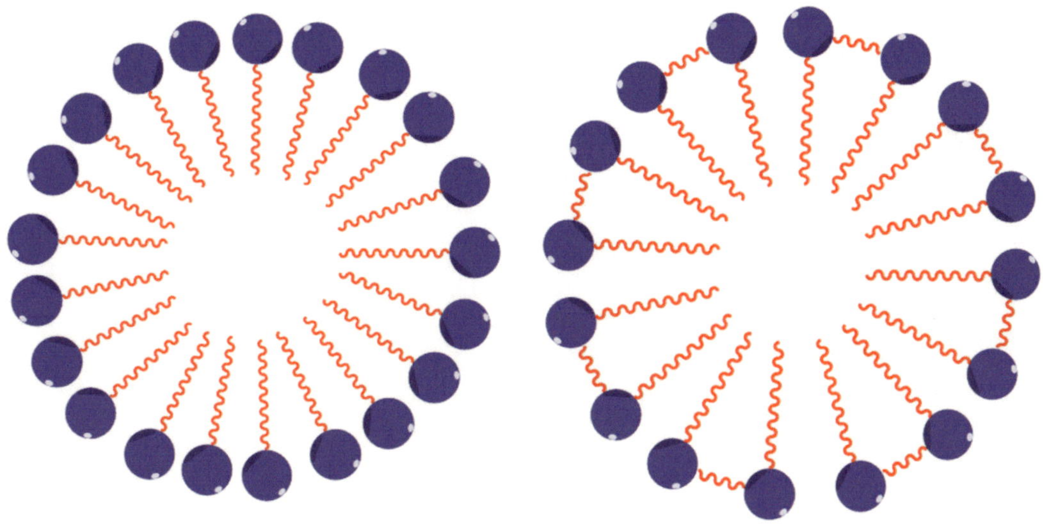

Figure 18. Micelles made of monomeric and dimeric surfactants.

geminis of type 12-s-12 (s—number of carbon atoms in the spacer) surfactants is presented in **Figure 20** [111]. This effect is due to the hydrophobicity of the spacer. The short spacers are fully extended on the air-water interface, whereas long spacers are much more hydrophobic and flexible; therefore they begin to fold into the air.

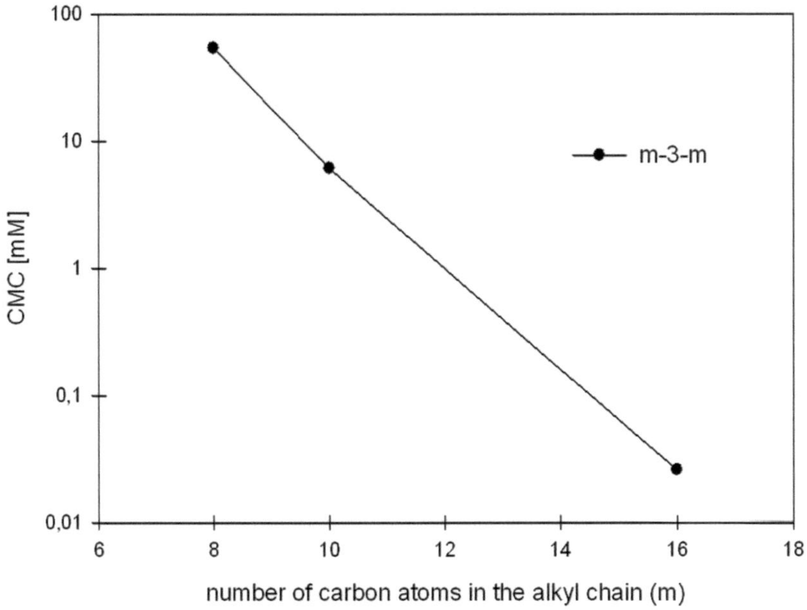

Figure 19. The relationship between the number of carbon atoms in the hydrophobic chain and the CMC values.

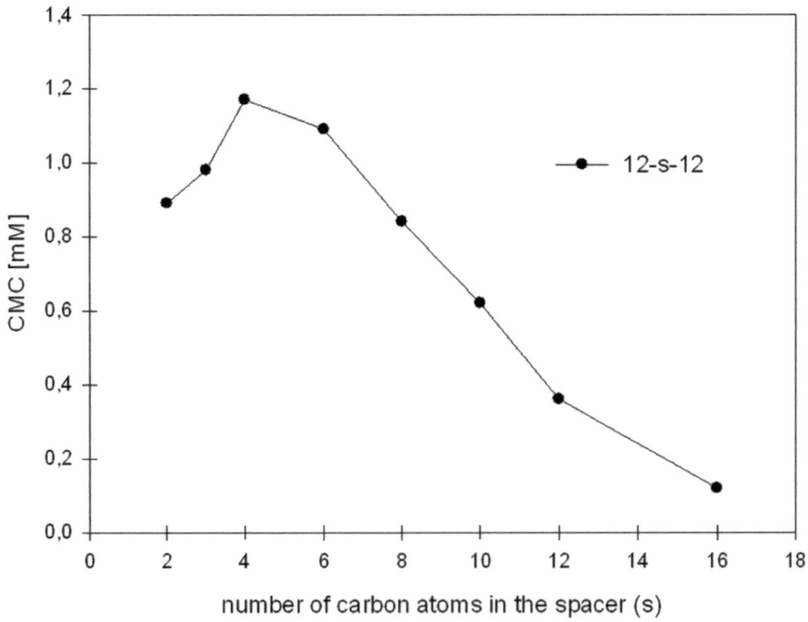

Figure 20. Relationship between the number of carbon atom in the spacer(s) and the CMC value.

The introduction of a polar group like oxygen to the spacer causes an increase of CMC. For gemini surfactant 12-5-12, the CMC value is 1.03 mM [34] and increases to 1.35 mM with spacer containing oxygen atom [114]. The multiplication of oxygen units in the spacers

12—$CH_2CH_2CH_2CH_2$—12	0.98 mM
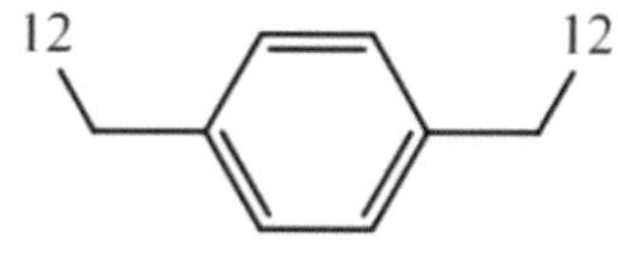	1.52 mM
16—$CH_2CH_2CH_2CH_2$—16	0.042 mM
16—$CH_2C{\equiv}CCH_2$—16	0.048 mM
12—$CH_2CH_2CH_2CH_2$—12	0.99 mM
12—$CH_2CH{=}CHCH_2$—12	2.0 mM

Table 1. CMC values for gemini surfactants with a rigid spacer and their analogues with a flexible one.

exerts a similar effect, that is, an increase of the CMC value. For gemini surfactant 16-$CH_2CH_2OCH_2CH_2$-16, CMC is 0.004 mM, whilst for compound with triple oxygen units, that is, 16-$CH_2(CH_2OCH_2)_3CH_2$-16, CMC value increases to 0.02 mM [115]. The presence of rigid spacers (unsaturated bond and benzene ring) also shifts the CMC to higher values. Some examples of CMC are presented in **Table 1** [103, 114, 116–118].

The structure of the head groups, not only the number of them, affects aggregation behaviour [118, 119]. The exchange of methyl groups at charged nitrogen atom to ethyl ones in 12-4-12 decreases the CMC value from 0.99 to 0.59 mM [120]. The relationship between critical micelle concentration and the type of groups linked to quaternary nitrogen atom is presented in **Figure 21**.

In contrast to the influence of hydrophilicity of the spacer on CMC, the increase of the hydrophilicity of the substituent significantly decreases CMC values. An example is shown in **Figure 22**.

The exchange of one methyl group to hydroxyethyl group at nitrogen atom in 12-4-12 reduces CMC value 10 times, whereas the exchange of both methyl groups to hydroxyethyl substituents lowers CMC almost 1000 times in comparison to the starting compound [121, 122].

Temperature is also a factor which affects the aggregation behaviour. An increase of temperature in the beginning causes the decrease of CMC to minimum around 25°C, and then with further increase of temperature, CMC becomes higher [101]. These effects are directly related to hydration and dehydration of alkyl chain that are sensitive to temperature. The correlation between temperature and CMC values for 12-4-12 is presented in **Figure 23** [105].

The shape of micelles may differ in a wide range, which mainly depends on the structure of surfactants. The most popular methods to estimate the shape of aggregates are dynamic light scattering, small-angle neutron scattering and NMR self-diffusion coefficients [101, 118, 123, 124].

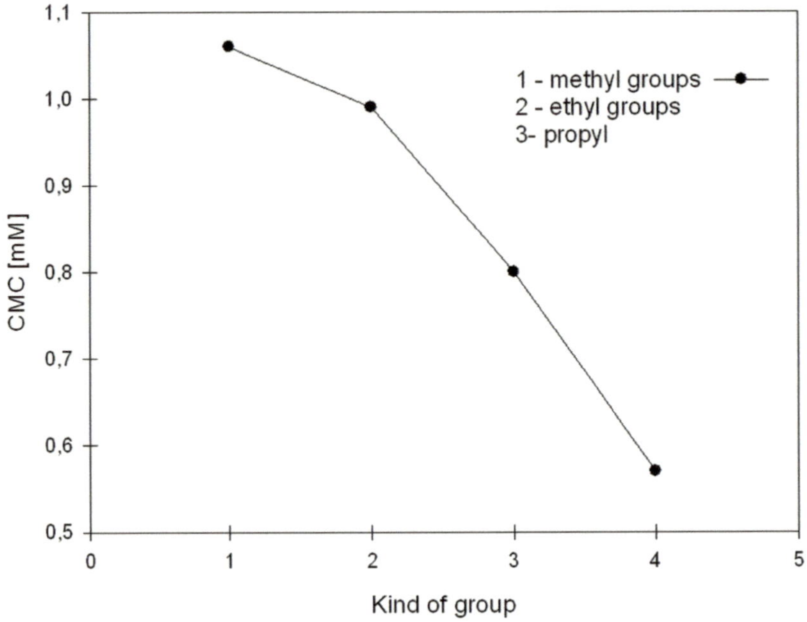

Figure 21. The relationship between CMC values and the kind of group in the hydrophilic part of the surfactant.

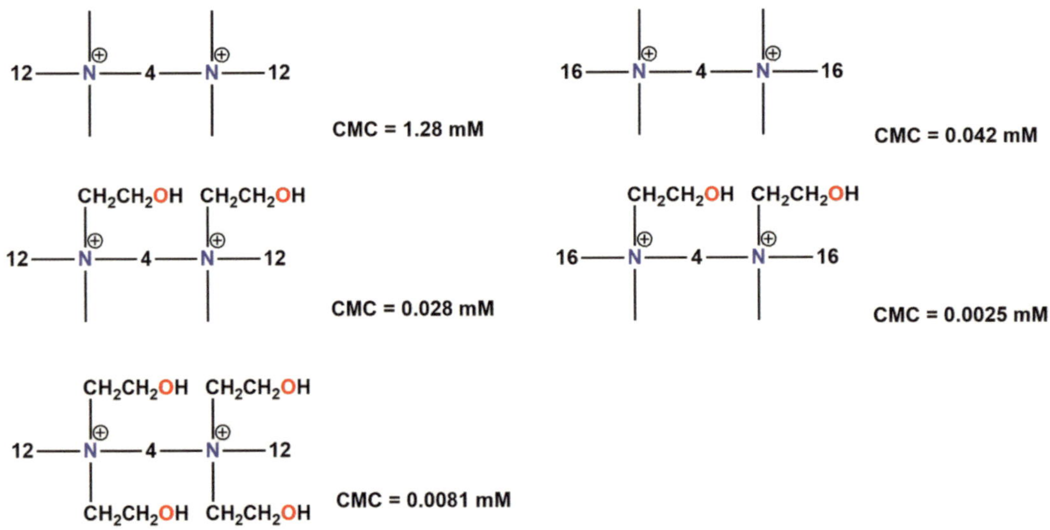

Figure 22. The CMC value for gemini surfactants modified by adding ethoxyl groups.

It was noticed that gemini surfactants with short spacers usually form cylindrical micelles, the one with medium spacers form spherical micelles and those compounds with long spacers form mainly vesicles [111]. However, gemini surfactant 12-2-12 has been shown to form spherical shape micelles [113]. The assessment of the micelle shape is somewhat difficult because it depends not only on the structure of surfactant. The significant influences on geometry and

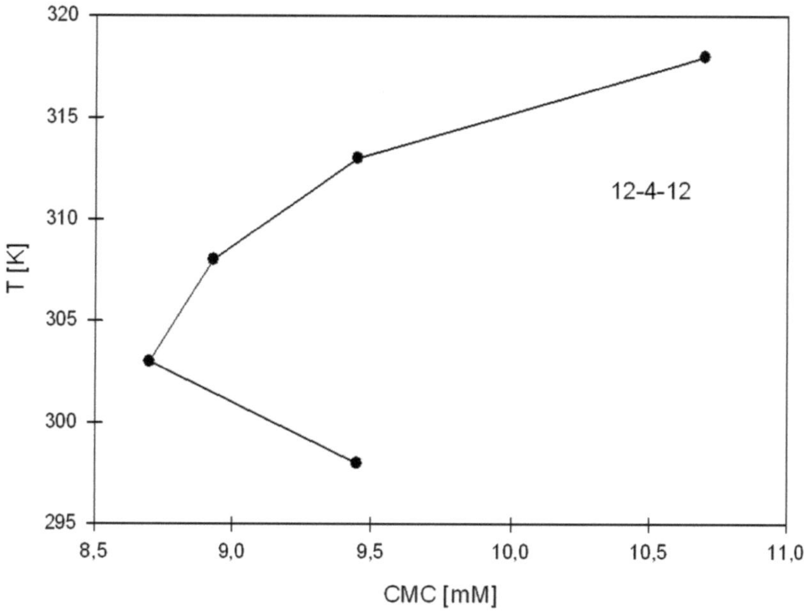

Figure 23. Relationship between temperature and CMC of 12-4-12.

structure of micelles are temperature, concentration, solution condition and ionic strength [115]. Geometrical construction of the surfactant aggregates can be determined by calculating critical packing parameter (P) [120]:

$$P = V_{hydrophobic}/(a_0 * l_0) \tag{1}$$

where $V_{hydrophobic}$ is the volume of hydrophobic chain (for gemini surfactants $V_{hydrophobic} = 2$ V) and l_0 is the length of hydrophobic chain (for gemini l_0 = 2l). They are estimated by using Tanford's expression:

$$V = (27.4 + 26.9m) * 10^{-3}[nm] \tag{2}$$

$$l_0 = (0.15 + 0.1265m)\,[nm] \tag{3}$$

where m is the carbon atom number of a single hydrophobic chain and a_0 is the average packing area of the hydrophilic head group by a single surfactant molecule, usually for gemini surfactants a_0 = 2 [120]. $0 < P > 1/3$ indicates spherical micelles, $1/3 < P > ½$—cylindrical, $½ < P > 1$—vesicles or lamellar and $P > 1$—inverse micelles in nonpolar media [101, 120]. Unfortunately, very often calculated shape varies from those estimated based on experimental methods. Transmission electron microscopy (TEM) is very often used to attain a direct visualization of micelles. Using a precise bar, measuring the size of micelles is possible [120].

Other aggregation parameters can be calculated from surface tension measurements. One of the most important is the aggregation number (N_A) which is the number of surfactant monomers obligatory for micelle formation [106]. The bigger the gemini surfactant (longer alkyl chains, longer spacer, etc.), the lower the N_A [108, 123]. It is in a good agreement with the relationship between the structure of surfactant and its CMC value. π_{CMC} is an effectiveness defined as a difference between surface tension of a pure water and the surface tension of a solution of the surfactant at the CMC. This value can be used to compare surfactants within one series. The lowest value of π_{CMC} belongs to the lowest surface-active analogue [112, 125]. The amounts of surfactant molecules adsorbed at the surface Γ_{max} are estimated from the slopes of straight lines in the plot of surface tension vs. logarithmic concentration drawn in the concentration region below the CMC according to Gibbs adsorption isotherm:

$$\Gamma_{max} = 1/3 \times 2.303nRT(d\gamma/d\log C) \tag{4}$$

The number of ionic species (n) at the interface varies with the surfactant concentration in the solution [125]. The minimum surface area per molecule (A_{min}) can be calculated from the equation:

$$A_{min} = 1/N\,\Gamma_{max} \tag{5}$$

where N is an Avogadro number [124]. A_{min} increases with increasing the length of the spacer and the length of the hydrophobic chains [105, 111].

Free Gibbs energy of micellization (ΔG_{mic}) gives information about the nature of the aggregation process. The energy for gemini surfactants is calculated from an equation proposed by Zana [126]:

$$\Delta G^{\circ}mic = 2RT\left(1/2 + \beta\right)\ln CMC - RT\ln 2 \tag{6}$$

B is the counterion binding parameter which gives the average number of counterions per surfactant ion in the micelle and can be estimated from the ratio of the slopes of conductometry measurements (conductivity vs. concentration) [127]. Negative values of ΔG_{mic} indicate that the process of micellization is spontaneous. ΔG_{mic} increases in the negative direction by increasing hydrophobic character [128]. **Figure 24** presents the relationship between the length of the spacer and ΔG_{mic} [113]. S = 0 represents a monomeric cationic surfactant, DTAB. It is noticed that ΔG_{mic} for gemini surfactant is much lower than for DTAB which means that forming micelles by dimeric salts is more favourable.

Using values of ΔG_{mic} at different temperatures, other thermodynamic parameters can be calculated: entropy (ΔS_{mic}), enthalpy (ΔH_{mic}) of micellization and free energy of adsorption (ΔG_{ads}) by following equations:

$$\Delta S_{mic} = -d(\Delta G_{mic}/\Delta T) \tag{7}$$

$$\Delta H_{mic} = \Delta G_{mic} + T\Delta S_{mic} \tag{8}$$

$$\Delta G_{ads} = \Delta G_{mic} - 6.023 * 10^{-2} * \Pi_{CMC} * A_{min} \tag{9}$$

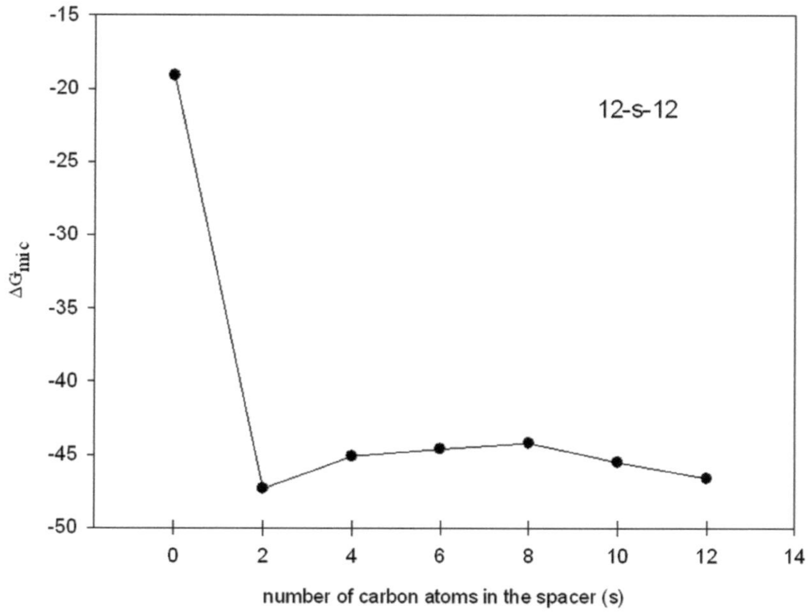

Figure 24. The relationship between the Gibbs free energy of micellization and the number of carbon atom is in the spacer(s).

Positive value of entropy indicates that the process of micellization is favoured. $\Delta H_{mic} < 0$ indicates an exothermic process whereas $\Delta H_{mic} > 0$ an endothermic. A negative value of ΔG_{ads} means that process of adsorption is spontaneous and usually increases by increasing temperature and the length of hydrophobic chain [106, 112, 113, 120, 128]. Moreover, if the value is more negative than $\Delta G_{mic,}$ the molecules of surfactants tend to adsorb at the air-water interface until complete surface coverage and afterwards micelles are formed [128].

6. Biological activity

6.1. Antimicrobial activity

Microorganisms are essential for a large number of metabolic and biotechnology processes. However, they are also responsible for diseases and demises as well as biodeterioration of technical materials like wood, paper, textiles, paints, stonework and steel. To reduce this considerable risk, the chemical compounds with biocidal activity—microbiocides—have been usually used. Microbiocides include some phenols and their derivatives, organic and inorganic halogen compounds, oxidizing substances, quaternary ammonium compounds, alcohols, aldehydes and organic and inorganic acids [9, 64, 129–132]. The most important group of microbiocides is quaternary ammonium compounds (QAC) because of their wide spectrum of biocidal activity, the safety of applications and low costs. Quaternary ammonium salts belong to lytic membrane-active microbiocides [133–135].

Mechanism of their biocidal activity begins with adsorption of quaternary ammonium cation on negatively charged cell surface. Subsequently long hydrocarbon chains can diffuse through the bilayer of the cell, which increases the hydrophobicity of the bacterial cell membrane and provokes disruption of the cytoplasmatic membrane. Damage of the membrane results in the release of potassium ions and other low molecular weight cytoplasmatic constituents, finally leading to the death of the microorganism cell [136–138]. Biocidal activity of the microbiocide is usually performed with minimum inhibitory concentration (MIC), that is, minimal concentration of compound which inhibits the growth of microorganism. MIC are affected by several factors like concentration of microbiocide, time of the contact, pH, temperature, the presence of organic matter or other compounds. Moreover MIC strongly depends on the nature, numbers, location and condition of the microorganism [139].

MIC could be determined by the broth or agar dilution method [140, 141] and expressed in the concentration units. Antimicrobial activity can be also expressed as a zone of inhibition by diffusion method. This method has many limitations, which mainly depend on diffusion ability of microbiocide. According to Klančnik, there are no straight correlations between these two methods [142].

The biocidal activity of gemini surfactants depends on the type of microorganisms. Gram-positive bacteria are more sensitive than the Gram-negative bacteria to ammonium microbiocide. This is due to morphology of the cell membranes. Gram-positive bacteria cell membranes are composed of peptidoglycan layers, which could be easily penetrated by surfactant, whereas Gram-negative cell membranes are mainly composed of lipopolysaccharides and proteins that restrict the entrance of microbiocides [137]. In general, the sensitivity of the microorganisms to gemini alkylammonium microbiocide decreases in the order: Gram-positive bacteria > fungi > Gram-negative bacteria [143]. Biocidal activity also depends on the strain of the microorganism. The environmental strains are more resistant than laboratory strains [144].

Structure of the microbiocide is the most important factor affecting antimicrobial activity. Gemini alkylammonium salts are much better microbiocides than their monomeric analogues. MIC values of geminis are usually 17–70 folds lower than MIC of single analogue QAC. For example, MICs against *Staphylococcus aureus* are 0.0036 [μM] for gemini [12-6-12] and 0.252 [μM] for dodecyltrimethylammonium bromide (DTAB) [144]. This is due to the fact that gemini surfactants posses not only double positive-charged nitrogen atoms but also two long lipophilic substituents. The adsorption on the microorganism cell wall and subsequent penetration of the bilayer is more efficient [138]. It could be said that antimicrobial activity increased with numbers of quaternary ammonium cations in the molecule, but Paniak et al. showed that number of charged nitrogens was not key determinant of bioactivity of ammonium surfactants [145].

Conventional gemini alkylammonium salts could be modified by the change of number of carbon atoms in the substituent or in the spacer. Compounds, which have 10–14 carbon atoms in the substituent, are more active against bacteria than others (**Figure 25**) [146]. The shorter substituents are too short to effectively penetrate the membrane. In turn the long substituents have a tendency to coil upwards loosing the ability to penetrate a cell wall. This is consistent

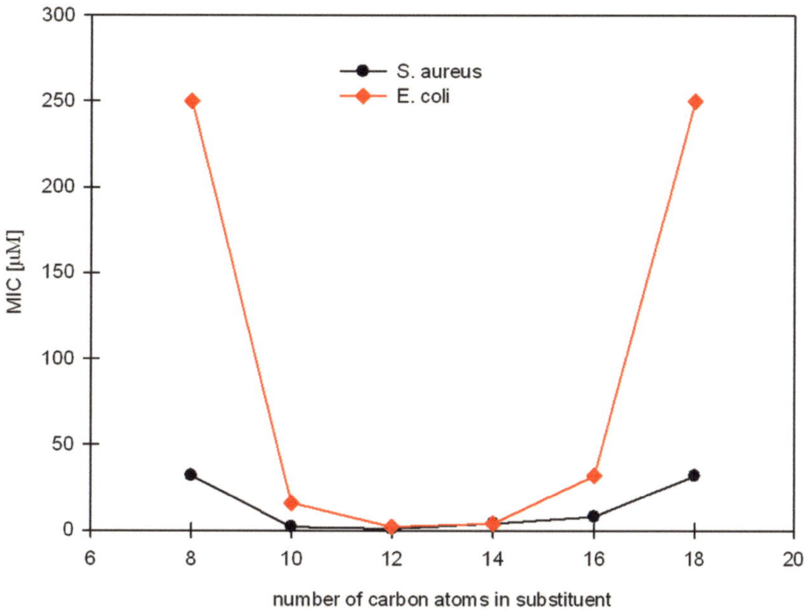

Figure 25. The relationship between MIC against *S. aureus* and *Escherichia coli* and a number of carbon atoms in the substituents in ethylene-1,2-bis(*N*,*N*-dimethyl-*N*-alkylammonium bromides).

with the parabolic relationship of MIC vs. the length of the substituent for monomeric QAC [129, 138].

The antimicrobial activity depends not only on the length of the substituent but also is dependent on the length of spacer [76, 137, 143, 147, 148]. In general, the longer the spacer, the better the antimicrobial activity (**Figure 26**) [77]. The longer spacers allow to better adjust of geminis to cell surface.

Tatsumi et al. compared antimicrobial activity of gemini surfactants with flexible and rigid spacers. In the case of surfactants with fourth carbon atom in the spacer, more effective are compounds with unsaturated bond in the linker [149]. Another possibility of stiffening of spacer is to introduce a ring. Martín et al. showed that the nature of the ring (aromatic or saturated) does not influence the antimicrobial activity of gemini surfactants [150].

The antimicrobial activity of gemini alkylammonium salts strongly depends on their hydrophilic-lipophilic balance (HLB), according to the equation $\log 1/MIC = a + b\log P + C[\log P]^2$, where P is octanol-water coefficient, which characterizes HLB of the molecule [138, 151, 152].

Antimicrobial activity of geminis with hydrophilic spacers modified by ester groups [148, 153–155], ether groups [156], amide groups [157, 158], amine group [145], phosphoryl group [159] and their antimicrobial has been frequently studied. It is important to note that there is no simple relationship between different types of hydrophilic groups in the spacer and antimicrobial activity of gemini surfactants. For the same type of the hydrophilic spacer, antimicrobial activity of geminis depends on the length of the alkyl substituent [145, 154].

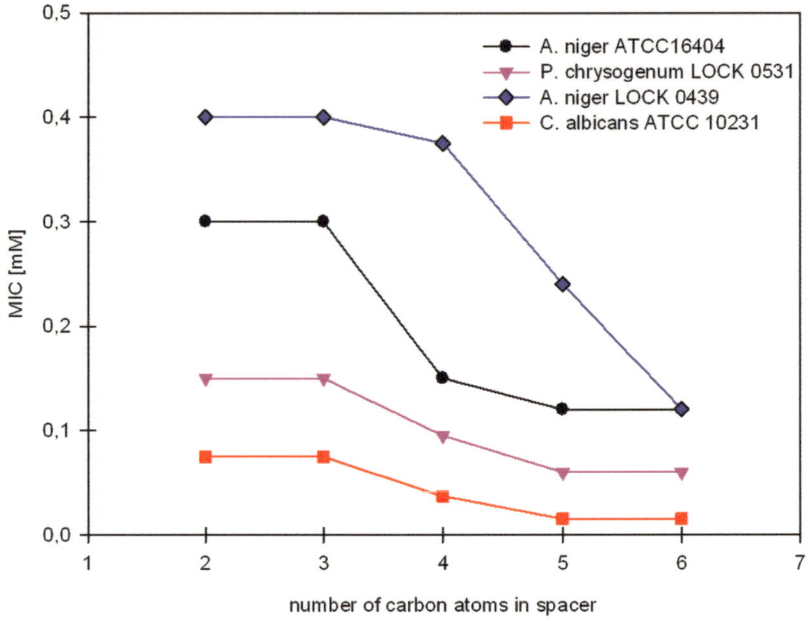

Figure 26. The relationship between MIC and a number of carbon atoms in spacer of polimethylene-α,ω,-bis(*N*,*N*-dimethyl-*N*-dodecylammonium bromides).

Gemini surfactants with higher HLB could be also obtained by introduction ester groups [160–162], amide group [163–165] or hydroxyl group [166, 167] to alkyl chain. These compounds possess usually better antifungal activity than corresponding classical gemini surfactant.

In general, the increase of hydrophilicity causes the better antimicrobial efficacy of gemini surfactants [164, 168–170].

Gemini alkylammoniums with two hydroxyethyl groups, [12-4-12] diethanol (DEA) show higher antimicrobial activity than monohydroxyethyl derivative [12-4-12] monoethanol (MEA). The latter one in turn is better than compounds without hydroxyl groups [12-4-12]. The same trend is observed for monomeric analogues with two hydroxyethyl groups DTAB-DEA, one DTAB-MEA and one DTAB (**Figure 27**) [170].

6.2. Biodegradability

The susceptibility of gemini surfactants to biodegradation is the objects of many tests. The alteration of chemical structure of a substance changing its properties is defined as primary biodegradation, whereas mineralization to carbon dioxide, mineral salts and biomass is an ultimate biodegradation [171]. Surfactants are called to be easily biodegradable if at least 60% biodegradation occurred during 28 days [171, 172].

Martin et al. studied biodegradation of gemini surfactants with phenyl or cyclohexyl ring in the spacer. Whereas monomeric analogues are biodegradable (especially one with phenyl ring), gemini surfactants show no biodegradability [150]. Due to excellent antimicrobial activity,

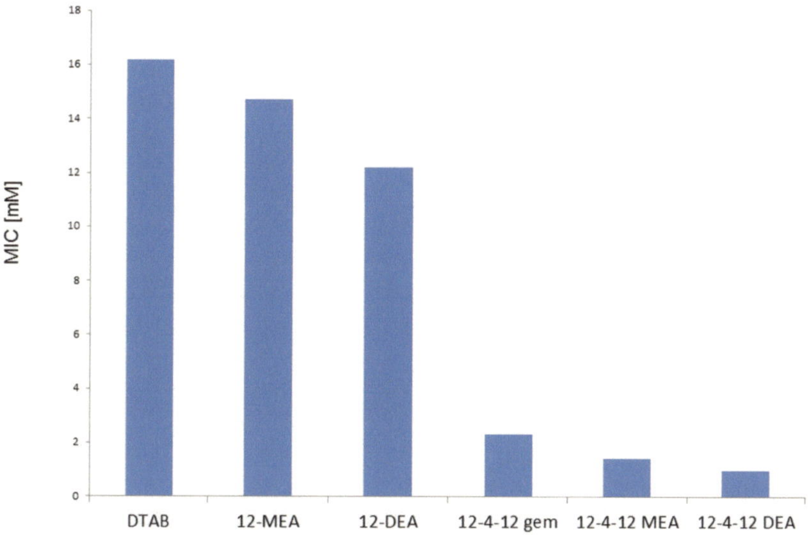

Figure 27. Antimicrobial activity of hydroxylated surfactants against *Bacillus subtilis* (MEA monoethanol, DEA diethanol).

gemini alkylammonium surfactants are considered as hard (or no) biodegradable. The high biological activity of cationic gemini surfactants might have a negative impact on their biodegradation. Modification of spacer to increase hydrophilicity of the surfactant molecule does not significantly change biodegradability of geminis [171]. Gemini surfactants with sugar substituents, like gemini alkyldeoxy-D-glucitolammonium salts, show a susceptibility to biodegradation in the range 20–32%. The degree of biodegradability depends on the length of the alkyl chain. The longer the hydrocarbon substituent or spacer, the lower the biodegradability [172]. Amino acid-based gemini surfactants are biodegradable to even 60%; however, their single analogues degrade rapidly [173]. Modification of the structure (spacer or substituent) with easily hydrolysed groups can significantly affect biodegradability. The widely described easily biodegradable gemini surfactants are that with ester bonds [125, 148, 174–177] (**Table 2**) [176, 178].

Another possibility to enhance the biodegradability of cationic surfactants is the use of immobilized consortium of microorganisms in Ca-alginate beads. This way allows the biodegradation of QAC up to 100% [179]. This method was applied to monomeric cationic surfactants; however, it is very possible that this would work also for gemini surfactants.

The crucial points to biodegrade chemical compounds, not only cationic surfactants, are the concerted activity of two or more groups of bacteria to fulfil enzymatic capabilities [181]. Moreover, there are no simple relationships between biocidal activity of gemini surfactants and their biodegradability.

6.3. Hemolycity

Gemini alkylammonium surfactants possess amphiphilic character and can interact with various surface, also with the membrane of erythrocytes. Łuczyński et al. report that the

Compound		Biodegradation (%)	Reference
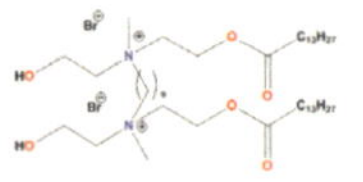	s = 2	80	[179]
	s = 6	75	
	s = 12	71	
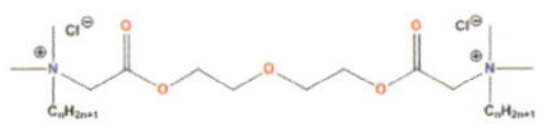	n = 12	59.85	[177]
	n = 14	51.65	
	n = 16	52.19	

Table 2. Biodegradability of gemini surfactants with ester bond in the substituent or in the spacer.

hydrocarbon chains of the gemini surfactants penetrate the hydrophobic lipid bilayer of the erythrocyte membrane, which causes weakness of the interaction between the lipid molecules, leading to lysis of the cell [56]. The haemolytic activity of the surfactants is usually expressed as HC_{50}, that is, concentration that induces the haemolysis of 50% of the total number of erythrocytes [164], and it depends strongly on structure of surfactants. Koziróg et al. notice that gemini [12-6-12] did not exhibit haemolytic activity at MIC against *Candida albicans*, whilst monomeric surfactant DTAB at the same MIC caused a slight haemolysis of erythrocyte [144]. Similar conclusions have been described for geminis with ester group in the spacer [153, 180] and for amino acid-based gemini surfactants [181]. Łuczyński et al. have shown that haemolytic activity depends on the alkyl chain length, whilst compounds with 10 and 12 carbon atoms exhibit the highest haemolytic activity (the lowest HC_{50}) (**Figure 28**). Surfactants with shorter alkyl chain induce haemolysis only at very high concentration. Also single-chain analogue shows haemolytic activity comparable with gemini with the same length of the alkyl chain [56].

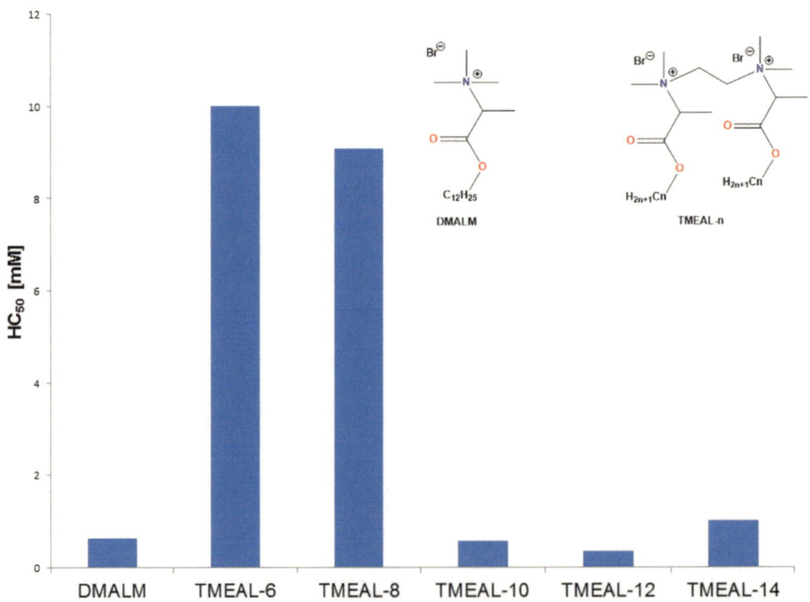

Figure 28. Comparison of HC_{50} of gemini surfactant (TMEAL-n) and its monomer analogue (DMALM).

Zhou has shown that HC_{50} of geminis is the highest for decyl, dodecyl and tetradecyl substituents that correspond to their high antimicrobial activity [182].

Hoque et al. studied haemolytic activity of amide-based gemini surfactants with different lengths of spacer. He found that the increase of the spacer length causes the increase of haemolicity (**Figure 29**) [164]. It is a result of an increasing hydrophobicity of gemini surfactants investigated. The similar results have been described for amino acid-based gemini surfactants; haemolytic power is higher for the compounds with more hydrophobic content, that is, with longer spacer and alkyl chain lengths [173, 181].

6.4. Cytotoxicity

Gemini alkylammonium surfactants are tested as nonviral gene vectors, so their cytotoxicity has to be studied. It is usually specified by the IC_{50} value i.e. the concentration of the compound (in μM) that attenuates the living cell survival to 50% [183]. IC_{50} depends on the structure of surfactants. For gemini surfactant with fixed length of spacer, cytotoxicity decreases as the length of alkyl chain increases (**Figure 30**). Moreover, for gemini surfactant with fixed length of alkyl substituent, IC_{50} decreases as number of methylene groups in spacer increases from 2 to 8 and then increases as number of methylene groups in spacer increases up to 12. These changes are very similar to those observed for MIC values. IC_{50} values show that monomeric surfactants are more cytotoxic than gemini ones, for example, IC_{50} of CTAB are 10.3 and 8.0 μM towards C6 and HEK293 cells, whilst IC_{50} of 16-4-16 are 3.5 and 4.1 μM, respectively [137]. Cytotoxicity also depends on the structure of head groups. Chauhan et al. find that pyridinium-gemini surfactant possesses lower cytotoxicity towards BV2 and C6 glioma cells than conventional gemini surfactants [14-2-14] [183].

Due to the structure of gemini alkylammonium surfactants and their ability to penetrate biological membranes, they can be potentially used as skin permeation enhancers. Almeida et al. studied cytotoxicity towards NCTC 2544 cell line, a human skin keratinocyte cell line of several dicationic gemini surfactants, and compared with a commercial single-tail surfactant

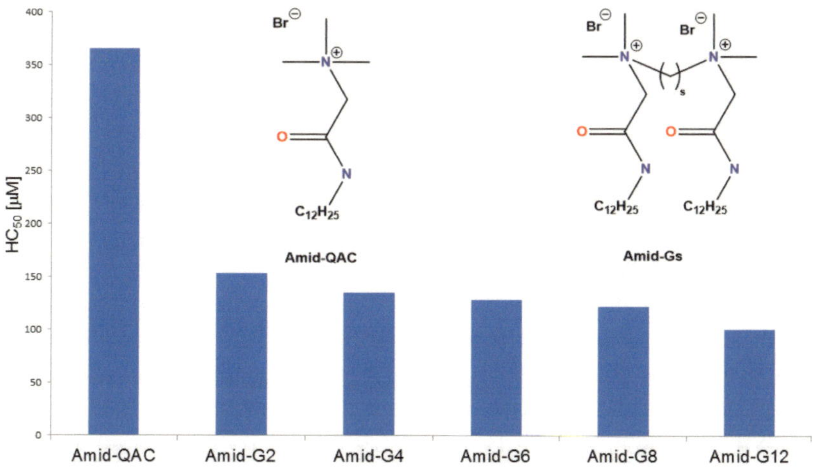

Figure 29. Comparison of HC_{50} of gemini surfactant (AMID-Gs) and its monomer analogue Amid-QAC.

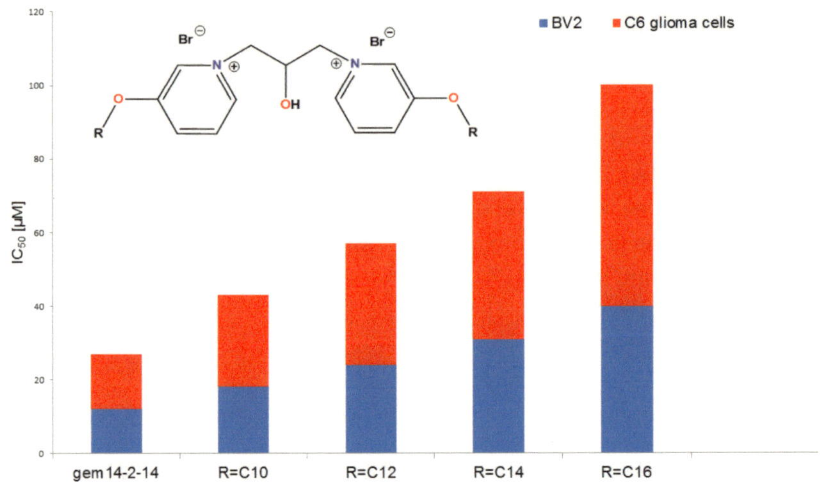

Figure 30. Cytotoxicity of gemini surfactants against BV2 and C6 glioma cells [185].

(DTAB) [184]. For the lower concentrations tested (up to 10 mM), none of gemini surfactant reveals a significant cytotoxicity upon the cellular line. However, over 25-mM toxicity is observed for some of them. Because gemini surfactants are more effective in disrupting the membrane than the single-tail counterpart, it means that a low amount of gemini, below threshold toxicity, may be needed to achieve the same effect of a significantly higher amount of DTAB. Silva et al. have shown that gemini surfactants are promising candidates, directed at permeation enhancing of hydrophilic drugs. They possess similar cytotoxic profiles and are even a little more effective than Azone (the most effective permeation enhancer for ketoprofen) [185].

6.5. Aquatic toxicity

The increasing use of gemini alkylammonium surfactants entails the need to define their biological profile, especially toxicity to aquatic organisms. For this purpose selected model organisms highly sensitive to pollution are used. Usually determined parameter is IC_{50} i.e. concentration of surfactants to immobilize 50% of organisms. Garcia et al. studied aquatic toxicity to *Daphnia magna* of several gemini surfactants with dodecyl substituent and different spacers (**Figure 31**) (dodecyltrimethylammonium bromide, DTAB; 1,6-hexamethylene-bis(*N*-dodecyl-*N,N*-dimethy lammonium) dibromide, 12-6-12; 3-oxa-1,5-pentamethylene-bis(*N*-dodecyl-*N,N*-dimethylammo nium) dichloride, 12-O-12; 3-azamethyl-1,5-bis(*N*-dodecyl-*N,N*-dimethylammonium) dibromide, 3N-12; 1,4-bis-[*N*-(1-dodecyl)-*N,N*-dimethylammoniummethyl]benzene dibromide, QSB2-12; and 1,6-hexamethylene-bis(*N*-dodecyl-*N*-hydroxyethyl-*N*-methylammonium) dibromide, G6-MOH-12). They find that aquatic toxicity decreases with increasing the hydrophilicity of the surfactant molecule. The structure of the spacer, rigid (benzene ring) or flexible (alkyl chain), has no significant effect on the acute toxicity to *D. magna*. Comparing the acute toxicity of gemini surfactants with that of monomeric surfactants DTAB (IC_{50} = 0.35 mg/l), dimeric surfactants are less toxic than monomeric surfactants [171]. Similar results were observed for amino acid-based

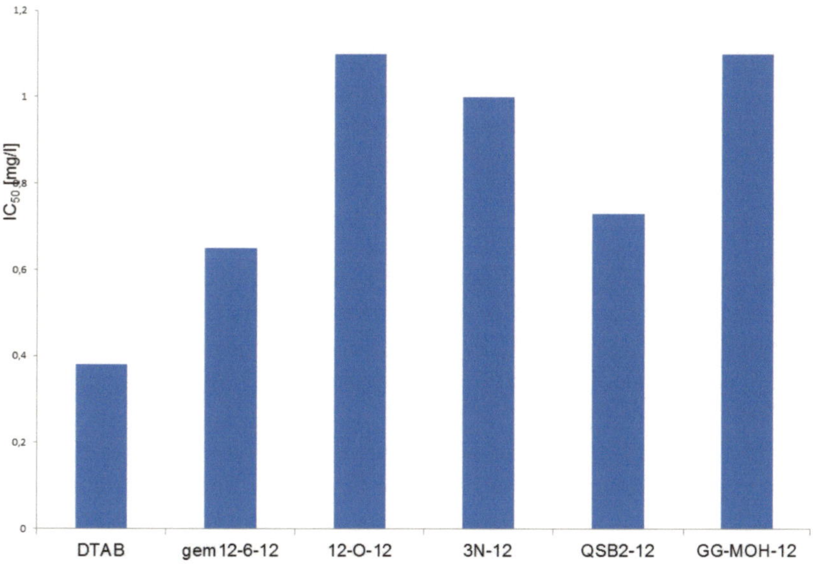

Figure 31. Aquatic toxicity to *D. magna* of DTAB and gemini surfactant with hydrophilicity of the surfactant molecule.

gemini surfactants [173]. These compounds are less toxic to freshwater *D. magna* and seawater *Photobacterium phosphoreum* than conventional monomeric ammonium salts [181].

7. Anticorrosion activity

Corrosion is a process of deterioration (degradation) of materials' properties due to the interactions between a surface and an environment [186], which leads to changes in the material properties because of a disintegration of the structure of the material. The process destroys surfaces of the metals (iron, aluminium and copper) but also non-metallic materials (concrete, wood, glass and paper) [187, 188]. Usually a term "corrosion" is booked for the deterioration of metals, and according to a definition given by the American Section of the International Association for Testing Materials (ASTM), it is the chemical or electrochemical reaction between a material, usually a metal, and its environment that produces a deterioration of the material and its properties [189]. The problem of corrosion affects many areas of industries, oil and gas [190], electronic [191], food, paint, coating [192], marine, chemical [193], automotive and in daily life [176], by destroying metallic equipment, pipelines, vessels, storage tanks [190], heaters and electrical power lines [191] and leading to scale results in reduced heat transfer, loss of production capacity and energy loss [194]. Corrosion is induced by acids that are extensively used in industry [195]. Organic acids are used for preparation of chemicals, drugs, fibbers and other processes [196], whereas mineral acids (HCL, H_2S, H_2SO_4 or H_3PO_4) are used for cleaning, acidification and pickling [190, 197]. Corrosion is a costly and dangerous process, which plays an important role in the field of economic and safety [194, 195]. The damages caused by corrosion can be estimated using different methods. The most popular are gravimetric (weight loss measurements) and electrochemical (potentiometry and electrochemical impedance spectroscopy) [198]. The rarer methods include spectroscopic (UV-VIS) [84], volumetric (amount of released hydrogen), analytical (assay of metal ions) or radiography (using radiation) [186, 188, 199, 200]. The morphology of destroyed metal surface is analysed using microscopes: scanning electron microscope (SEM) or atomic force microscope (AFM) [199].

In order to reduce the corrosion of metallic materials, several methods have been applied: electrochemical protection (anodic and cathodic), coatings (metallic and non-metallic) and corrosion inhibitors [187, 201–205]. Among them, the use of organic corrosion inhibitors, especially cationic gemini surfactants, is the most efficient and practical method [192], particularly to control acid-induced corrosion [84]. Corrosion inhibitors are chemical substances which, when added, in a small amount, to the corrosive environment, significantly decreases the corrosion rate of metals [194]. The general mechanism of action of organic corrosion inhibitors is based on adsorption of molecules of inhibitor onto a metal surface by displacing water molecules and forming a protective film [191, 206]. The adsorption process can be physical (electrostatic interaction), chemical (donor-acceptor interaction) or mixed [200]. The process is influenced by the molecular structure of inhibitor (functional groups, aromaticity and electron density at donor atoms), surface charge of metal and type of electrolyte. Compounds with heteroatoms (N, O, S, P) [84] and π groups (multiple bonds, benzene ring) [207] have been found to be more efficient due to donation of a lone pair of electrons to a free orbital of the metal making them stronger adsorbed [190]. The order of corrosion inhibition is the reverse order of the electronegativity of the heteroatoms [208]:

$$O < N < S < P \qquad (10)$$

It was noticed that in acid environment, heteroatoms are protonated which favours the physical adsorption and has increased the interest of quaternary ammonium salts (QAS) as corrosion inhibitors [128]. Cationic gemini surfactants are more efficient than monomeric QAS. It is related with lower values of CMC which is a key from the point of view of corrosion. Cationic surfactants reach the highest inhibition efficiency around CMC [200]. The corrosion rate (CR) of steel in 0.5 M HCl with addition of monomeric quaternary ammonium salts tetradecyl trimethyl ammonium bromide (TTAB) is higher than for dimeric analogue (1,4-butan-bis(-tetradecyl dimethyl ammonium bromide) (14-4-14) [192] (**Figure 32**).

Two positively charged nitrogen atoms are better adsorbed onto the metal surface due to electrostatic interactions between cations and the negatively charged surface of metal which provides better protection [186, 209, 210]. The size and molecular weight of organic inhibitors have an impact on the effectiveness of action as corrosion inhibitors [209]. Increasing the length of the aliphatic chains increases the inhibition efficiency [192, 200, 211, 212]. The relationship of m-6-m surfactants (C = 1 mM) is presented in **Figure 33** [200].

Another important factor is the length of the spacer. Surfactants with longer hydrocarbon spacer are more effective corrosion inhibitors [176, 199, 213–215]. As an example, inhibition efficiency for gemini surfactants $(C_{12}H_{25})_3N^+(CH_2)_nN^+(C_{12}H_{25})_3$ (C = 5 mM) is presented in **Figure 34** [215].

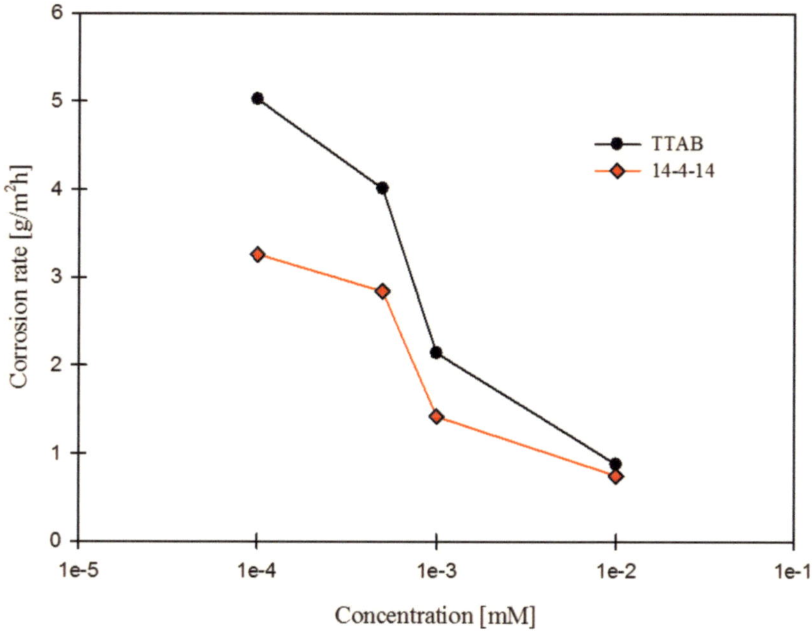

Figure 32. Corrosion rate of steel in acid in the presence of monomeric and dimeric quaternary ammonium slats.

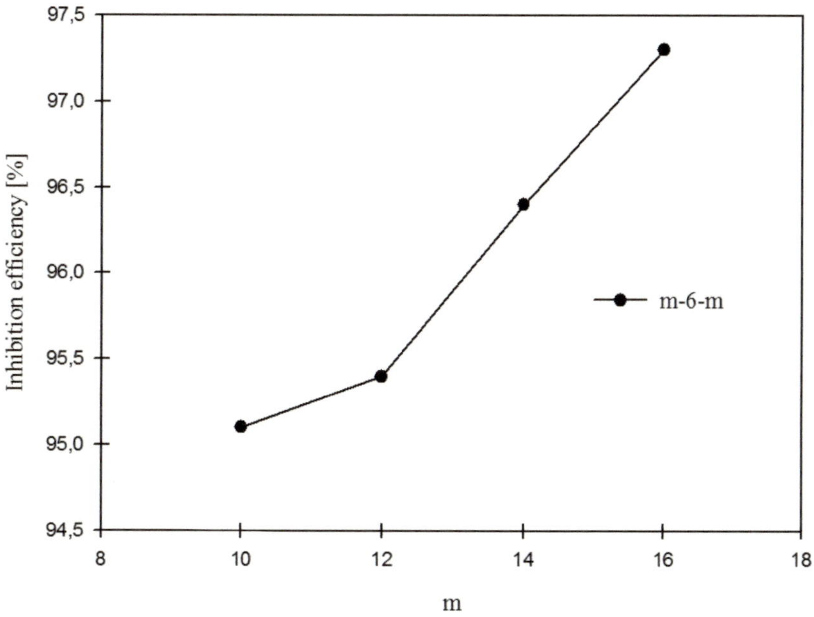

Figure 33. The relationship between the length of the alkyl chains (m) and inhibition efficiency of aluminium in hydrochloric acid.

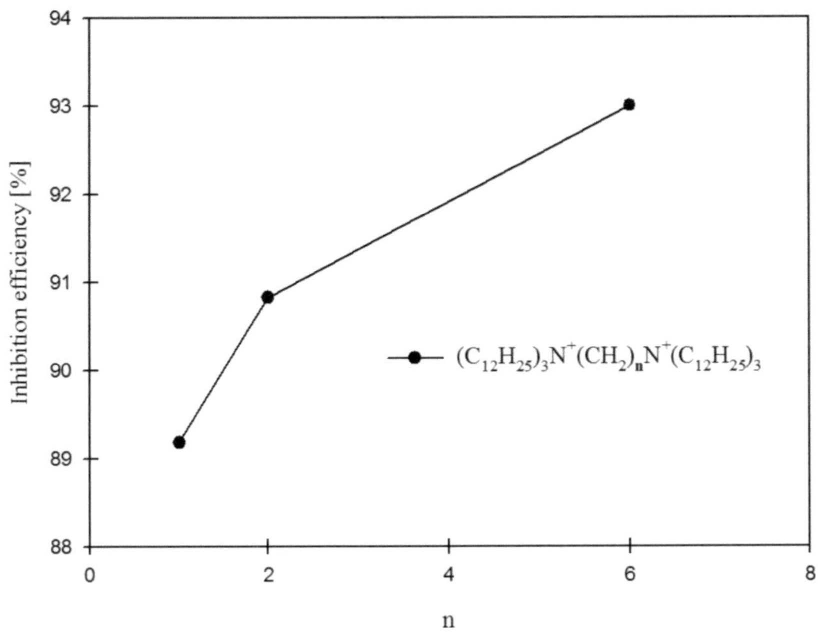

Figure 34. Inhibition efficiency of inhibition corrosion of carbon steel in 1 M HCl.

Inhibition efficiency is related to corrosion-resistance properties of the metals. The adsorption of gemini surfactant molecules changes it by increasing the values of resistance of the metal which makes the material more resistant to corrosion [198]. **Figure 35** presents the values of resistance of $(C_{12}H_{25})_3N^+(CH_2)_nN^+(C_{12}H_{25})_3$ and a value of blank sample, after immersion in acid without addition of inhibitors [215].

Introducing heteroatoms into a molecule promotes the inhibition behaviour. Due to lone electron pars, which can additionally interact with free metal orbitals, the adsorption is stronger and inhibition efficiency higher. Exchanging ethyl groups to ethoxyl groups in gemini surfactants with rigid spacer (**Figure 36**) increases the efficiency from 91.64 to 95.63% (C = 10 mM, carbon steel, 1 M HCl) [210].

The standard energy of adsorption ($\Delta G^0{}_{ads}$) gives information about the type of adsorption. Values up to −20 kJ/mol are related to the electrostatic interaction (physical adsorption), whereas more negative than −40 kJ/mol indicate chemisorption takes place. Negative values of $\Delta G^0{}_{ads}$ mean that the process of adsorption is spontaneous [209]. The energy of adsorption first decreases with increasing the length of the spacer and after reaching maximum starts decreasing which is related to the free energy of micellization (the same relation) [113]. Increasing the length of the alkyl chain increases the values of the $\Delta G^0{}_{ads}$ [199, 200]. The standard enthalpy of adsorption ($\Delta H^0{}_{ads}$) provides valuable information about the mechanism of the corrosion inhibition. Chemisorption is attributed to an endothermic process ($\Delta H^0{}_{ads} > 0$), whilst exothermic adsorption is represented by values lower than 0 [186] and is related to physical or mixed adsorption [215]. Another thermodynamic parameter which gives information of the adsorption process is entropy ($\Delta S^0{}_{ads}$). Positive values are attributed to the increase of

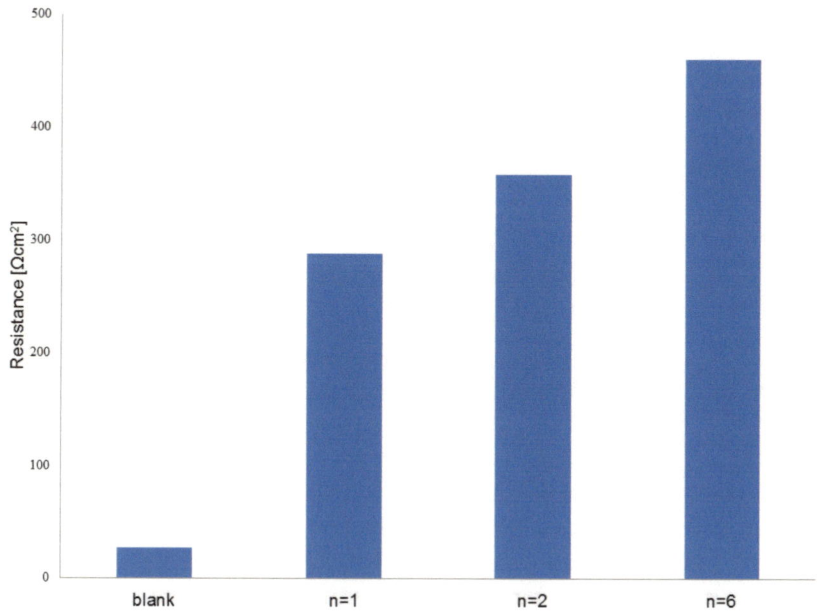

Figure 35. The values of resistance for $(C_{12}H_{25})_3N^+(CH_2)_nN^+(C_{12}H_{25})_3$.

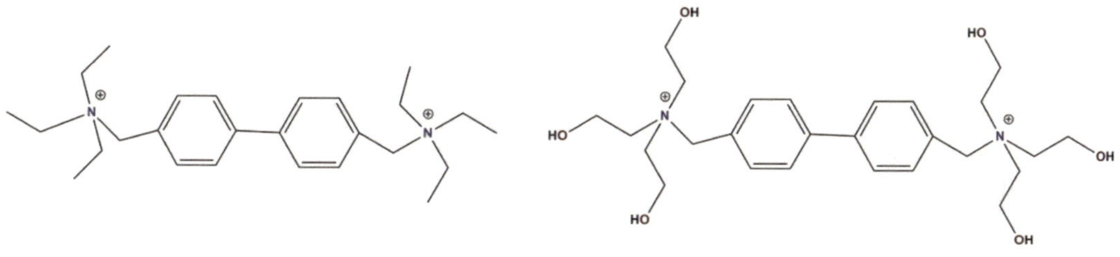

Figure 36. The structures of gemini surfactants with rigid spacer.

disorder due to the dissolution of metal and the adsorption of only one molecule of inhibitor by desorption of more water molecules [195].

Gemini surfactants which are used as commercial agents are not used alone. Special formulations are prepared based on synergistic effect. Thanks to that, the level of protection is higher, and very often due to synergism effect, less amount of surfactant is needed to protect the metal surface [193, 216–219]. The biggest consumption of corrosion inhibitors based on cationic gemini surfactants belongs to petrochemical industry. Using the special formulations leads to decreasing corrosion rate and protection against deterioration during long time. Some formulations are already used in industries. Some of gemini surfactants are already patented as multifunctional corrosion inhibitors of ferrous metals that transported or stored crude oil and liquid fuels by the presence of acidic pollutants, sulphur compounds and water and equipment and pipes used in cooling systems that use water with a high concentration of divalent ions such as calcium and magnesium, which are the main cause of producing pitting corrosion in this environment [220] and also for inhibiting corrosion and biofouling of metallic surfaces in contact with corrosive fluids in gas- and oil-field applications [221]. All of them contain heteroatoms and π groups and exhibit good inhibition efficiency (more than 90%).

8. Special applications

The unique physicochemical and biological properties of gemini surfactants designate them to many applications in industry and pharmaceutical and biomedical branch where the safety profile of products must be optimized.

8.1. Nanoscience and nanotechnology

8.1.1. Gene therapy and bioimaging

Gemini alkylammonium surfactants are applied to introducing genes into cells, due to their ability to interact with DNA [222–224]. This interaction must be strong enough to overcome the biologic membrane barrier and weak enough to release DNA in the right place in the cell. The gemini surfactant is shown to bind and compact DNA efficiently and form a "lipoplex". The lipoplex can penetrate the outer membranes of many cell types, to appear in the cytoplasm encapsulated within endosomes. Escape from the endosome may be controlled by changes in the aggregation behaviour of the lipoplex as the pH decreases. DNA may be released from the lipoplex before entry into the nucleus, where the new gene can be expressed with high efficiency. Some gemini surfactants with sugar substituent, peptide moiety [225, 226] or cholesterol-based diquaternary ammonium gemini surfactant [227] were tested as a gene transfection vectors. It was recently shown that hydroxyethylated gemini surfactants [228], fluorinated bispyridinium gemini surfactants [229] and geminis derived from cysteine [230, 231] can be also used for this application.

Bioimaging is a very useful technique in the cancer diagnosis where the stable fluorescent marker is necessary. It has been recently shown that geminis like 12-2-12 and 12-6-12 are good stabilizers for model genetic material constructed from DNA and polysaccharide-based chitosan on nanoemulsion core containing IR-780 indocyanine as fluorescent marker [232].

8.1.2. Drug nanocarriers

Gemini surfactants can very easily change their morphological structures upon pH, temperature and salts [233–239].

The reversible transition from micelles to other structures, especially to vesicles by changing pH, is very useful for drug delivery. Li et al. showed that gemini amino acid surfactants, where pH is the key driving force to control the aggregation behaviours, can be applied to build colloidal systems for delivering hydrophobic drugs or nutrition [240].

Similarly, Ref. [241] showed that geminis with morpholinium moieties exhibit high solubilization capacity towards a thymolphthalein as well as indomethacin, an inflammatory drug, exceeding that of reference amphiphiles.

8.1.3. Nanoparticles

Nanoparticles (NPs) have a lot of applications in medicine, physics, optics and electronics. The size and morphology of nanoparticles determine to high extent their properties and applications. These parameters can be mainly regulated by surfactants which act as soft templates or nanocontainers. The preparation of gold, silver and gold-silver alloy nanoparticles by seed-mediated method using gemini surfactant has been described by Tiwari et al. [242]. The obtained NPs were stable and were characterized by UV-vis, XPS, TEM, energy dispersive spectroscopy (EDS) and zeta potential techniques. The orientation of gemini surfactant molecules on the metal NPs has been determined by twisted intramolecular charge transfer (TICT).

A very interesting synthetic approach was developed by Wang et al. [243] for creating versatile hollow Au nanostructures. The reduction of Au(III) by ascorbic acid with the use of hexamethylene-1,6-bis(*N*-dodecyl-*N*,*N*-dimethylammonium bromide) (C12C6C12Br2) as a template agent leads to vesicle, capsule-like and tube-like aggregates which act as soft templates for hollow Au nanostructures upon further reduction of Au(I) to Au(0) by $NaBH_4$. Gemini surfactant plays a crucial role in formation of the final structure. The electrostatic repulsion between head groups of gemini surfactant is greatly weakened as Au(III) is converted to Au(I), which is in favour of the construction of vesicle, capsule-like and tube-like aggregates. This method allows to prepare nanostructures of gold potentially useful for many applications.

The industrial scale production of monodispersed gold nanorods (AuNRs) has been described by Xu et al. [244]. By using gemini surfactants, the cost of the synthesis of high-quality AuNRs can be reduced by 90%. Moreover, varying the concentration of the surfactant, the shape of AuNRs can be tailored from straight nanorods to "dog bones".

A special group of nanoparticles, quantum dots (QDs) [245], like lead telluride [246] hydrophobic quantum dots CdSe/ZnS [247] with strictly defined size and morphology are usually prepared with auxiliary of gemini surfactants.

8.1.4. Supramolecular solvents

Supramolecular solvents (SUPRAS) are nanostructured liquids made up of surfactant aggregates synthesized through a self-assembly process. This kind of solvent is mainly assigned to microextraction methods. Feizi et al. [248] applied a new gemini-based SUPRAS for the determination of methylparaben (MP), ethylparaben (EP) and propylparaben (PP) in cosmetics, beverages and water samples on the basis of pecation and Van der Waals interactions into the SUPRAS. The gemini-based SUPRAS followed by HPLC-UV has been found to have excellent detection sensitivity with a limit of detection (LOD, S/N = 3) of 0.5 mg/L for EP and PP and 0.7 mg/L for MP.

8.1.5. Interactions with proteins

Interactions between proteins and gemini surfactants derived from amino acids have also been investigated. This type of studies can help to understand the action of surfactants as denaturants and solubilizing agents for proteins that is important in medical and cosmetic branch. Gemini surfactants from glutamic acid exhibit different interactions with haemoglobin than their corresponding single-chain homolog. The gemini surfactants showed lower denaturing ability to haemoglobin, probably due to their bigger size, and the denaturation degree decreased when the spacer length increased. It was also observed that when the gemini surfactants content are low, the secondary structure of haemoglobin can be stabilized [249]. Takeda et al. reported the protective effect of gemini surfactants on thermal denaturation of BSA. The gemini surfactant studied by these authors consists of two glutamic acids as polar heads and a lysine as spacer. For this gemini surfactant, the protection of the recovery of the helicity of BSA appeared at lower concentration comparing to SDS due to the higher hydrophobicity of these compounds [250].

8.2. Technology

8.2.1. Solubilization

Gemini surfactants are very good solubilization agents [251]. Polycyclic aromatic hydrocarbons (PAHs) like anthracene, naphthalene, fluorene or pyrene [252, 253], which are organic pollutants, can be easily removed from water solution by the use of gemini surfactants. It significantly reduces the risk to the environment caused by these compounds [254]. Gemini surfactants are better for solubilization of PAH than their monomeric analogues. After mixing them together, values of molar solubilization ratio (MSR) are higher (**Table 3**) [252].

Gemini surfactants are also efficient as solubilization agents of organic dyes (Quinizan, Sudan I, orange OT) which are used to colour textiles, waxes or oils [6, 255, 256]. Cationic surfactants promote the adsorption of solubilized dye to the surface, especially textile fiber surface which carries a negative charge [255].

Solubilization power of gemini surfactants increases with the elongation of the alkyl chain length [255, 257, 258] and elongation of the spacer length [257, 259] which is related to a larger size of micelles.

8.2.2. Dispersion

Another potential application of gemini surfactants due to their ability to form micelles is the capacity to disperse insoluble in water particles and form stable colloids. Carbon nanotubes (CNT) have unique electrical, optical and mechanical properties, and due to that, they are used as medical sensors, electronics and compatible materials [260]. However, because of strong Van der Waals interactions, the bundles are insoluble in water and common organic solvents

	CMC (mM)	MSR
16-6-16	0.001	0.2110
CTAB	0.776	0.1236
16-6-16 + CTAB	0.0015	0.371

Table 3. Molar solubilization of naphthalene of gemini surfactant, cationic surfactant and gemini-conventional mixtures.

which limit their potential applications [261, 262]. Cationic surfactants are widely used to disperse CNT in water even at low concentration giving stable solutions for long time [263]. Gold nanoparticles, because of their properties [264, 265], also have various potential applications in different areas, but to make them useful, forming stable nanofluids is required. It can be reached by using gemini surfactants as a stabilizer to prepare stable gold/oil nanofluids [264]. It has also been shown that gemini surfactants can effectively disperse hydrogels to form supramolecular, three-dimensional micellar-hybridized network [266–268]. The formation of a spatial network of well-dispersed molecules is very significant for biomedical and optoelectronic applications.

8.2.3. Enhanced oil recovery

Traditional oil extraction methods produce depleted reservoirs that contain about 20–40% of trapped oil [269]. The remaining oil is trapped in porous media, due to the viscous, surface and interfacial forces, which results in poor displacement efficiency [270]. The implementation of advanced methods or their combinations to enable the recovery of residual oil is called enhanced oil recovery (EOR). Some techniques can be distinguished: thermal steam flooding (for heavy and extra heavy crude oil) [271], miscible gas flooding (for light, concentrated and volatile oil reservoirs) and chemical flooding (for medium or light reservoirs) [269]. Chemical flooding is one of the successful methods, especially the use of surfactants [272]. They are added into the flooding solution and improve the properties of reservoir fluids, to make them more conductive to extraction [273]. Tuning the capillary forces of the trapped oil and to achieve a complete miscibility, interfacial tension has to be reduced to the

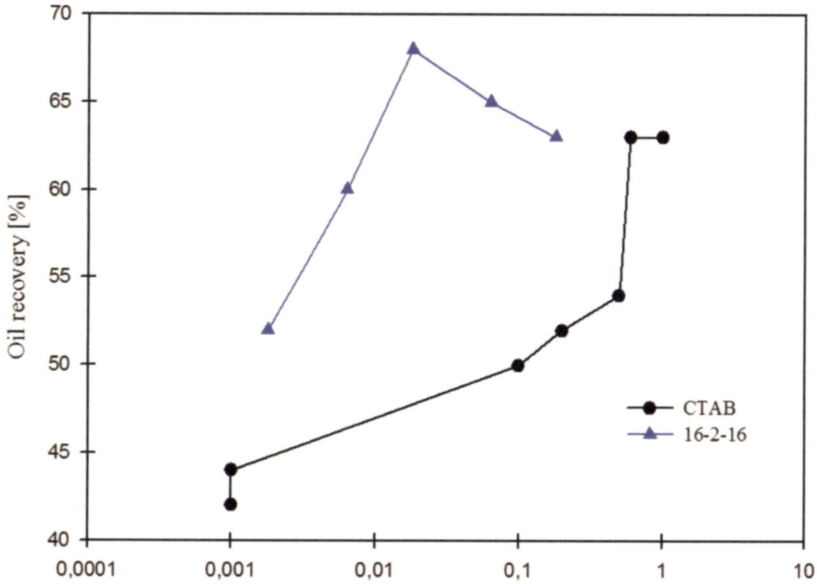

Figure 37. Mimic oil recovery of CTAB and 16-2-16 aqueous solutions.

lowest possible value [274]. Due to their excellent surface-active properties, cationic gemini surfactants are great at lowering surface tension and changing the wettability [273]. Solutions of cationic surfactants, both monomeric (CTAB) and dimeric (16-2-16), were tested as mimic flooding solution (oil, n-dodecane; porous material, silica gel powder). It was noticed that the best results were achieved around CMC values, for CTAB 1 mM and for [16-2-16] 0.018 mM [273].

The highest oil recovery of 16-2-16 (68%) was reached at 0.018 mM whereas for CTAB (63%) at 0.6 mM (**Figure 37**) [273]. The tested gemini surfactant allows to achieve similar percent of oil recovery at lower concentration which makes the process more efficient [272, 275].

Author details

Bogumil E. Brycki*, Iwona H. Kowalczyk, Adrianna Szulc, Olga Kaczerewska and Marta Pakiet

*Address all correspondence to: brycki@amu.edu.pl

Faculty of Chemistry, Laboratory of Microbiocides Chemistry, Adam Mickiewicz University, Poznan, Poland

References

[1] CAS Databases. CAS Databases [Internet]. Available from: http://www.cas.org/

[2] Myers D. Surfactant Science and Technology. 3rd ed. Hoboken, N.J: J. Wiley; 2006. 380 p

[3] Rhein LD, editor. Surfactants in Personal Care Products and Decorative Cosmetics. 3rd ed. Boca Raton: CRC Press; 2007. 480 p. (Surfactant science series)

[4] Karsa DR. Surfactants in Polymers, Coatings, Inks, and Adhesives. 1st ed. Blackwell; Bosa Roca, USA, 2003. (Sheffield Annual Surfactants Review)

[5] Schramm LL, Stasiuk EN, Marangoni DG. Surfactants and their applications. Annual Reports on the Progress of Chemistry, Section C: Physical Chemistry. 2003;**99**:3-48

[6] Tehrani-Bagha A, Holmberg K. Solubilization of hydrophobic dyes in surfactant solutions. Materials. 2013 Feb 21;**6**(2):580–608

[7] Siedenbiedel F, Tiller JC. Antimicrobial polymers in solution and on surfaces: overview and functional principles. Polymers. 2012 Jan 9;**4**(4):46-71

[8] Rahman PKSM, editor. Microbiotechnology Based Surfactants and Their Applications [Internet]. Frontiers Media SA; 2016 [cited 2017 Jan 20]. (Frontiers Research Topics). Available from: http://www.frontiersin.org/books/Microbiotechnology_Based_Surfactants_and_Their_Applications/804

[9] Russell H& A, editor. Principles and Practice of Disinfection, Preservation, and Sterilization. 5th ed. Chichester, West Sussex: John Wiley & Sons; 2012

[10] Zhou C, Wang F, Chen H, Li M, Qiao F, Liu Z, et al. Selective antimicrobial activities and action mechanism of micelles self-assembled by cationic oligomeric surfactants. ACS Applied Materials & Interfaces. 2016 Feb 17;**8**(6):4242-4249

[11] John VT, Simmons B, McPherson GL, Bose A. Recent developments in materials synthesis in surfactant systems. Current Opinion in Colloid & Interface Science. 2002;**7**(5):288-295

[12] Shinde PV, Kategaonkar AH, Shingate BB, Shingare MS. Surfactant catalyzed convenient and greener synthesis of tetrahydrobenzo[*a*]xanthene-11-ones at ambient temperature. Beilstein Journal of Organic Chemistry 2011 Jan 13;**7**:53-58

[13] Akbar JR. Pharmaceutical Applications of Gemini Surfactants. 2011 [cited 2017 Jan 20]; Available from: https://uwspace.uwaterloo.ca/handle/10012/5700

[14] Sivaramakrishnan CN. The use of surfactants in the finishing of technical textiles. In: Advances in the Dyeing and Finishing of Technical Textiles [Internet]. Elsevier; 2013 [cited 2017 Jan 20]. pp. 199-235. Available from: http://linkinghub.elsevier.com/retrieve/pii/B9780857094339500092

[15] Le Marechal AM, Križanec B, Vajnhandl S, Valh JV. Textile finishing industry as an important source of organic pollutants. In: Organic Pollutants Ten Years After the Stockholm Convention-Environmental and Analytical Update [Internet]. Rijeka, Croatia: InTech; 2012 [cited 2017 Jan 20]. Available from: http://www.intechopen.com/source/pdfs/29368/InTech-Textile_finishing_industry_as_an_important_source_of_organic_pollutants.pdf

[16] Castro MJL, Ojeda C, Cirelli AF. Advances in surfactants for agrochemicals. Environmental Chemistry Letters. 2014 Mar;**12**(1):85-95

[17] Shen AQ, Gleason B, McKinley GH, Stone HA. Fiber coating with surfactant solutions. Physics of Fluids. 2002 Nov;**14**(11):4055-4068

[18] Ma J, Tang J, Cheng Q, Zhang H, Shinya N, Qin LC. Effects of surfactants on spinning carbon nanotube fibers by an electrophoretic method. Science and Technology of Advanced Materials. 2010 Dec;**11**(6):065005

[19] Lin Y, Qiao Y, Cheng X, Yan Y, Li Z, Huang J. Hydrotropic salt promotes anionic surfactant self-assembly into vesicles and ultralong fibers. Journal of Colloid and Interface Science. 2012 Mar;**369**(1):238-244

[20] Pritchard G, editor. Plastics Additives [Internet]. Dordrecht: Springer Netherlands; 1998 [cited 2017 Jan 20]. (Brewis D, Briggs D, editors. Polymer Science and Technology Series; Vol. 1). Available from: http://link.springer.com/10.1007/978-94-011-5862-6

[21] Kralova I, Sjöblom J. Surfactants used in food industry: a review. Journal of Dispersion Science and Technology. 2009 Sep 30;**30**(9):1363-1383

[22] Nitschke M, Silva SS. Recent food applications of microbial surfactants. Critical Reviews in Food Science and Nutrition. 2016 Jul **20**

[23] Raffa P, Broekhuis AA, Picchioni F. Polymeric surfactants for enhanced oil recovery: A review. Journal of Petroleum Science & Engineering. 2016 Sep;**145**:723-733

[24] Maithufi MN, Joubert DJ, Klumperman B. Application of Gemini surfactants as diesel fuel wax dispersants. Energy & Fuels. 2011 Jan 20;**25**(1):162-171

[25] Kumar N, Tyagi R. Industrial applications of dimeric surfactants: A review. Journal of Dispersion Science and Technology. 2014 Feb;**35**(2):205-214

[26] Athas JC, Jun K, McCafferty C, Owoseni O, John VT, Raghavan SR. An effective dispersant for oil spills based on food-grade amphiphiles. Langmuir. 2014 Aug 12;**30**(31): 9285-9294

[27] Song D. Development of high efficient and low toxic oil spill dispersants based on sorbitol derivants nonionic surfactants and glycolipid biosurfactants. Journal of Environmental Protection. 2013;**04**(01):16-22

[28] Lopes LRB, Soares VLP, Barcellos MTC, Mansur CRE. Desenvolvimento de surfatantes para aplicação na indústria de explosivos. Polímeros. 2014 Aug;**24**(4):474-477

[29] Tornero V, Hanke G. Chemical contaminants entering the marine environment from sea-based sources: A review with a focus on European seas. Marine Pollution Bulletin. 2016 Nov;**112**(1–2):17-38

[30] Brown P, Butts CP, Eastoe J. Stimuli-responsive surfactants. Soft Matter. 2013;**9**(8):2365

[31] Transparency Market Research. Surfactants Market. Global Industry Analysis, Size, Share, Growth, Trends and Forecast, 2015–2023 [Internet]. Available from: http://www.transparencymarketresearch.com/

[32] Li J, Wang X, Zhang T, Wang C, Huang Z, Luo X, et al. A review on phospholipids and their main applications in drug delivery systems. Asian Journal of Pharmaceutical Sciences. 2015 Apr;**10**(2):81-98

[33] Transparency Market Research. Surfactants Market Rising at 4.20% CAGR from 2015–2023 due to Rising Demand for Detergents and Personal Care Products [Internet]. [cited 2017 Jan 16]. Available from: https://globenewswire.com/news-release/2016/01/29/805884/0/en/Surfactants-Market-Rising-at-4-20-CAGR-from-2015–2023-due-to-Rising-Demand--for-Detergents-and-Personal-Care-Products-Transparency-Market-Research.html

[34] Zana R, Xia J. Gemini Surfactants: Synthesis, Interfacial and Solution-Phase Behavior, and Applications [Internet]. New York: Marcel Dekker; 2004 [cited 2017 Jan 20]. Available from: http://www.chemlibnetbase.com/ejournals/books/book_km.asp?id=2399

[35] Patial P, Chandel M. Synthesis, Characterization & Evaluation of Cationic Gemini Surfactants—Synthesis of Surfactants. Verlag, Germany: Lambert Academic Publishing; 2016

[36] Ping M, Lu L, Cheng ZY. Synthesis and Properties of Gemini Surfactant (Chinese Edition). Chemical Industry Press; Beijing, China, 2014

[37] Damen M, Cristóbal-Lecina E, Sanmartí GC, van Dongen SFM, García Rodríguez CL, Dolbnya IP, et al. Structure–delivery relationships of lysine-based Gemini surfactants and their lipoplexes. Soft Matter. 2014;**10**(31):5702-5714

[38] Kwaśniewska D, Staszak K, Wieczorek D, Zieliński R. Synthesis and Interfacial Activity of Novel Heterogemini Sulfobetaines in Aqueous Solution. Journal of Surfactants and Detergents. 2015 May;**18**(3):477-486

[39] Wang L, Zhang Y, Ding L, Liu J, Zhao B, Deng Q, et al. Synthesis and physiochemical properties of novel Gemini surfactants with phenyl-1,4-bis(carbamoylmethyl) spacer. RSC Advances. 2015;**5**(91):74764-74773

[40] Badea I, Poorghorban M, Das U, Alaidi O, Chitanda J, Michel D, et al. Characterization of the host-guest complex of a curcumin analog with β-cyclodextrin and β-cyclodextrin-gemini surfactant and evaluation of its anticancer activity. International Journal of Nanomedicine. 2015 Jan;**503**

[41] Ilies MA, Seitz WA, Johnson BH, Ezell EL, Miller AL, Thompson EB, et al. Lipophilic pyrylium salts in the synthesis of efficient pyridinium-based cationic lipids, Gemini surfactants, and lipophilic oligomers for gene delivery. Journal of Medicinal Chemistry. 2006 Jun;**49**(13):3872-3887

[42] Kateb ME, Givenchy ET, Baklouti A, Guittard F. Synthesis and surface properties of semi-fluorinated Gemini surfactants with two reactive bromo pendant groups. Journal of Colloid and Interface Science. 2011 May;**357**(1):129-134

[43] Jia W, Rao X, Song Z, Shang S. Microwave-assisted synthesis and properties of a novel cationic gemini surfactant with the hydrophenanthrene structure. Journal of Surfactants and Detergents. 2009 Aug;**12**(3):261-267

[44] Tehrani-Bagha AR, Oskarsson H, van Ginkel CG, Holmberg K. Cationic ester-containing Gemini surfactants: Chemical hydrolysis and biodegradation. Journal of Colloid and Interface Science. 2007 Aug;**312**(2):444-452

[45] Bunton CA, Robinson LB, Schaak J, Stam MF. Catalysis of nucleophilic substitutions by micelles of dicationic detergents. Journal of Organic Chemistry. 1971;**36**(16):2346-2350

[46] Devinsky F, Masárová L, Lacko I. Surface activity and micelle formation of some new bisquaternary ammonium salts. Journal of Colloid and Interface Science. 1985;**105**(1):235-239

[47] Zana R. Dimeric (Gemini) surfactants: Effect of the spacer group on the association behavior in aqueous solution. Journal of Colloid and Interface Science. 2002 Apr;**248**(2):203-220

[48] Esumi K, Taguma K, Koide Y. Aqueous properties of multichain quaternary cationic surfactants. Langmuir. 1996;**12**(16):4039-4041

[49] Zhu Y, Masuyama A, Okahara M. Preparation and surface active properties of amphipathic compounds with two sulfate groups and two lipophilic alkyl chains. Journal of the American Oil Chemists' Society. 1990;**67**(7):459-463

[50] Zana R, Benrraou M, Rueff R. Alkanediyl-. alpha.,. omega.-bis (dimethylalkylammonium bromide) surfactants. 1. Effect of the spacer chain length on the critical micelle concentration and micelle ionization degree. Langmuir. 1991;**7**(6):1072-1075

[51] De S, Aswal VK, Goyal PS, Bhattacharya S. Novel gemini micelles from dimeric surfactants with oxyethylene spacer chain. Small angle neutron scattering and fluorescence studies. Journal of Physical Chemistry. 1998;**102**:6152-6160

[52] Lu T, Huang J. Synthesis and properties of novel gemini surfactant with short spacer. Chinese Science Bulletin. 2007 Oct;**52**(19):2618-2620

[53] Brycki B, Drgas M, Bielawska M, Zdziennicka A, Jańczuk B. Synthesis, spectroscopic studies, aggregation and surface behavior of hexamethylene-1,6-bis(N,N-dimethyl-N-dodecylammonium bromide). Journal of Molecular Liquids 2016 Sep;**221**:1086-1096

[54] Liao B, Li Y, Li Y. Synthesis of alpha- chloroacetates of ethylene glycol and its oligomers. Acta Polymerica Sinica. 1992;**1**(2):34-41

[55] Gao Z, Tai S, Zhang Q, Zhao Y, Lu B, Ge Y, et al. Synthesis and surface activity of bisquaternary ammonium salt Gemini surfactants with ester bond. Wuhan University Journal of Natural Sciences. 2008;**13**(2):227-231

[56] Łuczyński J, Frąckowiak R, Włoch A, Kleszczyńska H, Witek S. Gemini ester quat surfactants and their biological activity. Cellular and Molecular Biology Letters [Internet]. 2013 Jan 1 [cited 2017 Jan 20];**18**(1). Available from: http://www.degruyter.com/view/j/cmble.2013.18.issue-1/s11658-012-0041-4/s11658-012-0041-4.xml

[57] Morán MC, Pinazo A, Pérez L, Clapés P, Angelet M, García MT, et al. "Green" amino acid-based surfactants. Green Chemistry 2004;**6**(5):233-240

[58] Goreti Silva S, Fernandes RF, Marques EF, do Vale MLC. Serine-based bis-quat Gemini surfactants: Synthesis and micellization properties. European Journal of Organic Chemistry. 2012 Jan;**2012**(2):345-352

[59] Clapés P, Rosa Infante M. Amino acid-based surfactants: Enzymatic synthesis, Properties and Potential Applications. Biocatalysis and Biotransformation. 2002 Jan;**20**(4):215-233

[60] Colomer A, Pinazo A, Manresa MA, Vinardell MP, Mitjans M, Infante MR, et al. Cationic surfactants derived from lysine: Effects of their structure and charge type on antimicrobial and hemolytic activities. Journal of Medicinal Chemistry. 2011 Feb 24;**54**(4):989-1002

[61] Tan H, Xiao H. Synthesis and antimicrobial characterization of novel l-lysine Gemini surfactants pended with reactive groups. Tetrahedron Letters. 2008 Mar;**49**(11):1759-1761

[62] Gomes P, Araújo MJ, Marques EF, Falcão S, Brito RO. Straightforward method for the preparation of lysine-based double-chained anionic surfactants. Synthetic Communications. 2008 May 23;**38**(12):2025-2036

[63] Yoshimura T, Sakato A, Tsuchiya K, Ohkubo T, Sakai H, Abe M, et al. Adsorption and aggregation properties of amino acid-based N-alkyl cysteine monomeric and -dialkyl cystine Gemini surfactants. Journal of Colloid and Interface Science. 2007 Apr;**308**(2):466-473

[64] Seredyuk V, Alami E, Nyden M, Holmberg K, Peresypkin AV, Menger FM. Adsorption of zwitterionic Gemini surfactants at the air–water and solid–water interfaces. Colloids and Surfaces A: Physicochemical and Engineering Aspects. 2002;**203**(1):245-258

[65] Peresypkin AV, Menger FM. Zwitterionic geminis. Coacervate formation from a single organic compound. Organic Letters. 1999 Nov;**1**(9):1347-1350

[66] Yoshimura T, Ichinokawa T, Kaji M, Esumi K. Synthesis and surface-active properties of sulfobetaine-type zwitterionic Gemini surfactants. Colloids and Surfaces A: Physicochemical and Engineering Aspects. 2006 Feb;**273**(1–3):208-212

[67] Yoshimura T, Nyuta K, Esumi K. Zwitterionic heterogemini surfactants containing ammonium and carboxylate headgroups. 1. Adsorption and micellization. Langmuir. 2005;**21**(7):2682-2688

[68] Xie Z, Feng Y. Synthesis and properties of alkylbetaine zwitterionic Gemini surfactants. Journal of Surfactants and Detergents. 2010 Jan;**13**(1):51-57

[69] Quagliotto P, Viscardi G, Barolo C, Barni E, Bellinvia S, Fisicaro E, et al. Gemini pyridinium surfactants: Synthesis and conductometric study of a novel class of amphiphiles. Journal of Organic Chemistry. 2003 Oct;**68**(20):7651-7660

[70] Zhou L, Jiang X, Li Y, Chen Z, Hu X. Synthesis and properties of a novel class of gemini pyridinium surfactants. Langmuir. 2007 Nov;**23**(23):11404-11408

[71] Shi L, Lundberg D, Musaev DG, Menger FM. Annulene Gemini surfactants: Structure and self-assembly. Angewandte Chemie, International Edition. 2007 Aug 3;**46**(31):5889-5891

[72] Yamaguchi I, Gobara Y, Sato M. One-pot synthesis of N-substituted diaza[12] annulenes. Organic Letters. 2006 Sep;**8**(19):4279-4281.

[73] Alami EO, Holmberg K. HeteroGemini surfactants. Advances in Colloid and Interface Science 2003;**100**:13-46

[74] Jaeger DA, Li B, Clark TJ. Cleavable double-chain surfactants with one cationic and one anionic head group that form vesicles. Langmuir 1996;**12**:4314-4316

[75] Xu Q, Wang L, Xing F. Synthesis and properties of dissymmetric Gemini surfactants. Journal of Surfactants and Detergents. 2011 Jan;**14**(1):85-90

[76] Muslim AA, Ayyash D, Gujral SS, Mekhail GM, Rao PPN, Wettig SD. Synthesis and characterization of asymmetrical Gemini surfactants. Physical Chemistry Chemical Physics. 2017;**19**(3):1953-1962

[77] Horikoshi S, Matsuzaki S, Mitani T, Serpone N. Microwave frequency effects on dielectric properties of some common solvents and on microwave-assisted syntheses:

2-Allylphenol and the C12–C2–C12 Gemini surfactant. Radiation Physics and Chemistry. 2012 Dec;**81**(12):1885-1895

[78] Brycki B, Kowalczyk I, Kozirog A. Synthesis, molecular structure, spectral properties and antifungal activity of polymethylene-α,ω-bis(N,N- dimethyl-N-dodecylammonium bromides). Molecules. 2011 Jan 5;**16**(12):319-335

[79] Chen L, Xie H, Li Y, Yu W. Applications of cationic gemini surfactant in preparing multi-walled carbon nanotube contained nanofluids. Colloids and Surfaces A: Physicochemical and Engineering Aspects. 2008 Dec;**330**(2–3):176-179

[80] Landry JM, Marangoni DG, Lumsden MD, Berno R. 1D and 2D NMR investigations of the micelle-formation process in 8-phenyloctanoate micelles. Canadian Journal of Chemistry 2007;**85**:202-207

[81] Wuthrich KI. Nmr of Proteins and Nucleic Acids. New York, USA: Wiley VCH; 1986.

[82] Roscigno P, Asaro F, Pellizer G, Ortona O, Paduano L. Complex formation between poly (vinylpyrrolidone) and sodium decyl sulfate studied through NMR. Langmuir. 2003;**19** (23):9638-9644

[83] Han P, He Y, Chen C, Yu H, Liu F, Yang H, et al. Study on synergistic mechanism of inhibitor mixture based on electron transfer behavior. Scientific Reports 2016 Sep 27;**6**:33252

[84] Mobin M, Aslam R, Zehra S, Ahmad M. Bio-/environment-friendly cationic gemini surfactant as novel corrosion inhibitor for mild steel in 1 M HCl solution. Journal of Surfactants and Detergents. 2017 Jan;**20**(1):57-74

[85] Yoshimura T, Nagata Y, Esumi K. Interactions of quaternary ammonium salt-type Gemini surfactants with sodium poly(styrene sulfonate). Journal of Colloid and Interface Science. 2004 Jul;**275**(2):618-622

[86] Buse J, Badea I, Verrall RE, El-Aneed A. Tandem mass spectrometric analysis of the novel gemini surfactant nanoparticle families G12-s and G18:1-s. Spectroscopy Letters. 2010 Aug 17;**43**(6):447-457

[87] Aimé C, Plet B, Manet S, Schmitter JM, Huc I, Oda R, et al. Competing gas-phase substitution and elimination reactions of Gemini surfactants with anionic counterions by mass spectrometry. Density functional theory correlations with their bolaform halide salt models. Journal of Physical Chemistry B. 2008 Nov 20;**112**(46):14435-14445

[88] Wang S, Zhao K. Dielectric analysis for the spherical and rodlike micelle aggregates formed from a gemini surfactant: Driving forces of micellization and stability of micelles. Langmuir. 2016 Aug 2;**32**(30):7530-7540

[89] Wilkes AJ, Waraven G, Talbot JM. HPLC analysis of quaternary ammonium surfactants with the evaporative light scattering detector. JAOCS. 1992;**69**(7)

[90] Schmidtchen FP, Oswald H. RP-HPLC separation of highly charged quaternary ammonium salts. Journal of Liquid Chromatography. 1986 Apr;**9**(5):993-1002

[91] Chen H, Wang C, Ye J, Zhou H, Lu L, Yang Z. Synthesis and properties of a lacquer wax-based quaternary ammonium gemini surfactant. Molecules. 2014 Mar 24;**19**(3):3596-3606

[92] Wee VT, Kennedy JM. Determination of trace levels of quaternary ammonium compounds in river water by liquid chromatography with conductometric detection. Analytical Chemistry. 1982;**54**(9):1631-1633

[93] Prince SJ, McLaury HJ, Allen LV, McLaury P. Analysis of benzalkonium chloride and its homologs: HPLC versus HPCE. Journal of Pharmaceutical and Biomedical Analysis. 1999;**19**(6):877-882

[94] Vincent G, Kopferschmitt-Kubler MC, Mirabel P, Pauli G, Millet M. Sampling and analysis of quaternary ammonium compounds (QACs) traces in indoor atmosphere. Environmental Monitoring and Assessment. 2007 Oct;**133**(1–3):25-30

[95] Epton SR. A rapid method of analysis for certain surface-active agents. Nature. 1947;**6**(4075):795-796

[96] ASTM International D1681-05. Standart Test Method for Synthetic Anionic Active Ingredient in Detergents by Cationic Titration Procedure. ASTM International; West Conshohocken, USA, 2005

[97] ASTM International D3049-89. Standard Test Method for Synthetic Anionic Ingredient by Cationic Titration. ASTM International; West Conshohocken, USA, 2003

[98] BS 3762-3.1:1990, ISO 2271. Analysis of Formulated Detergents. Quantitative Test Methods. Method for Determination of Anionic-Active Matter Content. BSI; London, United Kingdom, 1989

[99] DIN 38409-23:2010-12. German Standard Methods for the Examination of Water, Waste Water and Sludge—Parameters Characterizing Effects and Substances (Group H)—Part 23: Determination of Bismuth Active Substances (H 23). Berlin, Germany, 2010

[100] Dietrich O. Handbook of Surfactant Analysis. London, United Kingdom: John Wiley & Sons; 2000

[101] Rosen MJ, Kunjappu JJ. Surfactants and Interfacial Phenomena. 4th ed. New Jersy: John Wiley & Sons Interscience; 2012

[102] Lai L, Mei P, Wu XM, Chen L, Liu Y. Interfacial dynamic properties and dilational rheology of mixed anionic and cationic Gemini surfactant systems at air–water interface. Colloids and Surfaces A: Physicochemical and Engineering Aspects 2016 Nov;**509**:341-350

[103] Menger FM, Keiper JS, Azov V. Gemini Surfactants with Acetylenic Spacers. Langmuir. 2000 Mar;**16**(5):2062-2067

[104] Garcia MT, Campos E, Sanchez-Leal J, Comelles F. Structure-activity relationships for sorption of alkyl trimethyl ammonium compounds on activated sludge. Tenside, Surfactants, Detergents. 2004;**41**(5):235-239

[105] Hajy Alimohammadi M, Javadian S, Gharibi H, Tehrani-Bagha AR, Alavijeh MR, Kakaei K. Aggregation behavior and intermicellar interactions of cationic Gemini surfactants: Effects of alkyl chain, spacer lengths and temperature. Journal of Chemical Thermodynamics. 2012 Jan;**44**(1):107-115

[106] Singh V, Tyagi R. Unique micellization and CMC aspects of Gemini surfactant: An overview. Journal of Dispersion Science and Technology. 2014 Nov 2;**35**(12):1774-1792

[107] Laschewsky A, Wattebled L, Arotçaréna M, Habib-Jiwan JL, Rakotoaly RH. Synthesis and properties of cationic oligomeric surfactants. Langmuir. 2005;**21**(16):7170-7179

[108] Yoshimura T, Chiba N, Matsuoka K. Supra-long chain surfactants with double or triple quaternary ammonium headgroups. Journal of Colloid and Interface Science. 2012 May;**374**(1):157-163

[109] In M, Bec V, Aguerre-Chariol O, Zana R. Quaternary ammonium bromide surfactant oligomers in aqueous solution: Self-association and microstructure. Langmuir. 2000 Jan;**16**(1):141-148

[110] Zana R. Alkanediyl-α,ω-bis(dimethylalkylammonium bromide) surfactants. Journal of Colloid and Interface Science. 2002 Feb;**246**(1):182-190

[111] Wettig SD, Verrall RE. Thermodynamic studies of aqueous m–s–m Gemini surfactant systems. Journal of Colloid and Interface Science. 2001 Mar;**235**(2):310-316

[112] Verma SK, Ghosh KK. Micellar and surface properties of some monomeric surfactants and a Gemini cationic surfactant. Journal of Surfactants and Detergents. 2011 Jul;**14**(3):347-352

[113] Grosmaire L, Chorro M, Chorro C, Partyka S, Zana R. Alkanediyl-α,ω-bis(dimethylalkylammonium bromide) surfactants. Journal of Colloid and Interface Science. 2002 Feb;**246**(1):175-181

[114] Wang X, Wang J, Wang Y, Yan H, Li P, Thomas RK. Effect of the nature of the spacer on the aggregation properties of Gemini surfactants in an aqueous solution. Langmuir. 2004 Jan;**20**(1):53-56

[115] Menger FM, Keiper JS. Gemini surfactants. Angewandte Chemie, International Edition. 2000;**39**:1906-1920

[116] Laschewsky A, Lunkenheimer K, Rakotoaly RH, Wattebled L. Spacer effects in dimeric cationic surfactants. Colloid & Polymer Science. 2005 Feb;**283**(5):469-479

[117] Zhang Z, Wang H, Zheng P, Shen W. Effect of spacer rigidity on the aggregations of ester containing Gemini surfactants in aqueous solutions: A study of density and fluorescence. Colloids and Surfaces A: Physicochemical and Engineering Aspects. 2013 Mar;**421**:193-200

[118] Teresa Garcia M, Kaczerewska O, Ribosa I, Brycki B, Materna P, Drgas M. Hydrophilicity and flexibility of the spacer as critical parameters on the aggregation behavior of long alkyl chain cationic gemini surfactants in aqueous solution. Journal of Molecular Liquids [Internet]. 2017 Jan [cited 2017 Jan 20]; Available from: http://linkinghub.elsevier.com/retrieve/pii/S0167732216317494

[119] Zhang Q, Gao Z, Xu F, Tai S. Effect of hydrocarbon structure of the headgroup on the thermodynamic properties of micellization of cationic gemini surfactants: An electrical conductivity study. Journal of Colloid and Interface Science. 2012 Apr;**371** (1):73-81

[120] Li B, Zhang Q, Xia Y, Gao Z. Surface properties and aggregation behavior of cationic gemini surfactants with dipropylammonium head-groups. Colloids and Surfaces A: Physicochemical and Engineering Aspects. 2015 Apr;**470**:211-217

[121] Borse MS, Devi S. Importance of head group polarity in controlling aggregation properties of cationic gemini surfactants. Advances in Colloid and Interface Science. 2006 Nov;**123–126**:387-399

[122] Kumar B, Tikariha D, Ghosh KK, Barbero N, Quagliotto P. Effect of polymers and temperature on critical micelle concentration of some gemini and monomeric surfactants. Journal of Chemical Thermodynamics 2013 Jul;**62**:178-185

[123] Borse M, Sharma V, Aswal VK, Pokhriyal NK, Joshi JV, Goyal PS, et al. Small angle neutron scattering and viscosity studies of micellar solutions of bis-cationic surfactants containing hydroxyethyl methyl quaternary ammonium head groups. Physical Chemistry Chemical Physics. 2004;**6**(13):3508

[124] Pisárčik M, Jampílek J, Devínsky F, Drábiková J, Tkacz J, Opravil T. Gemini surfactants with polymethylene spacer: Supramolecular structures at solid surface and aggregation in aqueous solution. Journal of Surfactants and Detergents. 2016 May;**19**(3):477-486

[125] Tawfik SM, Abd-Elaal AA, Shaban SM, Roshdy AA. Surface, thermodynamic and biological activities of some synthesized Gemini quaternary ammonium salts based on polyethylene glycol. Journal of Industrial and Engineering Chemistry. 2015 Oct;**30**: 112-119

[126] Zana R. Critical micellization concentration of surfactants in aqueous solution and free energy of micellization. Langmuir. 1996;**12**(5):1208-1211

[127] Cornellas A, Perez L, Comelles F, Ribosa I, Manresa A, Garcia MT. Self-aggregation and antimicrobial activity of imidazolium and pyridinium based ionic liquids in aqueous solution. Journal of Colloid and Interface Science. 2011 Mar;**355**(1):164-171

[128] Aiad I, El-Sukkary MM, Soliman EA, El-Awady MY, Shaban SM. Characterization, surface properties and biological activity of new prepared cationic surfactants. Journal of Industrial and Engineering Chemistry. 2014 Jul;**20**(4):1633-1640

[129] Block SS. Disinfection, Sterilization and Preservations 5th ed. Philadelphia: Lippincott Williams & Wilkins; 2001

[130] Manivannan G. Disinfection and Decontamination; Principles, Applications and Related Issues. Boca Raton USA: CRC Press Taylor & Francis Group; 2008

[131] Paulus W. Directory of Microbiocides for the Protection of Material. A Handbook. Dordrecht, The Netherlands: Springer; 2005

[132] Cross J, Singer J. Cationic Surfactants, Analytical and Biological Evaluation. New York, USA: Marcel Dekker; 1994

[133] Lambert PA. Mechanisms of action of microbicides. In: Principles and Practise of Disinfection, Preservation, and Sterilization. 5th ed. Chichester, West Sussex: John Wiley & Sons; 2012

[134] Millard JS. Mechanisms of bacterial resistance to microbicides. In: Principles and Practice of Disinfection, Preservation and Sterilization. 5th ed. Chichester, West Sussex: John Wiley & Sons; 2012

[135] Chapman JS. Biocide resistance mechanisms. International Biodeterioration and Biodegradation. 2003;**51**(2):133-138

[136] Brycki B, Szulc A. Gemini alkyldeoxy-D-glucitolammonium salts as modern surfactants and microbiocides: Synthesis, antimicrobial and surface activity, biodegradation. Johnson SJ, editor. PLoS ONE. 2014 Jan 8;**9**(1):e84936

[137] Zhang S, Ding S, Yu J, Chen X, Lei Q, Fang W. Antibacterial activity, in vitro cytotoxicity, and cell cycle arrest of Gemini quaternary ammonium surfactants. Langmuir. 2015 Nov 10;**31**(44):12161-12169

[138] Brycki B. Gemini alkylammonium salts as biodeterioration inhibitors. Polish Journal of Microbiology. 2010;**59**:227-231

[139] Russell AD. Biocide use and antibiotic resistance: the relevance of laboratory findings to clinical and environmental situations. Lancet Infectious Diseases. 2003;**3**(12):794-803

[140] Jorgensen JH, Ferraro MJ. Antimicrobial susceptibility testing: A review of general principles and contemporary practices. Clinical Infectious Diseases. 2009 Dec;**49**(11):1749-1755

[141] Tyagi S, Tyagi VK. Novel cationic Gemini surfactants and methods for determination of their antimicrobial activity–review. Tenside, Surfactants, Detergents. 2014;**51**:379-386

[142] Klančnik A, Piskernik S, Jeršek B, Možina SS. Evaluation of diffusion and dilution methods to determine the antibacterial activity of plant extracts. Journal of Microbiological Methods. 2010 May;**81**(2):121-126

[143] Kuperkar K, Modi J, Patel K. Surface-active properties and antimicrobial study of conventional cationic and synthesized symmetrical Gemini surfactants. Journal of Surfactants and Detergents. 2012 Jan;**15**(1):107-115

[144] Koziróg A, Brycki B. Monomeric and gemini surfactants as antimicrobial agents—influence on environmental and reference strains. Acta Biochimica Polonica. 2015;**62**(4): 879-883

[145] Paniak TJ, Jennings MC, Shanahan PC, Joyce MD, Santiago CN, Wuest WM, et al. The antimicrobial activity of mono-, bis-, tris-, and tetracationic amphiphiles derived from simple polyamine platforms. Bioorganic & Medicinal Chemistry Letters. 2014 Dec;**24** (24):5824-5828

[146] Black JW, Jennings MC, Azarewicz J, Paniak TJ, Grenier MC, Wuest WM, et al. TMEDA-derived biscationic amphiphiles: An economical preparation of potent antibacterial agents. Bioorganic & Medicinal Chemistry Letters. 2014 Jan;**24**(1):99-102

[147] Laatiris A, El Achouri M, Rosa Infante M, Bensouda Y. Antibacterial activity, structure and CMC relationships of alkanediyl α,ω-bis(dimethylammonium bromide) surfactants. Microbiological Research. 2008 Nov;**163**(6):645-650

[148] Banno T, Toshima K, Kawada K, Matsumura S. Synthesis and properties of Gemini-type cationic surfactants containing carbonate linkages in the linker moiety directed toward green and sustainable chemistry. Journal of Surfactants and Detergents. 2009 Aug;**12** (3):249-259

[149] Tatsumi T, Zhang W, Nakatsuji Y, Miyake K, Matsushima K, Tanaka M, et al. Preparation, surface-active properties, and antimicrobial activities of bis (alkylammonium) dichlorides having a butenylen or a butynylene spacer. Journal of Surfactants and Detergents. 2001;**4**(3):271-277

[150] Martín VI, de la Haba RR, Ventosa A, Congiu E, Ortega-Calvo JJ, Moyá ML. Colloidal and biological properties of cationic single-chain and dimeric surfactants. Colloids and Surfaces. B, Biointerfaces. 2014 Feb;**114**:247-254

[151] Hansch C, Fujita T. p-σ-π Analysis. A method for the correlation of biological activity and chemical structure. Journal of the American Chemical Society. 1964;**86**(8):1616-1626

[152] Hansch C, Clayton JM. Lipophilic character and biological activity of drugs II: The parabolic case. Journal of Pharmaceutical Sciences. 1973;**62**(1):1-21

[153] Fatma N, Panda M, Kabir-ud D, Beg M. Ester-bonded cationic gemini surfactants: Assessment of their cytotoxicity and antimicrobial activity. Journal of Molecular Liquids. 2016 Oct;**222**:390-394

[154] Guo S, Sun X, Zou Q, Zhang J, Ni H. Antibacterial activities of five cationic Gemini surfactants with ethylene glycol bisacetyl spacers. Journal of Surfactants and Detergents. 2014 Nov;**17**(6):1089-1097

[155] Zhu H, Hu Z, Ma X, Wang J, Cao D. Synthesis, surface and antimicrobial activities of cationic Gemini surfactants with semi-rigid spacers. Journal of Surfactants and Detergents. 2016 Mar;**19**(2):265-274.

[156] Li H, Yu C, Chen R, Li J, Li J. Novel ionic liquid-type Gemini surfactants: Synthesis, surface property and antimicrobial activity. Colloids and Surfaces A: Physicochemical and Engineering Aspects. 2012 Feb;**395**:116-124

[157] Diz M, Manresa A, Pinazo A, Erra P, Infante M. Synthesis, surface active properties and antimicrobial activity of new bis quaternary ammonium compounds. Journal of the Chemical Society, Perkin Transactions 2. 1994;(8):1871-1876

[158] Murguía MC, Machuca LM, Lurá MC, Cabrera MI, Grau RJ. Synthesis and properties of novel antifungal Gemini compounds derived from N-acetyl diethanolamines. Journal of Surfactants and Detergents. 2008 Sep;**11**(3):223-230

[159] Labena A, Hegazy MA, Horn H, Müller E. Cationic Gemini surfactant as a corrosion inhibitor and a biocide for high salinity sulfidogenic bacteria originating from an oil-field water tank. Journal of Surfactants and Detergents. 2014 May;**17**(3):419-431

[160] Obłąk E, Piecuch A, Krasowska A, Łuczyński J. Antifungal activity of gemini quaternary ammonium salts. Microbiological Research. 2013 Dec;**168**(10):630-638

[161] Obłąk E, Piecuch A, Guz-Regner K, Dworniczek E. Antibacterial activity of gemini quaternary ammonium salts. FEMS Microbiology Letters. 2014 Jan;**350**(2):190-198

[162] Obłąk E, Piecuch A, Dworniczek E, Olejniczak T. The influence of biodegradable gemini surfactants, N,N′-bis(1-decyloxy-1-oxypronan-2-yl)-N,N,N′,N′-tetramethylpropane-1,3-diammonium dibromide and N,N′-bis(1-dodecyloxy-1-oxopronan-2-yl)-N,N,N′,N′-tetramethane-1,2-diammonium dibromide, on fungal biofilm and adhesion. Journal of Oleo Science. 2015;**64**:527-537

[163] Badr EE, Kandeel EM, El-Sadek BM. Novel gemini cationic surfactants based on N,N-dimethyl fatty hydrazine and 1,3-dibromopropane; synthesis, mechanism of action, and cytotoxicities. Journal of Oleo Science 2010;**59**:647-652

[164] Hoque J, Akkapeddi P, Yarlagadda V, Uppu DSSM, Kumar P, Haldar J. Cleavable cationic antibacterial amphiphiles: Synthesis, mechanism of action, and cytotoxicities. Langmuir. 2012 Aug 21;**28**(33):12225-12234

[165] Ghumare AK, Pawar BV, Bhagwat SS. Synthesis and antibacterial activity of novel amido-amine-based cationic Gemini surfactants. Journal of Surfactants and Detergents. 2013 Jan;**16**(1):85-93

[166] Banno T, Kawada K, Matsumura S. Creation of novel green and sustainable Gemini-type cationics containing carbonate linkages. Journal of Surfactants and Detergents. 2010 Oct;**13**(4):387-398

[167] Ding Z, Fang S. Synthesis, surface and antimicrobial activities of novel cationic Gemini surfactants. Journal of Surfactants and Detergents. 2015 Nov;**18**(6):1051-1057

[168] Węgrzyńska J, Chlebicki J, Maliszewska I. Preparation, surface-active properties and antimicrobial activities of bis(ester quaternary ammonium) salts. Journal of Surfactants and Detergents. 2007 May 8;**10**(2):109-116

[169] Caillier L, Taffin de Givenchy E, Levy R, Vandenberghe Y, Geribaldi S, Guittard F. Polymerizable semi-fluorinated gemini surfactants designed for antimicrobial materials. Journal of Colloid and Interface Science. 2009 Apr;**332**(1):201-207

[170] Sharma V, Borse M, Devi S, Dave K, Pohnerkar J, Prajapati A. Oil solubilization capacity, liquid crystalline properties, and antibacterial activity of alkanolamine-based novel cationic surfactants. Journal of Dispersion Science and Technology. 2005 Jul;**26** (4):421-427

[171] Garcia MT, Kaczerewska O, Ribosa I, Brycki B, Materna P, Drgas M. Biodegradability and aquatic toxicity of quaternary ammonium-based gemini surfactants: Effect of the spacer on their ecological properties. Chemosphere 2016 Jul;**154**:155-160

[172] Brycki B, Waligórska M, Szulc A. The biodegradation of monomeric and dimeric alkylammonium surfactants. Journal of Hazardous Materials. 2014 Sep;**280**:797-815

[173] Colomer A, Pinazo A, García MT, Mitjans M, Vinardell MP, Infante MR, et al. pH-Sensitive surfactants from lysine: Assessment of their cytotoxicity and environmental behavior. Langmuir. 2012 Apr 10;**28**(14):5900-5912

[174] Tehrani-Bagha AR, Holmberg K, van Ginkel CG, Kean M. Cationic gemini surfactants with cleavable spacer: Chemical hydrolysis, biodegradation, and toxicity. Journal of Colloid and Interface Science 2015 Jul;**449**:72-79

[175] Tehrani-Bagha AR, Holmberg K. Cationic ester-containing Gemini surfactants: Physical–chemical properties. Langmuir. 2010 Jun 15;**26**(12):9276-9282

[176] Tawfik SM, Abd-Elaal AA, Aiad I. Three gemini cationic surfactants as biodegradable corrosion inhibitors for carbon steel in HCl solution. Research on Chemical Intermediates. 2016 Feb;**42**(2):1101-1123

[177] Akram M, Anwar S, Ansari F, Bhat IA, Kabir-ud D. Bio-physicochemical analysis of ethylene oxide-linked diester-functionalized green cationic gemini surfactants. RSC Advances. 2016;**6**(26):21697-21705

[178] Akram M, Ansari F, Bhat IA, Chaturvedi SK, Khan RH, Kabir-ud D. An ester-functionalized cationic gemini surfactant mediated structural transitions of porcine serum albumin (PSA) via binding interaction. Colloids and Surfaces A: Physicochemical and Engineering Aspects. 2017 Mar;**516**:245-253

[179] Bergero MF, Liffourrena AS, Opizzo BA, Fochesatto AS, Lucchesi GI. Immobilization of a microbial consortium on Ca-alginate enhances degradation of cationic surfactants in flasks and bioreactor. International Biodeterioration & Biodegradation 2017 Feb;**117**:39-44

[180] Akram M, Bhat IA, Kabir-ud D. Effect of salt counterions on the physicochemical characteristics of novel green surfactant, ethane-1,2-diyl bis(N,N-dimethyl-N-tetradecylammoniumacetoxy) dichloride. Colloids and Surfaces A: Physicochemical and Engineering Aspects. 2016 Mar;**493**:32-40

[181] Pérez L, Garcia MT, Ribosa I, Vinardell MP, Manresa A, Infante MR. Biological properties of arginine-based gemini cationic surfactants. Environmental Toxicology & Chemistry. 2002;**21**(6):1279-1285

[182] Zhou F, Maeda T, Nagamune H, Kourai H. Synthesis and antimicrobial characteristics of novel biocides, 1, 1′-(decanedioyl) bis (4-methy1-4-alkylpiperazinium iodide) s with a Gemini structure. Biocontrol Science. 2004;**9**(3):61-67

[183] Chauhan V, Singh S, Kaur T, Kaur G. Self-assembly and biophysical properties of Gemini 3-alkyloxypyridinium amphiphiles with a hydroxyl-substituted spacer. Langmuir. 2015 Mar 17;**31**(10):2956-2966

[184] Almeida JAS, Faneca H, Carvalho RA, Marques EF, Pais AACC. Dicationic alkylammo nium bromide Gemini surfactants. Membrane perturbation and skin irritation. Fraternali F, editor. PLoS ONE. 2011 Nov 10;6(11):e26965

[185] Silva SMC, Sousa JJS, Marques EF, Pais AACC, Michniak-Kohn BB. Structure activity relationships in alkylammonium C12-Gemini surfactants used as dermal permeation enhancers. The AAPS Journal. 2013 Oct;**15**(4):1119-1127

[186] McCafferty E. Introduction to Corrosion Science. Springer; London, United Kingdom, 2010. 575 p

[187] Bardal E. Corrosion and Protection. London; New York: Springer; 2004. 315 p

[188] Sastri VS, Ghali E, Elboujdaini M. Corrosion Prevention and Protection; Practical Solutions. England: John Wiley & Sons Ltd.; 2007

[189] ASTM International. ASTM G15-02. Standard Terminology Relating to Corrosion and Corrosion Testing. ASTM International; West Conshohocken, USA, 2002.

[190] Hegazy MA, Abd El-Rehim SS, Badr EA, Kamel WM, Youssif AH. Mono-, di- and tetracationic surfactants as carbon steel corrosion inhibitors. Journal of Surfactants and Detergents. 2015 Nov;**18**(6):1033–1042

[191] Cao K, Sun HY, Hou BR. Corrosion inhibition of Gemini surfactant for copper in 3.5% NaCl. Advances in Materials Research. 2014 Jun;**936**:1125-1131

[192] Asefi D, Arami M, Mahmoodi NM. Comparing chain length effect of single chain and Gemini surfactants on corrosion inhibition of steel in acid. ECS Transactions. 2011;**35** (17):89-101

[193] Hegazy MA, El-Tabei AS. Synthesis, surface properties, synergism parameter and inhibitive performance of novel cationic Gemini surfactant on carbon steel corrosion in 1 M HCl solution. Journal of Surfactants and Detergents. 2013 Mar;**16**(2):221-232

[194] Migahed MA, Rashwan SM, Kamel MM, Habib RE. Synthesis, characterization of polyaspartic acid-glycine adduct and evaluation of their performance as scale and corrosion inhibitor in desalination water plants. Journal of Molecular Liquids 2016 Dec;**224**:849-858

[195] Hegazy MA, Rashwan SM, Kamel MM, El Kotb MS. Synthesis, surface properties and inhibition behavior of novel cationic gemini surfactant for corrosion of carbon steel tubes in acidic solution. Journal of Molecular Liquids 2015 Nov;**211**:126-134

[196] Mobin M, Masroor S. Adsorption and corrosion inhibition behavior of schiff base-based cationic Gemini surfactant on mild steel in formic acid. Journal of Dispersion Science and Technology. 2014 Apr 3;**35**(4):535-543

[197] Hegazy MA, El-Etre AY, El-Shafaie M, Berry KM. Novel cationic surfactants for corrosion inhibition of carbon steel pipelines in oil and gas wells applications. Journal of Molecular Liquids 2016 Feb;**214**:347-356

[198] Aiad I, Riya MA, Tawfik SM, Abousehly MA. Protection of carbon steel against corrosion in hydrochloric acid solution by some synthesized cationic surfactants. Protection of Metals and Physical Chemistry of Surfaces. 2016 Mar;**52**(2):339-347

[199] Mobin M, Masroor S. Cationic gemini surfactants as novel corrosion inhibitor for mild steel in 1M HCl. International Journal of Electrochemical Science. 2012;**7**:6920-6940

[200] Zhang Q, Gao Z, Xu F, Zou X. Adsorption and corrosion inhibitive properties of gemini surfactants in the series of hexanediyl-1,6-bis-(diethyl alkyl ammonium bromide) on aluminium in hydrochloric acid solution. Colloids and Surfaces A: Physicochemical and Engineering Aspects. 2011 May;**380**(1–3):191-200

[201] Ash M, Ash I. Handbook of Corrosion Inhibitors [Internet]. Endicott, N.Y.: Synapse Information Resources; 2011 [cited 2017 Jan 20]. Available from: http://search.ebscohost.com/login.aspx?direct=true&scope=site&db=nlebk&db=nlabk&AN=470439

[202] von Baeckmann W, Schwenk W, Prinz W, von Baeckmann W, editors. Handbook of Cathodic Corrosion Protection: Theory and Practice of Electrochemical Protection Processes. 3rd ed. Houston, Tex: Gulf Pub. Co; 1997. 567 p

[203] Grundmeier G, Schmidt W, Stratmann M. Corrosion protection by organic coatings: electrochemical mechanism and novel methods of investigation. Electrochimica Acta. 2000;**45**(15):2515-2533

[204] Riggs OL, Locke CE, Hamner NE. Anodic Protection Theory and Practice in the Prevention of Corrosion. Boston, MA: Springer US; 1981

[205] Wojciechowski J, Szubert K, Peipmann R, Fritz M, Schmidt U, Bund A, et al. Anticorrosive properties of silane coatings deposited on anodised aluminium. Electrochimica Acta 2016 Dec;**220**:1-10

[206] Sun YM, Chen HL. A study of corrosion inhibition of carbon steel in hydrochloric acid using BIMGCS12-3. Advances in Materials Research. 2012 Jan;**427**:3-6

[207] Hegazy MA, Azzam EMS, Kandil NG, Badawi AM, Sami RM. Corrosion inhibition of carbon steel pipelines by some new amphoteric and di-cationic surfactants in acidic

solution by chemical and electrochemical methods. Journal of Surfactants and Detergents. 2016 Jul;**19**(4):861-871

[208] Abdallah M, Eltass HM, Hegazy MA, Ahmed H. Adsorption and inhibition effect of novel cationic surfactant for pipelines carbon steel in acidic solution. Protection of Metals and Physical Chemistry of Surfaces. 2016 Jul;**52**(4):721-730

[209] Hegazy MA, Abdallah M, Awad MK, Rezk M. Three novel di-quaternary ammonium salts as corrosion inhibitors for API X65 steel pipeline in acidic solution. Part I: Experimental results. Corrosion Science. 2014 Apr;**81**:54-64

[210] Hegazy MA, Atlam FM. Three novel bolaamphiphiles as corrosion inhibitors for carbon steel in hydrochloric acid: Experimental and computational studies. Journal of Molecular Liquids. 2016 Jun;**218**:649-662

[211] Mahdavian M, Tehrani-Bagha AR, Holmberg K. Comparison of a cationic Gemini surfactant and the corresponding monomeric surfactant for corrosion protection of mild steel in hydrochloric acid. Journal of Surfactants and Detergents. 2011 Oct;**14**(4):605-613

[212] Nessim IM, Hamdy A, Osman MM, Shalaby MN. Inhibitory effect of some cationic Gemini surfactants for carbon steel in sea water. International Journal of Chemistry. 2012;**02**.

[213] Mobin M, Masroor S. Alkanediyl-α, ω-bis (dimethyl cetylammonium bromide) gemini surfactants as novel corrosion inhibitors for mild steel in formic acid. Materials Research. 2012 Dec;**15**(6):837-847

[214] Qiu LG, Xie AJ, Shen YH. Understanding the effect of the spacer length on adsorption of gemini surfactants onto steel surface in acid medium. Applied Surface Science. 2005 Jun;**246**(1–3):1-5

[215] Tawfik SM. Ionic liquids based gemini cationic surfactants as corrosion inhibitors for carbon steel in hydrochloric acid solution. Journal of Molecular Liquids. 2016 Apr;**216**:624-635

[216] Migahed MA, Hegazy MA, Al-Sabagh AM. Synergistic inhibition effect between Cu^{2+} and cationic gemini surfactant on the corrosion of downhole tubing steel during secondary oil recovery of old wells. Corrosion Science. 2012 Aug;**61**:10-18

[217] Qiu LG, Wu Y, Wang YM, Jiang X. Synergistic effect between cationic gemini surfactant and chloride ion for the corrosion inhibition of steel in sulphuric acid. Corrosion Science. 2008 Feb;**50**(2):576-582

[218] Wu ZY, Fang Z, Qiu LG, Wu Y, Li ZQ, Xu T, et al. Synergistic inhibition between the gemini surfactant and bromide ion for steel corrosion in sulphuric acid. Journal of Applied Electrochemistry. 2009 Jun;**39**(6):779-784

[219] Zhao J, Duan H, Jiang R. Synergistic corrosion inhibition effect of quinoline quaternary ammonium salt and Gemini surfactant in H_2S and CO_2 saturated brine solution. Corrosion Science 2015 Feb;**91**:108-119

[220] Altamirano RH, Cervantes VYM, Rivera LSZ, Conde HIB, Ramirez SL. Gemini Surfactants, Process of Manufacture and Use as Multifunctional Corrosion Inhibitors. US 9023785 B2, 2015

[221] Henry KM, Hicks KD. Bis-quaternary ammonium salt corrosion inhibitors. US 8999315 B2, 2015

[222] Ahmed T, Kamel AO, Wettig SD. Interactions between DNA and Gemini surfactant: Impact on gene therapy: Part I. Nanomedicine. 2016 Feb;**11**(3):289-306

[223] Pisárčik M, Devínsky F. Surface tension study of cationic gemini surfactants binding to DNA. Central European Journal of Chemistry. 2014 May;**12**(5):577-585

[224] Geoffroy M, Faure D, Oda R, Bassani DM, Baigl D. Photocontrol of genomic DNA conformation by using a photosensitive Gemini surfactant: Binding affinity versus reversibility. ChemBioChem. 2008 Oct 13;**9**(15):2382-2385

[225] Kirby AJ, Camilleri P, Engberts JBFN, Feiters MC, Nolte RJM, Söderman O, et al. Gemini surfactants: New synthetic vectors for gene transfection. Angewandte Chemie, International Edition. 2003 Apr 4;**42**(13):1448-1457

[226] Al-Dulaymi MA, Chitanda JM, Mohammed-Saeid W, Araghi HY, Verrall RE, Grochulski P, et al. Di-peptide-modified Gemini surfactants as gene delivery vectors: Exploring the role of the alkyl tail in their physicochemical behavior and biological activity. The AAPS Journal. 2016 Sep;**18**(5):1168-1181

[227] Kim BK. Synthesis and optimization of cholesterol-based diquaternary ammonium Gemini surfactant (Chol-GS) as a new gene delivery vector. Journal of Microbiology and Biotechnology. 2011;**21**(1):93-99

[228] Zakharova LY, Gabdrakhmanov DR, Ibragimova AR, Vasilieva EA, Nizameev IR, Kadirov MK, et al. Structural, biocomplexation and gene delivery properties of hydroxyethylated gemini surfactants with varied spacer length. Colloids and Surfaces. B, Biointerfaces. 2016 Apr;**140**:269-277

[229] Fisicaro E, Compari C, Bacciottini F, Contardi L, Pongiluppi E, Barbero N, et al. Nonviral gene-delivery by highly fluorinated gemini bispyridinium surfactant-based DNA nanoparticles. Journal of Colloid and Interface Science. 2017 Feb;**487**:182-191

[230] McGregor C, Perrin C, Monck M, Camilleri P, Kirby AJ. Rational approaches to the design of cationic Gemini surfactants for gene delivery. Journal of the American Chemical Society. 2001 Jul;**123**(26):6215-6220

[231] Camilleri P, Kremer A, Edwards AJ, Jennings KH, Jenkins O, Marshall I, et al. A novel class of cationic gemini surfactants showing efficient in vitro gene transfection properties. Chemical Communications. 2000;(14):1253-1254

[232] Bazylińska U, Saczko J. Nanoemulsion-templated polyelectrolyte multifunctional nanocapsules for DNA entrapment and bioimaging. Colloids and Surfaces. B, Biointerfaces. 2016 Jan;**137**:191-202

[233] Wang L, Zhang C, Xie H, Sun W, Chen X, Wang X, et al. Calcium alginate gel capsules loaded with inhibitor for corrosion protection of downhole tube in oilfields. Corrosion Science. 2015 Jan;**90**:296-304

[234] Minkenberg CB, Li F, van Rijn P, Florusse L, Boekhoven J, Stuart MCA, et al. Responsive vesicles from dynamic covalent surfactants. Angewandte Chemie, International Edition. 2011 Apr 4;**50**(15):3421-3424

[235] Minkenberg CB, Homan B, Boekhoven J, Norder B, Koper GJM, Eelkema R, et al. Responsive wormlike micelles from dynamic covalent surfactants. Langmuir. 2012 Sep 25;**28**(38):13570-13576

[236] Feng Y, Chu Z. pH-Tunable wormlike micelles based on an ultra-long-chain "pseudo" gemini surfactant. Soft Matter. 2015;**11**(23):4614-4620

[237] Wang X, Zhang Z, Cao Y, Hao J. Ionogels of pseudogemini supra-amphiphiles in ethylammonium nitrate: Structures and properties. Journal of Colloid and Interface Science. 2017 Apr;**491**:64-71

[238] Akay G, Hassan-Raeisi A, Tuncaboylu DC, Orakdogen N, Abdurrahmanoglu S, Oppermann W, et al. Self-healing hydrogels formed in catanionic surfactant solutions. Soft Matter. 2013;**9**(7):2254

[239] Menger FM, Peresypkin AV, Caran KL, Apkarian RP. A sponge morphology in an elementary coacervate. Langmuir. 2000 Nov;**16**(24):9113-9116

[240] Lv J, Qiao W, Li Z. Vesicles from pH-regulated reversible gemini amino-acid surfactants as nanocapsules for delivery. Colloids and Surfaces. B, Biointerfaces. 2016 Oct;**146**: 523-531

[241] Mirgorodskaya AB, Ya Zakharova L, Khairutdinova EI, Lukashenko SS, Sinyashin OG. Supramolecular systems based on gemini surfactants for enhancing solubility of spectral probes and drugs in aqueous solution. Colloids and Surfaces A: Physicochemical and Engineering Aspects. 2016 Dec;**510**:33-42

[242] Tiwari AK, Gangopadhyay S, Chang CH, Pande S, Saha SK. Study on metal nanoparticles synthesis and orientation of gemini surfactant molecules used as stabilizer. Journal of Colloid and Interface Science. 2015 May;**445**:76-83

[243] Wang W, Han Y, Tian M, Fan Y, Tang Y, Gao M, et al. Cationic Gemini surfactant-assisted synthesis of hollow Au nanostructures by stepwise reductions. ACS Applied Materials & Interfaces. 2013 Jun 26;**5**(12):5709-5716

[244] Xu Y, Zhao Y, Chen L, Wang X, Sun J, Wu H, et al. Large-scale, low-cost synthesis of monodispersed gold nanorods using a gemini surfactant. Nanoscale. 2015;**7**(15):6790-6797

[245] Alejo T, Paulo PMR, Merchán MD, Garcia-Fernandez E, Costa SMB, Velázquez MM. Influence of 3D aggregation on the photoluminescence dynamics of CdSe quantum dot films. Journal of Luminescence. 2017 Mar;**183**:113-120

[246] Jamwal D, Rana D, Singh P, Pathak D, Kalia S, Thakur P, et al. Well-defined quantum dots and broadening of optical phonon line from hydrothermal method. RSC Advances. 2016;**6**(104):102010-102014

[247] Gaynanova GA, Vasilieva EA, Bekmukhametova AM, Nizameev IR, Kadirov MK, Zakharova LY, et al. Encapsulation of quantum dots in supramolecular systems based on amphiphilic compounds and polyelectrolytes. Russian Chemical Bulletin. 2016;**65** (1):151-157

[248] Feizi N, Yamini Y, Moradi M, Karimi M, Salamat Q, Amanzadeh H. A new generation of nano-structured supramolecular solvents based on propanol/gemini surfactant for liquid phase microextraction. Analytica Chimica Acta. 2017 Feb;**953**:1-9

[249] Wang Y, Guo R, Xi J. Comparative studies of interactions of hemoglobin with single-chain and with gemini surfactants. Journal of Colloid and Interface Science. 2009 Mar;**331**(2):470-475

[250] Li Y, Wang X, Wang Y. Comparative studies on interactions of bovine serum albumin with cationic Gemini and single-chain surfactants. Journal of Physical Chemistry B. 2006 Apr;**110**(16):8499-8505

[251] Ansari WH, Fatma N, Panda M, Kabir-ud D. Solubilization of polycyclic aromatic hydrocarbons by novel biodegradable cationic gemini surfactant ethane-1,2-diyl bis (N,N-dimethyl-N-hexadecylammoniumacetoxy) dichloride and its binary mixtures with conventional surfactants. Soft Matter. 2013;**9**(5):1478

[252] Panda M, Kabir-ud D. Study of surface and solution properties of gemini-conventional surfactant mixtures and their effects on solubilization of polycyclic aromatic hydrocarbons. Journal of Molecular Liquids. 2011 Sep;**163**(2):93-98

[253] Siddiqui H, Kamil M, Panda M, Kabir-ud D. Solubilization of phenanthrene and fluorene in equimolar binary mixtures of Gemini/conventional surfactants. Chinese Journal of Chemical Engineering. 2014 Sep;**22**(9):1009-1015

[254] Siddiqui H, Kamil M, Nazish F. Surface and solution properties of single and mixed gemini/conventional micelles on solubilization of polycyclic aromatic hydrocarbons. Indian Journal of Chemical Technology. 2015;**22**:194-200

[255] Tehrani-Bagha AR, Singh RG, Holmberg K. Solubilization of two organic dyes by cationic ester-containing gemini surfactants. Journal of Colloid and Interface Science. 2012 Jun;**376**(1):112-118

[256] Gharanjig K, Sadeghi-Kiakhani M, Tehrani-Bagha AR, Khosravi A, Menger FM. Solubility of two disperse dyes derived from N-alkyl and N-carboxylic acid naphthalimides in the presence of Gemini cationic surfactants. Journal of Surfactants and Detergents. 2011 Jul;**14**(3):381-389

[257] Lakra J, Tikariha D, Yadav T, Das S, Ghosh S, Satnami ML, et al. Mixed micellization of gemini and cationic surfactants: Physicochemical properties and solubilization of polycyclic aromatic hydrocarbons. Colloids and Surfaces A: Physicochemical and Engineering Aspects. 2014 Jun;**451**:56-65

[258] Wei J, Huang G, An C, Yu H. Investigation on the solubilization of polycyclic aromatic hydrocarbons in the presence of single and mixed Gemini surfactants. Journal of Hazardous Materials. 2011 Jun;**190**(1–3):840-847

[259] Wei J, Huang G, Wang S, Zhao S, Yao Y. Improved solubilities of PAHs by multicomponent Gemini surfactant systems with different spacer lengths. Colloids and Surfaces A: Physicochemical and Engineering Aspects. 2013 Apr;**423**:50-57

[260] Silva PR, Almeida VO, Machado GB, Benvenutti EV, Costa TMH, Gallas MR. Surfactant-based dispersant for multiwall carbon nanotubes to prepare ceramic composites by a sol–gel method. Langmuir. 2012 Jan 17;**28**(2):1447-1452

[261] Di Crescenzo A, Cambré S, Germani R, Di Profio P, Fontana A. Dispersion of SWCNTs with imidazolium-rich surfactants. Langmuir. 2014 Apr 15;**30**(14):3979-3987

[262] Poorgholami-Bejarpasi N, Sohrabi B. Role of surfactant structure in aqueous dispersions of carbon nanotubes. Fluid Phase Equilibria 2015 May;**394**:19-28

[263] Zhang S, Lu F, Zheng L. Dispersion of multiwalled carbon nanotubes (MWCNTs) by ionic liquid-based Gemini pyrrolidinium surfactants in aqueous solution. Colloid & Polymer Science. 2011 Nov;**289**(17–18):1815-1819

[264] Li D, Fang W, Wang H, Gao C, Zhang R, Cai K. Gold/oil nanofluids stabilized by a Gemini surfactant and their catalytic property. Industrial & Engineering Chemistry Research. 2013 Jun 19;**52**(24):8109-8113

[265] Xu J, Han X, Liu H, Hu Y. Synthesis of monodisperse gold nanoparticles stabilized by Gemini surfactant in reverse micelles. Journal of Dispersion Science and Technology. 2005 Jul;**26**(4):473-476

[266] Bi S, Peng H, Liao Y, Yang Y, Brycki B, Xie X. Microstructure and performances of pva dispersed liquid crystals containing gemini surfactant. Acta Polymerica Sinica. 2012;**16** (6):628-632

[267] Wang H, He L, Brycki BE, Kowalczyk IH, Kuliszewska E, Yang Y. Electrochemical characterization of the hydrophobic microenvironment with gemini surfactant micellar-hybridized supramolecular gels. Electrochimica Acta 2013;**90**:326-331

[268] Fu C, He D, Yu Y, Wu S, Dong C, Wang H. Fluorescent sensitization of gemini surfactant micellar-hybridized supramolecular hydrogels. Journal of Luminescence 2017;**181**:8-13

[269] Sakthivel S, Gardas RL, Sangwai JS. Effect of alkyl ammonium ionic liquids on the interfacial tension of the crude oil–water system and their use for the enhanced oil recovery using ionic liquid-polymer flooding. Energy & Fuels. 2016 Mar 17;**30**(3):2514-2623

[270] Pal N, Babu K, Mandal A. Surface tension, dynamic light scattering and rheological studies of a new polymeric surfactant for application in enhanced oil recovery. Journal of Petroleum Science & Engineering. 2016 Oct;**146**:591-600

[271] Malkin AY, Khadzhiev SN. On the rheology of oil (review). Petroleum Chemistry. 2016 Jul;**56**(7):541-551

[272] Zhou M, Zhao J, Hu X. Synthesis of bis[N,N′-(alkylamideethyl)ethyl] triethylenediamine bromide surfactants and their oilfield application investigation. Journal of Surfactants and Detergents. 2012 May;**15**(3):309-315

[273] Bi ZC, Qi LY, Liao WS. Dynamic surface properties, wettability and mimic oil recovery of ethanediyl-α, β-bis (cetyldimethylammonium bromide) on dodecane modified silica powder. Journal of Materials Science. 2005;**40**(11):2783-2788

[274] Nguele R, Sasaki K, Salim HS-A, Sugai Y. Physicochemical and microemulsion properties of dimeric quaternary ammonium salts with trimethylene spacer for enhanced oil recovery. Colloid & Polymer Science. 2015 Dec;**293**(12):3487-3497

[275] Qiu LG, Cheng MJ, Xie AJ, Shen YH. Study on the viscosity of cationic gemini surfactant–nonionic polymer complex in water. Journal of Colloid and Interface Science. 2004 Oct;**278**(1):40-43

Chapter 5

The Versatile Dioctadecyldimethylammonium Bromide

Ana Maria Carmona-Ribeiro

Additional information is available at the end of the chapter

http://dx.doi.org/10.5772/68020

Abstract

Dioctadecyldimethylammonium bromide (DODAB) is a quaternary ammonium surfactant (Quat) with interesting properties and applications. In this chapter, DODAB characteristics as compared to other Quats emphasize its self-assembly in aqueous solutions and the novel applications involving this useful cationic lipid so easily combined with biomolecules and interfaces to yield a wide range of novel uses in many fields such as delivery of drugs, vaccines and genes, design of nanoparticles, modification of interfaces, and many others yet to come.

Keywords: quaternary ammonium surfactants, self-assembly in water, cationic lipid in novel applications

1. The quaternary ammonium surfactants (Quats)

The quaternary ammonium surfactants or "Quats" encompass many individual chemicals [1, 2]. They are present in thousands of end-use formulations, many of which are blends of various Quats [1]. Common uses include disinfection, detergency, fabric softening, antistatic, and wood preservation [2]. The chemical structure determines their chemical behavior and utility. Quats will be strongly cationic due to their quaternary and positively charged nitrogen able to attach to surfaces, both organic and inorganic [3]. With remarkable chemical stability, they can exhibit long-lasting biocidal effects [4]. They attract anions, for example, soaps, detergents and hard water constituents, for example, carbonate and sulfate [5]. They are attracted by negatively charged cells such as bacteria or fungus and become attached to them eventually causing their cytoplasmic membrane to leak with membrane damages leading to antimicrobial effects [6–9]. Certain Quats will biodegrade and the biodegradability decreases with increase in their alkyl chain length [10, 11]. The degradation takes place via partitioning to sludge and processing by biodegradation. The complex Quats biodegradation occurs in several steps and depends on the Quat chemical structure, Quat interactions with the sludge determining adsorption and desorption, microorganisms present in the sludge and the presence or absence of anions; alkylammonium surfactants chemically modified with biological moieties such as carbohydrates, amide, aminoacids or peptides were better degraded [12]. From the point of view of Quats synthesis, compounds bearing more than one positive charge were readily obtained at economical cost from compounds with at least two tertiary amines that could be readily quaternized; some of these displayed potent antibacterial and antibiofilm activity and did not trigger bacterial resistance systems as those from methicillin-resistant Staphylococcus aureus (MRSA); mono-Quats and several bis-, tris- and tetra-Quats tested against bacteria within a few hundred generations yielded a lack of resistance for Quats of higher charge when compared to mono-Quats [13].

Quats chemical structure determines their self-assembly in water solution. The theory for the self-assembly of dilute surfactant solutions is well established and very successful [14, 15]. This theory applies also to Quats since their amphiphilic molecular nature includes polar and apolar regions in the same molecule. The theory relates the self-assembly in water solution with the geometric parameter v/al. The definition of v/al is given by v, the volume of the hydrocarbon region of the surfactant; a, the optimal head group area, and l, as the optimal hydrocarbon chain length related to its maximum extended length. One should notice that the nature and shape of the assemblies are intimately related to the v/al value. For instance, in the case of spherical micelles, $v/al < 1/3$ whereas for vesicles or bilayers, $½ < v/al \leq 1$. When bilayer vesicles are the desired structure, larger v is required as is the case of the double-chained surfactants. Single-chained surfactants and lower v are required for micellar structures. For example, a single-chained quaternary ammonium surfactant such as cetyltrimethylammonium bromide (CTAB) has a lower v than the corresponding double-chained quaternary ammonium surfactant. The self-assembly of CTAB and dioctadecyldimethylammonium bromide [DODAB] from calculations for their respective geometric parameters predicts, as indeed observed, CTAB molecules assembling as micelles and DODAB molecules assembling as bilayers in water solutions.

Not only the molecular geometry of the Quats determines their assembly in water solution: specific counterion effects also do [16]. Counterion adsorption and Stern layer effects change the optimal headgroup area a. In general, counterions will adsorb to some extent to the surfactant headgroups. Specific interactions of a nonelectrostatic origin like dehydration or hydration of the surfactants, conformational changes in the surfactant headgroup, size of the adsorbed counterion are important because they determine the thickness of the Stern layer and the actual surface potential. Specific counterions can change the lateral interactions between surfactants in a micelle, monolayer or bilayer. By means of the direct force measurement technique developed by Israelachvili [15] after depositing DODAB bilayers with the Langmuir-Blodgett technique on two molecularly smooth mica surfaces and bringing these surfaces together in an aqueous solution, the measurements of the interaction forces between the bilayers as a function of their separation a repulsive double-layer force are experienced. Fitting the measured double-layer force with theory allows the surface potential to be estimated, from which the binding affinity of the ions can be determined [15]. Apart from the repulsive double-layer interaction, the van der Waals interaction and possibly the ion-ion correlation interaction, which are both attractive, must be taken into account [17]. The interactions between bilayers of dihexadecyldimethylammonium acetate and bromide surfactants, which are soluble in water and adsorbed from solution as a bilayer onto the mica surfaces, were determined by Pashley and coworkers [18]. Marra employed the Langmuir-Blodgett deposition technique for an insoluble surfactant like DODAB so that the solution did not contain any aggregates and the binding of anions to the quaternary ammonium headgroups would not depend sensitively on the precise length of the hydrocarbon tails [16]. The anions investigated bound to the headgroups following a lyotropic series where the least hydrated, smallest anions bound with highest affinity [16]. Lateral interactions between DODAB adjacent molecules in a monolayer at the air-water interface and interactions between bilayers of DODAB surfactants exhibited a pronounced ion specificity. Large hydrated counterions like the fluoride, hydroxide, and acetate ions gave expanded monolayer compression isotherms. Fluoride, hydroxide, and acetate counterions did not bind to DODAB headgroups. Following the lyotropic series for anion sizes F-> C1-> Br-, the smaller the (hydrated) anion, the more contracted the monolayer [16]. For dioctadecyldimethylammonium (DODA) acetate, chloride or bromide, vesicle size and zeta-potentials were inversely related; an increase in the zeta-potential was accompanied by a decrease in vesicle size, in accordance with the self-assembly theory; DODA acetate bilayer vesicles had the largest, less tightly bound and more hydrated counterion and exhibited the smallest size in comparison with those obtained from the other DODA salts [19].

2. DODAB hybrid assemblies

DODAB remarkable interactive capability with opposite charges of silica particles [20–23], silicon wafers [24], polymeric particles [25–31]; polymer films [32–34], drugs [35–45], nucleic acids [31, 46], oligonucleotides [47–49], proteins [30, 50–54], peptides [9, 55–57], polyelectrolytes [8, 9, 36, 58, 59] and many other important surfaces, biological cells, molecules and nanostructures [60–67] is at the root of DODAB popularity in the literature spanning a huge variety of subjects. Today (December 10th, 2016) a search in American Chemical Society, PubMed and Scopus databases retrieved 104, 140 and 1208 documents, respectively, quoting DODAB. Therefore, this review just gives an overview on DODAB recent possibilities, and many others have already appeared or are yet to come.

The interaction between DODAB and solid surfaces like silicon wafers depends on the charge density of the solid surface, which depends on the nature and concentration of bound counterions and DODAB ability to displace them; the cation more tightly bound to the negatively charged surface solid surface should be Li+ that would be difficult to displace by the DODAB cation, in contrast to the loosely bound Cs+ with its large ion radius and low charge density. In summary, DODAB adsorption proceeded in accordance with charge density on the solid surface thus depending on nature and concentration of counterions and DODAB ability to displace them; increasing the ionic strength increases silanol dissociation, surface charge density, and DODAB adsorption [24]. The effect of monovalent salt nature and concentration over a range of low ionic strengths (0–10 mM LiCl, NaCl, KCl, or CsCl) and at two different pH values (6.3 and 10.0) on DODAB adsorption onto flat SiO_2 surfaces evaluated by in situ ellipsometry. This technique allowed precise evaluation of thin film thicknesses on very smooth solid surfaces such as those of silicon wafers. Thereby, DODAB adsorption isotherms of high affinity showed adsorption maxima consistent with bilayer deposition only around 10 mM monovalent salt at both pH values. In contrast, when pure water was the intervening medium, DODAB adsorption decreased substantially. The nature of counterion on the charged solid surface was also important to determine DODAB adsorption: at 10 mM CsCl or LiCl, the highest and the lowest affinity constants for DODAB adsorption onto SiO_2 were, respectively, obtained [24]. This was understandable from the fact that DODAB adsorption onto the solid surface required as a first step the displacement and cation exchange at the solid surface. DODAB adsorption consistently followed the expected facility of cation exchange at the surface required for DODAB adsorption. **Figure 1** illustrates the effect of counterion nature and

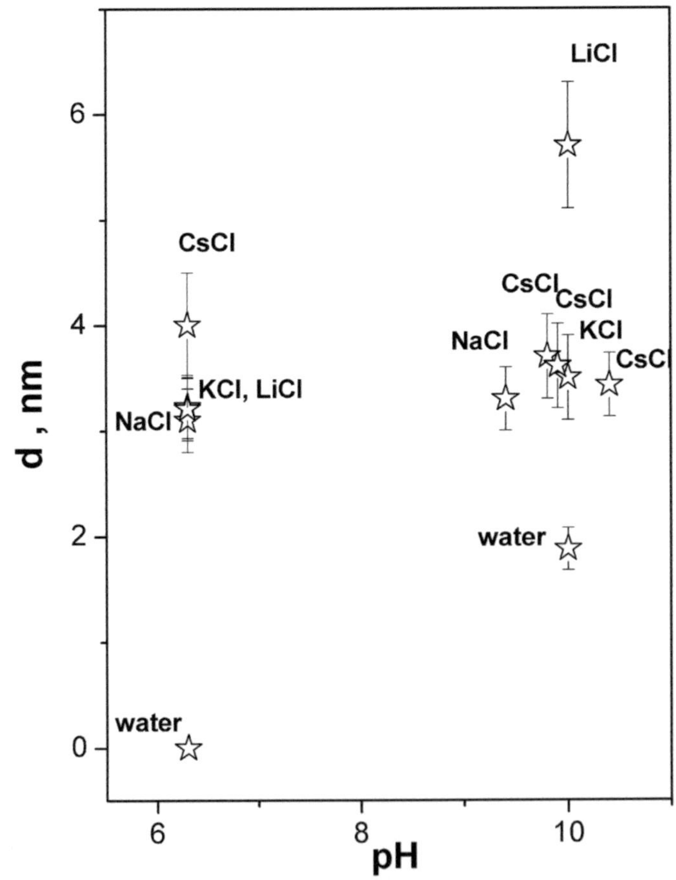

Figure 1. The effect of monovalent salt nature and concentration at 0 and 10 mM LiCl, NaCl, KCl, or CsCl and pH 6.3 or 10.0 on the thickness of the DODAB adsorbed layer deposited from bilayer fragments onto flat SiO_2 surfaces from in situ ellipsometry [24]. Reprinted with permission from Ref. [24]. Copyright (2006) American Chemical Society.

concentration on DODAB adsorption from bilayer fragments (BF) onto silicon wafers as determined from in situ ellipsometry measurements [24].

The changes of the electrostatic repulsion between adjacent DODAB molecules in a bilayer as the one due to interaction with counterions or oppositely charged inorganic or organic species can drastically change the nature of DODAB assemblies. For example, monovalent salt at a moderate concentration was reported to induce fusion of DODAB bilayer fragments [68–71] with induction of hydrophobic defects at the bilayer-water interface [72]. When the electrostatic repulsion is high as in pure water or in the presence of low concentrations of poorly bound counterions, interdigitation represents a way of relaxing the intermolecular repulsion in the bilayer; adhesion between DODAB bilayers due to interdigitation between DODAB molecules in the bilayer [26], molecular dynamic simulations [73], differential scanning calorimetry (DSC), and X-ray scattering in the subgel state [74] further supported DODAB tendency to display hydrophobic moieties in its assemblies for relaxation of the electrostatic repulsion.

Other interesting instances refer to the formation of catanionic bilayers from DODAB and anionic oleosiloxanes [75] or oleic acid [76]; DODAB membrane fragments and fatty-acid esters of cyclosiloxanes formed dense multibilayered vesicles; the transformation took place once the ester groups hydrolyzed to yield carboxyl groups yielding the anionic silicone surfactant in situ and the catanionic system with DODAB. The oleo-silica compound was obtained via hydrosilylation of methyl undec-10-enoate with 1,3,5,7-tetra-methylcyclo-tetrasiloxane (1). Flat DODAB/oleic acid bilayer sheets were obtained at about 1:1 molar ratios for DODAB/oleic acid binary dispersions; the relaxation of the electrostatic repulsion between DODAB molecules in the bilayer due to the incorporation of OA into DODAB bilayer decreased the membrane curvature and increased the aggregate size; introduction of the fatty acid around equimolar ratios led to flat DODAB/OA bilayer assemblies in the dispersions [76]. The electrostatic attraction between DODAB and anionic amphiphiles decreased the mean area per molecule, increased the geometric parameter v/al, and increased the aggregate size similarly to the fusogenic effects reported upon increasing counterion concentration [68–72, 75, 76].

Figure 2 shows cryo-transmission electron micrographs (cryo-TEM) of vitrified DODAB bilayer fragments obtained by sonication of DODAB in water [77], unilamellar vesicles of about 200–400 nm obtained by vaporization of a DODAB chloroform solution in water at 70 degrees centigrades (above the gel to liquid-crystalline phase transition temperature of the DODAB bilayer and above the chloroform boiling point) [78] and very large unilamellar DODAB vesicles from salt-induced fusion of DODAB bilayer fragments [68, 69].

Combinations of DODAB and dihexadecylphosphate (DHP) yielded miscible catanionic bilayers over a range of molar ratios, though DODAB and DHP miscibility in the bilayer domain was non-ideal [79]. For vesicles with DODAB as the predominant lipid, small sizes, high positive zeta potential, low main transition temperature, less angular structure, good stability, and high internal water compartment contrasted with similar properties determined for the DHP-rich vesicles; DODAB improved the bilayer fluidity of DHP vesicles both in the liquid-crystalline and in the rippled bilayer phases [79]. Interestingly, the reduction

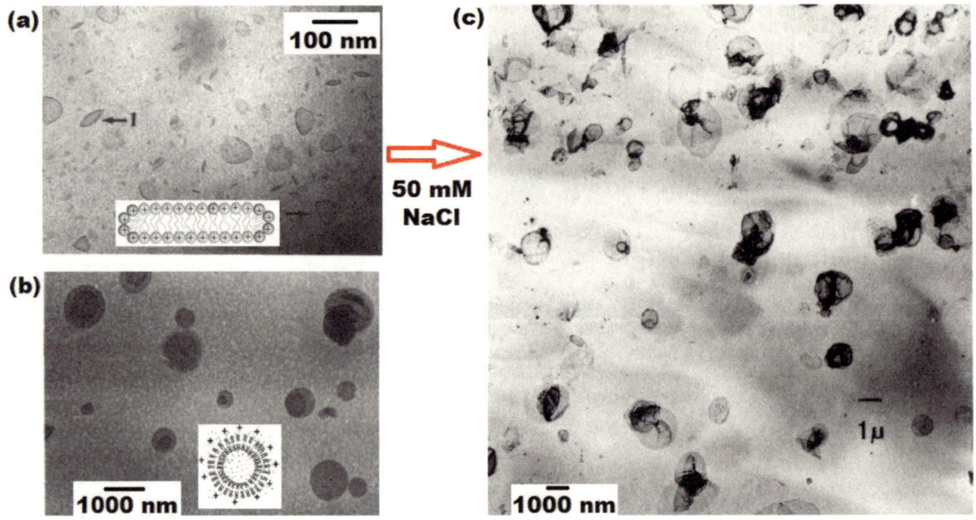

Figure 2. DODAB dispersions in water solutions obtained by different dispersion methods. (a) Cryo-transmission electron micrographs (cryo-TEM) of vitrified DODAB bilayer fragments obtained by sonication of DODAB in water [77]. Reprinted with permission from Ref. [77]. Copyright (1995) American Chemical Society. (b) Transmission electron microscopy of electronically stained large unilamellar DODAB vesicles (200–400 nm mean diameter) from vaporization of a DODAB chloroform solution in water at 70°C [78]. Reprinted from Ref. [78]. Copyright (1983) with permission of Elsevier. (c) Transmission electron microscopy of electronically stained and very large micrometric unilamellar DODAB vesicles obtained by NaCl-induced fusion of DODAB bilayer fragments [69]. Reprinted from Ref. [69]. Copyright (1986) with permission of Elsevier.

of positive charges on the DODAB/DHP vesicles improved also the survival of mammalian cells in culture [79]. These results might become important for future drug/gene delivery applications.

Cholesterol has been suggested to play a role in stable vesicle formation by adjusting the molecular packing of the vesicular bilayer. The Langmuir monolayer approach with infrared reflection-absorption spectroscopy (IRRAS) elucidated the effects of cholesterol on molecular packing of double-chained cationic surfactants [80]. Combining cholesterol with DXDAB monolayers at the air-water interface (X meaning the hydrocarbon chain length) reduced desorption of DXDAB with short alkyl chains, for example, ditetradecyldimethylammonium bromide or dihexadecyldimethylammonium bromide, into the water sub-phase and condensed the DXDAB monolayers [80]. For the DODAB monolayers, cholesterol had a dual effect inducing both order and disorder of the neighboring hydrocarbon chains; the flexible alkyl side-chain of cholesterol along with the corresponding portion of neighboring hydrocarbon chains formed a fluidic region, counteracting the conformational order induced by the sterol ring of cholesterol interacting with the alkyl chains [80].

The effect of varying the molar proportion of DODAB and neutral dipalmitoylphosphatidylcholine (DPPC) in DODAB/DPPC vesicles revealed a high bilayer and coloidal stability with good miscibility for the binary system and absence of phase separation at a molar proportion equal to 1 [81]. The demixing and crystallization of DODAB/DPPC binary lipid system were recently found to take place when DODAB or DPPC was dominant in the mixture (DPPC/DODAB = 1/2 or DPPC/DODAB = 2/1); when DODAB was no more than equimolar (e.g., DPPC/DODAB = 2/1 and 1/1), there was good miscibility in absence of DODAB crystallization [82]. At high or low DODAB, DPPC molar proportions, phase separation occurred upon cooling so that gel domains rich in DODAB phase-separated from DPPC-DODAB domains or DPPC domains. This phase separation for the gels would mean demixing and crystallization originating DODAB-rich and DPPC-rich tilted gel separated domains upon incubation at low temperatures [82].

Figure 3 illustrates the development of interdigitated regions in the DODAB bilayer as predicted from molecular dynamics simulation at two instants in time [73].

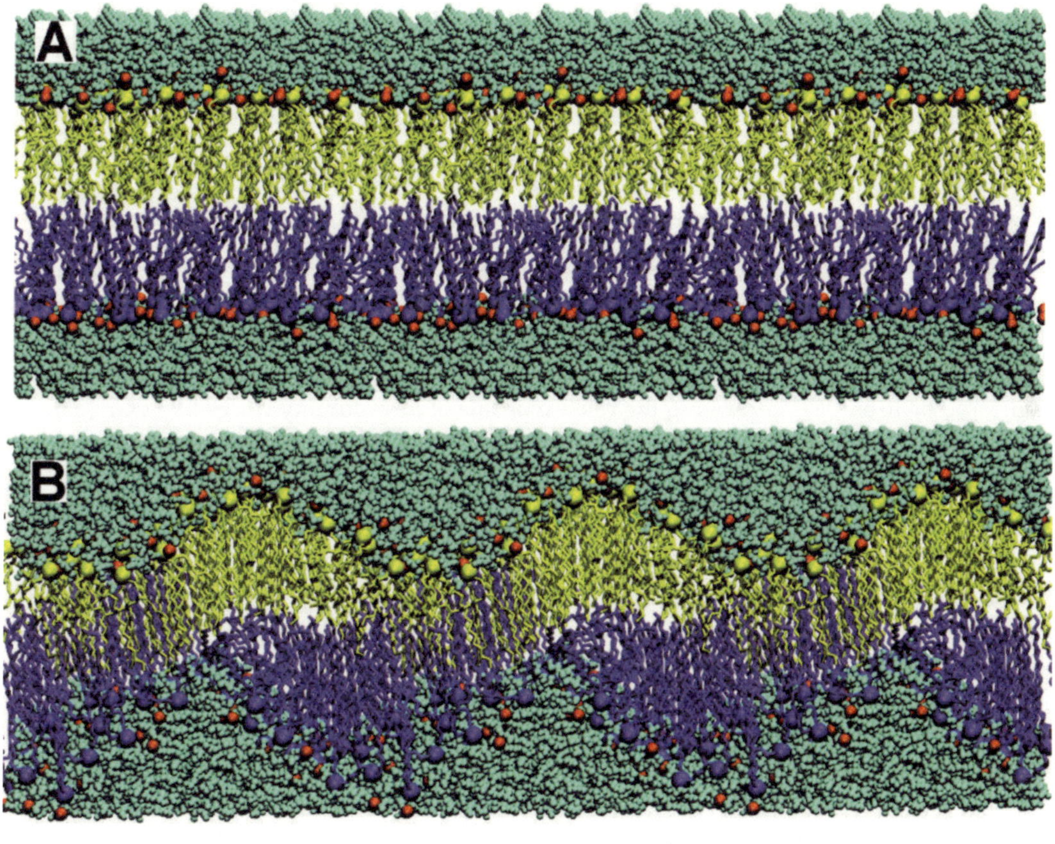

Figure 3. Molecular dynamics simulations of the DODAB bilayer at two different instants in time: 0 (A) and 90 ns (B); DODAB molecules assembled as a conventional (A) or as an interdigitated bilayer (B) where the hydrophilic quaternary ammonium heads were represented as spheres, similarly to the bromide ions; the water molecules are displayed as small spheres [73]. Reprinted with permission from Ref. [73]. Copyright (2010) American Chemical Society.

3. Novel applications for DODAB hybrid assemblies

Aqueous solubilization of water-insoluble materials is highly important for pharmaceuticals, detergency, emulsion polymerization, enhanced oil recovery, and textile dyeing. Among colloidal Self-assembled structures, micelles/vesicles are efficient solubilizers but the solubilization properties of bilayers of vesicles are superior [83, 84]. A series of double-chained surfactants, with increasing chain length (C12–18) mixed with single chained dodecylethyldimethylammonium bromide (DODABB) solubilized curcumin thanks to hydrophobic-hydrophobic and electrostatic interactions with preservation of curcumin antioxidant activity in food [85].

Aiming at the production of nanoparticles (NPs) for drug delivery, DODAB has been very useful to harmonize oppositely charged polysaccharides such as carboxymethylcellulose [58] or hyaluronic acid [86] with hydrophobic drugs such as amphotericin B [36], indomethacin [45], and tocoferol (vitamin E) [86]. Carboxymethylcellulose/DODAB/indomethacin NPs were prepared by direct injection of DODAB/indomethacin ethanol solution into a carboxymethylcellulose water solution [45]. Similarly, hyaluronate/soybean lecithin/DODAB/vitamin E NPs were prepared by direct injection of vitamin E/soybean lecithin/DODAB ethanol solution into hyaluronic acid water solution; further incorporation of these NPs in polymeric, bioadhesive films containing Aloe vera extract, hyaluronic acid, sodium alginate, polyethyleneoxide (PEO) and polyvinylalcohol (PVA) represented an innovative treatment for skin wounds [86].

A three-dimensional layer-by-layer (LbL) structure composed by xanthan and galactomannan biopolymers on DODAB liposome template created a LbL structure up to eight layers, evaluated using quartz crystal microbalance (QCM) and zeta potential analysis; these bilayer-coated NPs increased up to five times the sustained release of epidermal growth factor (EGF) and could be useful for improving the release profile of low-stability drugs like EGF [87].

The approach of combining important biomolecules such as proteins or nucleic acids with DODAB and further stabilizing the hybrids with hydrophilic polymers has been very useful for several biomedical and biotechnological applications. For instance, the delivery of DNA plasmids or small interference RNA (siRNA) to cells requires nanocarrier stability after in vivo administration though too strong stabilization can decrease the carrier efficiency; after characterizing DODAB/monoolein/pDNA or siRNA lipoplexes [88, 89], the nanocarriers were pegylated and tested for stability in serum and gene silencing in cultured cancer cells with promising results: pegylation avoided siRNA dissociation from the nanocarriers in human serum and improved transfection efficiency [90]. Stable lipoplexes of small size (100–160 nm) with a positive surface charge (>+45 mV) were readily internalized by human non-small cell lung carcinoma (H1299) cells and were efficient in promoting gene silencing. Monolein had a similar gene silencing ability as the commonly used helper lipid 1,2-dioleyl-3-phosphatidylethanolamine (DOPE), but with much lower cytotoxicity [91]. More recently, the same DODAB/monolein system was used to incorporate cell wall surface proteins (CWSP) from *Candida albicans* aiming at the production of an antigen delivery system (ADS) for a potential vaccine against candidiasis; the system facilitated antigen uptake by dendritic cells in vitro and induced higher levels of pro-inflammatory cytokines and opsonizing specific IgG antibodies in serum of mice immunized subcutaneously [92].

DODAB was also used to treat spores of *Bacillus subtilis* aiming at gene gun delivery of DNA plasmids in mice; DODAB treated spores allowed efficient plasmid adsorption and could be loaded into biolistic cartridges and efficiently delivered into mice for induction of specific cellular and antibody responses required for DNA vaccines in vivo [93].

For textile materials, sometimes modification of the wettability of hydrophobic surfaces is essential. For instance, DODAB adsorption to hydrophobic polypropylene (PP) thin films dramatically enhanced surface adsorption of different proteins from soybeans and represented a facile treatment to obtain PP-modified surfaces that were completely hydrophilic [94].

DODAB combinations with graphene enhanced adsorption of hydrophobic analytes and improved the design of novel sensors for phenolic compounds; graphene/DODAB films exhibited remarkable synergistic effects toward the oxidation of tetrabromobisphenol TBBPA, due to the greatly increased TBBPA accumulation in the film and magnitude of the peak currents detected by chronocoulometry [95]. In another interesting instance, immobilization of urease for urea biosensing was achieved employing a DODAB monolayer at the air-water interface and natural exopolysaccharides from microalgae in the aqueous subphase; both DODAB and polysaccharide provided an appropriate microenvironment for the enzyme, enhanced its adsorption in the monolayer and could be used for the production of films supported on solid substrates [96].

Interestingly, the anisotropic polymerization of DNA adsorbed to a DODAB monolayer at the air-water interface yielded a one-dimensionally assembled belt-shaped structure and a unimolecular thickness for the polymerized DNA; thereby, the polymerization could be regulated in the two-dimensionally confined medium of the Langmuir-Blodgett film [97].

In another instance, DODAB monolayers allowed to ascertain the nanostructure of assembled oligonucleotides; two oligonucleotides, a 19-mer bearing thrombin binding aptamer sequence and a 21-mer with human telomeric sequence were end-labeled with fluorescent groups and their fluorescence spectra and G-quadruplex folding at DODAB monolayer interface were reported for the first time. Thanks to film balance measurements (pressure-area isotherms), the fluorescence spectra recording using a fiber optic accessory interfaced with a spectrofluorimeter and the DODAB monolayer, the fluorescence energy transfer efficiency of monolayer adsorbed probes increased significantly in the presence of sodium or potassium ion in subphase, which indicated that the probes retained their cation binding properties when adsorbed at the DODAB monolayer interface [98].

In the fields of antimicrobials and adjuvants for vaccines, DODAB has also been playing important roles. Biocompatible NPs of poly (methylmethacrylate) (PMMA) were synthesized in the presence of DODAB and characterized by dynamic light scattering for sizing, polydispersity and zeta potential analysis, scanning electron microscopy (SEM) for morphology visualization, and plating plus colony-forming unities (CFU) counting for the determination of antimicrobial activity; there was a high permanent load of DODAB in the NPs, and a remarkable antimicrobial activity of PMMA/DODAB NPs, which was much higher than the one determined for DODAB itself [61]. PMMA particles loaded with DODAB were thus obtained from particle synthesis by emulsion polymerization in the presence of DODAB, a facile, fast, low-cost approach to obtaining highly efficient antimicrobial nanoparticles with a permanent DODAB load. Other hybrid DODAB assemblies with the antimicrobial peptide gramicidin (Gr) reunited the complementary antimicrobial properties of DODAB with those of the peptide [56]. DODAB dispersed as large closed bilayer vesicles (LV) or bilayer disks (BF) was added of gramicidin (Gr), which is an antimicrobial peptide assembling as channels in membranes, increasing their permeability toward cations and displaying high toxicity against mammalian cells; DODAB/Gr bilayers exhibited microbicidal action and reduced cytotoxicity against eukaryotic cells [56]. The novel formulations were characterized by dynamic light scattering for sizes an zeta-potentials, leakage from large vesicles induced by transmembrane gramicidin pores with dissipation of osmotic gradients, determination of lytic effects on bacteria and plating plus viable bacteria counting over a range of DODAB and/or Gr concentrations [56]. Gr dimers reconstituted functional channels in LV and the insertion of these channels in DODAB bilayer increased the charge density for LV but did not affect charge density of BF, with Gr at the BF borders. DODAB/Gr combinations diminished the high peptide toxicity against *Saccharomyces cerevisae* and had the advantage of broadening the spectrum of antimicrobial activity for the combination by inducing *Escherichia coli* and *Staphylococcus aureus* lysis and bacterial death. Thereby, the cytotoxicity of the peptide against eukariotic cells was reduced, and the spectrum of antimicrobial activity was broadened since DODAB and Gr displayed complementary activities [56]. More recently, the PMMA/DODAB and DODAB/Gr antimicrobial systems revealed potential uses in food microbiology for killing important food-borne pathogens such as *Escherichia coli, Salmonella enterica, Staphylococcus aureus* and *Listeria monocytogenes* [9].Nowadays, a large family of bacterial genes (generally termed quaternary ammonium genes) encode efflux pumps capable of expelling many Quat structures from bacterial cells, leading to a decrease in susceptibility to Quats [99]. Since bacteria will inevitably find ways of resisting the existing antibiotics and Quats, maybe hybrid assemblies of antimicrobials will prove strategical to overcome resistance. **Table 1** shows some schematic representations of DODAB combinations with gramicidin [56] or biocompatible PMMA polymer in PMMA/DODAB nanoparticles [61]. Their antimicrobial effects against food-borne bacteria were summarized on **Table 2** [9].

In vaccine development, adjuvants and immunostimulants have the important task of presenting antigens to the immune system eliciting an amplified and antigen-specific immune response. Among the adjuvants, DODAB is especially important due to its biomimetic hybrid nanostructures with an outer DODAB coating or an inner DODAB core, which join the advantages of particles and lipids and permit a robust control over size-dependent immune responses in vivo. Recently, hybrid nanomaterials based on DODAB with potential for combination with antigens and immunostimulants for vaccine development were reviewed [100]. For instance, in compositions with derivatives of the myco-bacterial cell wall component, the cord factor trehalose dimycolate (TDM), which is the most abundant glycolipid in the mycobacterial cell wall, DODAB yielded highly efficacious immunoadjuvant formulations

http://dx.doi.org/10.5772/68020

Assemblies	Schematic representation	References
DODAB BF	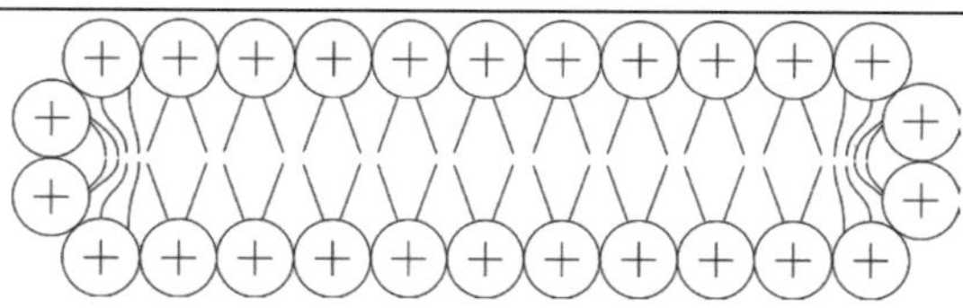	[71, 77]
DODAB BF/Gr	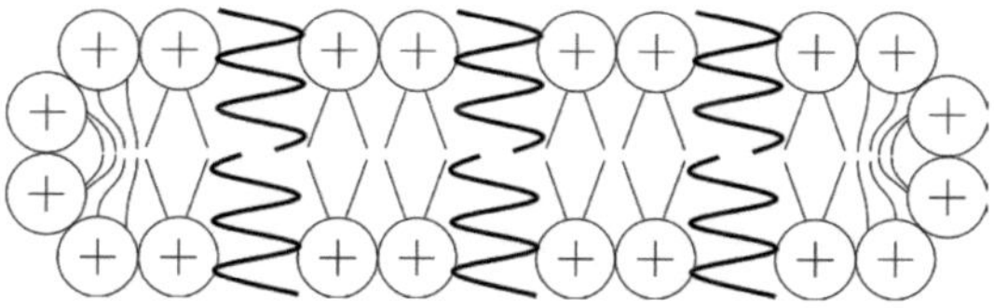	[9]
DODAB LV	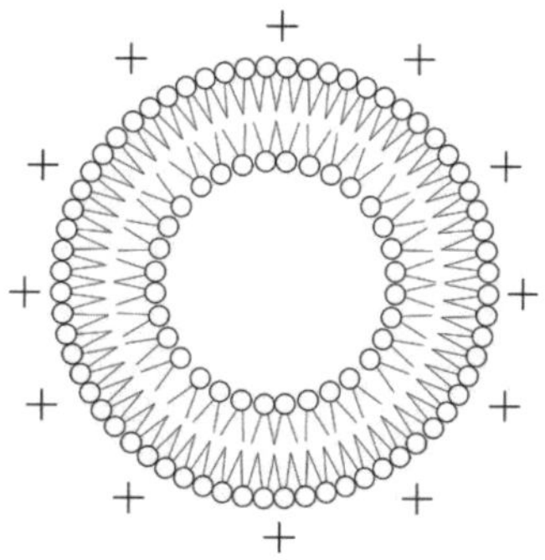	[71, 78]
DODAB LV/Gr	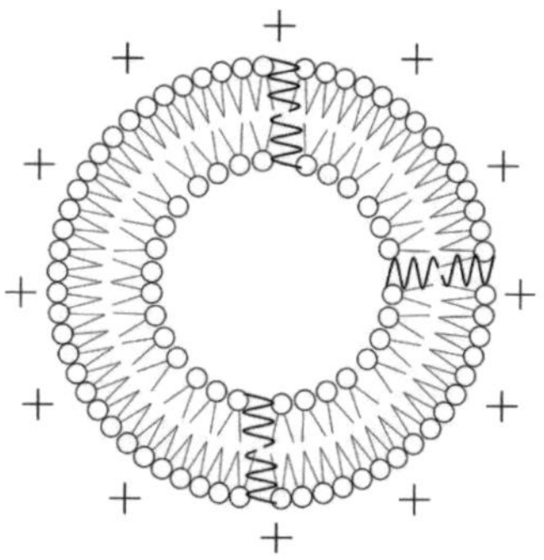	[55, 56]
PMMA/DODAB	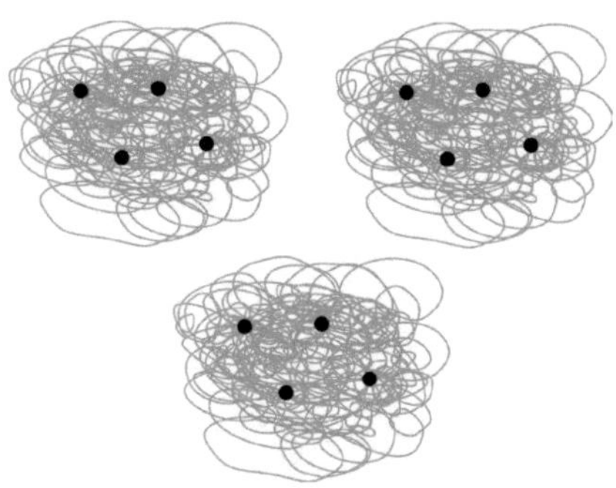	[33, 61]

Table 1. Some DODAB supramolecular assemblies: DODAB bilayer fragments (BF) or large closed vesicles (LV), antimicrobial peptide gramicidin D (Gr) and its assemblies with DODAB BF or DODAB LV and DODAB molecules in PMMA biocompatible polymer.

Assembly	MBC in mM; mg/mL/reduction in log(CFU/mL)			
	E. coli	*S. enterica*	*S. aureus*	*L. monocytogenes*
Gr	0.010; 0.019/0.3	0.010; 0.019/0.5	0.010; 0.019/2.1	0.005; 0.009/7.6
DODAB BF	0.063; 0.039/7.6	0.500; 0.316/1.3	0.063; 0.039/3.4	0.125; 0.079/7.8
DODAB BF/Gr	0.031; 0.019/7.5	0.250; 0.158/0.9	0.015; 0.010/3.8	0.125; 0.079/8.0
DODAB LV	0.015; 0.010/4.5	0.500; 0.316/0.7	0.015; 0.010/2.9	0.250; 0.158/5.7
DODAB LV/Gr	0.015; 0.010/4.6	0.500; 0.316/0.4	0.031; 0.019/2.7	0.063; 0.039/6.0
PMMA/DODAB NPs	–; 2.500/2.2	–; 1.250/0.1	–; 5.000/3.1	–; 5.000/1.5

Minimal bactericidal concentrations (MBC) (in mM; mg/mL) and log of viability reduction at MBC for the cationic assemblies were determined against important food-borne pathogens. For DODAB/Gr combinations, the molar ratio was [Gr] = 0.1 [DODAB]. Adapted from Ref. [9].

Table 2. Antimicrobial activity of DODAB and some of its hybrid assemblies with the antimicrobial peptide gramicidin (Gr) or the biocompatible polymer PMMA.

for tuberculosis vaccines able to induce cell-mediated immunoresponses against intracelular bacteria [101, 102]. In general, DODAB has been combined not only with antigens of interest but also with important immunostimulants such as oligonucleotides, glycolipids or lipopeptides [100].

DODAB-covered particles and DODAB bilayer fragments were often used as immunoadjuvants since DODAB can both adsorb onto several hydrophobic or hydrophilic particles and present antigens (Ag) to elicit amplified immunoresponses [65]. The electrostatic attraction drives the adsorption of a cationic DODAB bilayer onto oppositely charged polystyrene sulfate (PSS) nanoparticles (NPs) over a range of particle sizes [25, 27]. Adsorption isotherms and electrokinetic properties of the covered particles show the deposition of DODAB onto silica or PSS particles at maximal adsorption [21, 22, 25, 27, 28]. At maximal adsorption, the area per DODAB molecule adsorbed onto PSS particles is 0.286 nm^2, which is half of the usual area per monomer in DODAB monolayers at the air-water interface and suggests bilayer deposition onto the polystyrene surface; electrokinetic properties of the covered particles are very similar to those of DODAB vesicles [25]. The hydrodynamic diameter of particles in the particles/DODAB mixtures increases 9–10 nm. A tiny concentration of 10-micromolar is required for bilayer coverage of 10^9 particles (300 nm diameter) per mL at sub-toxic DODAB concentrations. DODAB toxicity against fibroblasts in cell culture becomes significant above 0.1 mM DODAB; there is 50% of cell death at 0.5 mM DODAB [103]. Lipid-covered NPs are useful for antigen presentation [30].

The mean molecular area of DODAB in a monolayer at the air-water interface is 0.6 nm^2 [70]. For particles with 300 nm of mean diameter, the bilayer coverage of total surface area on 5×10^9 particles/mL requires 10 μM DODAB only [30]. At this minute amount, the usual DODAB toxicity is not relevant. In contrast, DODAB vesicles used as immunoadjuvants over the millimolar range of DODAB concentrations may be toxic in vivo [52]. Antigen (Ag) adsorption to the PSS/DODAB assembly does not disturb the order of the particulate over a range of Ag concentrations; the PSS/DODAB system at 5 × 10^9 particles/mL accommodates well up to 25 μg/mL Ag with narrow size distributions for PSS/DODAB/Ag NPs over this range of Ag concentrations [30]. This homogeneity for the particle size in the dispersions yields low polydispersities determined by dynamic light scattering, inside the 0.05–0.10 range [30].

DODAB molecules ultrasonically dispersed in aqueous solution are nano-sized bilayer disks or bilayer fragments (BF); the electrostatic repulsion at low ionic strength keeps the BF stable in aqueous dispersions [39, 64]. DODAB BFs are antimicrobial agents [39, 43], carriers for hydrophobic drugs [104] and useful for the production of lipid-covered particles such as bilayer-coated sílica [22] or PSS [28]. DODAB BFs also present antigens to the immune system inducing cellular immune responses [54]. These open bilayers differ from their mother vesicles by especial features. They do not respond to osmotic gradients because they do not have an inner aqueous compartment. They have a discoidal shape with disks exhibiting one bilayer thickness and both faces available to display antigens [54]. They have domains of fluid and gel lipid phases [105]. They solubilize hydrophobic molecules sometimes in contrast to their mother vesicles that do not do so as in the case of amphotericin B [104]. DODAB BF interact with proteins, oligonucleotides or DNA via both the hydrophobic effect and the electrostatic attraction at low ionic strength. Bovine serum albumin (BSA) purified 18/14 kDa antigens from *Taenia crassiceps* cysticerci (18/14-Tcra) or a recombinant heat-shock protein (hsp-18 kDa) from *Mycobacterium leprae* adsorb on DODAB BF [54]. DODAB BF/Ag NPs are stable over a range of DODAB and Ag concentrations; these ranges vary with the Ag nature and are different for different antigens [54]. The production of cytokines by lymph nodes (LN) cells of immunized mice in culture is important to determine the nature of immune response induced by PSS/DODAB/Ag or DODAB BF/Ag. The mice immunized with antigen alone, adjuvant/antigen or adjuvant alone provide LN cells in culture that produce different cytokines depending on Ag and adjuvant nature [54]. A sandwich kit enzyme-linked immunosorbent assay (ELISA) determines the analytical concentrations of the cytokines produced after reestimulating the cells in culture. The cytokines profile is rather different from immunization with the parasite and the bacterium antigens [54]. The high levels of IL-12 and IFN-gamma induced by PSS/DODAB/Ag and DODAB BF/Ag when Ag is hsp-18kDa shows that these adjuvants are useful for the design of subunit vaccines against intracelular bacteria. IL-12 and IFN-gamma are the most important cytokines in innate responses to intracellular bacteria such as M. leprae or tuberculosis; when Ag is 18/14-Tcra, there is an enhancement in production of IL-10 and Il-13 by LN cells elicited by DODAB BF/Ag. These cytokines are typically associated with responses to allergens and parasites such as helminths and mediate differentiation of CD4+-T cells into Th2 cells [106]. On the other hand, the *Mycobacterium leprae* antigen carried by DODAB BF or PSS/DODAB adjuvants elicits low levels of these cytokines. Responses are indeed different for the helminthes and the bacteria antigens and antigen-specific as they should be [54, 106].

IL-10 exerts an inhibitory effect on macrophages and dendritic cells by decreasing the production of IL-12 and the expression of class II major histocompatibility complex (MHC) [106]. Macrophages and DCs also secrete IL-12 that induces T cells differentiation into Th1 and natural killer (NK) cells with increased IFN-gamma synthesis and cytotoxic activity. The adaptive immunity against intracellular bacteria is principally cell mediated and consists of activation of macrophages by CD4+T cells as well as killing of infected cells by CD8+ cytotoxic T lymphocytes (CTL). Naïve CD4+ T cells may differentiate into distinct subsets, such as Th1 and Th2 cells in response to different antigens.

Due to its chemical stability and low cost when compared to other natural or synthetic lipids, DODAB has been intensively investigated aiming at subunit vaccine design. Major problems of liposomal formulations based on DODAB are the high DODAB concentration (1–10 mM DODAB) and the large liposomes size [52, 106, 107]. Minimization of DODAB dose is required for administration in vivo. DODAB BF effectively present antigens at 0.1 mM DODAB only; supported DODAB bilayers on PSS or silica require even lower DODAB concentrations [22, 25, 30, 54]. The total surface area on the BF dispersion available for antigen association are much larger than the one for closed, large and sometimes multibilayered liposomes. Thus, the first advantage of DODAB BF, PSS/DODAB or silica/DODAB as adjuvants would be the low DODAB concentration required for Ag presentation. The second advantage of BF is the nanosize. Depending on sonication power and time plus composition of the dispersing medium that determine colloidal stability, DODAB BF/Ag complexes have a few tenths of nanometers in size (40–80 nm). This size is effective for antigen delivery to antigen-presenting cells (APCs), generating potent and combined humoral and CD8+ T cell immunity [109–111]. Over a range of low DODAB and antigen concentrations ([DODAB] ≤0.1 mM; 0.001–0.05 mg/mL antigen), adjuvant/antigen combinations were cationic, stable, homodisperse and immunogenic at low DODAB dose, low cost, low sizes for improved dendritic cells uptake, high chemical stability, prone to present several different antigens and displaying low or even absent cytotoxicity. They were remarkably superior to alum due to their ability to elicit the cellular Th1 immune response. Contrary to alum or DODAB LV (1–10 mM DODAB), local or systemic adverse effects in mice were completely absent over the 0.1–0.01 mM DODAB range. Silica/DODAB, PSS/DODAB, and DODAB BF are available over the sub-200 nm range of sizes thus presenting potential also for design of mucosal vaccines. The third advantage of BF is the absence of depots at the site of injection, an inflammatory reaction that is not always desirable [54]. These depots occur for DODAB large vesicles (LV) and appear due to inflammatory responses at the site of injection [107, 108]. Similar sizes for adjuvant and adjuvant-antigen complexes evidenced that the antigens readily adsorbed and stabilized the adjuvant; conversely, the adjuvant also stabilized the antigens preventing antigen-antigen aggregation as often observed for protein-protein interactions [30, 54].

An important component of the early innate immune response to viruses and bacteria is IL-12 that enhances the IFN-gamma production and the development of Th1 cells; IL-12 is involved in the combat of infections by cell-mediated immunity, for example, leishmaniasis [106]. Subunit vaccines against protozoa that survive within macrophages require as principal defense mechanism the cell-mediated immunity, particularly directed to macrophage activation by Th1 cell-derived cytokines. Immune responses to leishmaniasis against the parasite *Leishmania donovani* involve cell-mediated immune response of the Th1 type and CD4+ Th1 cells activation for killing phagocytosed parasites. Leishmania-specific Th1 CD4+ T cells produce IFN-gamma, that activates macrophages to kill intracellular parasites. On the other hand, the parasite activates Th2 cells increasing their production of Th2 cytokines that suppress the activity of macrophages and increase parasite survival

[106]. Similarly, during the liver stages of malaria, CD8+ T cells kill infected hepatocytes and induce the secretion of IFN-gamma activating the production of nitric oxide and other agents by the hepatocytes for killing the parasites. IL-12 stimulates IFN-gamma production inducing resistance to sporozoite challenge in rodents and non-human primates [106]. Il-12 also increases the cytotoxic activity of natural killer (NK) cells after viral infections thereby mediating the NK cell killing of virus-infected cells for combating the infection. Recombinant DNA vaccines expressing membrane and envelope of viral proteins may benefit from the DODAB BF or PSS/DODAB adjuvants, which can also carry DNA [31] or oligonucleotides [49].

DNA sequences containing unmethylated CpG dinucleotide generate danger signals that are recognized by the immune system; they are typical of bacteria and viruses but rare in vertebrates activating cells that express Toll-like receptor 9 and induce an innate immune response characterized by the production of Th1 cytokines [112]. Both CpG and DODAB improve Th1 responses against antigens when used separately. DODAB BF/CpG presenting ovalbumin (OVA) also enhanced Th1 immune responses [50]. DODAB BF/CpG/OVA also did not result in any observable depot effect at the site of prime suggesting their direct action on the antigen presenting cells (APC) of the draining LN. Only NPs can specifically target LN-resident cells [113]. The interstitial flow convects sub-100 nm NPs into the draining lymphatic vessels; NPs are not trapped in the tissue interstitium. Nano-sizes allow direct LN targeting without the use of specific ligands. In the LN, antigen-presenting cells (APCs) rapidly capture the NPs. A few reviews are available on DODAB applications for the development of novel hybrid assemblies useful as immunoadjuvants, gene or RNA carriers [114–118].

Acknowledgements

The author thanks Conselho Nacional de Desenvolvimento Científico e Tecnológico (CNPq) for financial support (CNPq302352/2014-7).

Author details

Ana Maria Carmona-Ribeiro

Address all correspondence to: amcr@usp.br

Biocolloids Laboratory, Departamento de Bioquímica, Instituto de Química, Universidade de São Paulo, São Paulo, SP, Brazil

References

[1] Block SS. Disinfection, Sterilization and Preservation (5th edition). Block SS (Ed.), Lippincott, Williams & Wilkins, Philadelphia, 2001. 1481 p.

[2] Attwood D, Florence AT, editors. Surfactant Systems: Their Chemistry, Pharmacy and Biology. 1st ed. London EC4P 4EE: Chapman & Hall, 1983. 794 p. Available from http://www.springer.com/br/book/9789400957770

[3] Carmona-Ribeiro AM. Lipid-based biomimetics in drug and vaccine delivery, In: Mukherjee A, editor. Biomimetics Learning from Nature. Olajinica: InTech: 2010. pp. 507-534, doi:10.5772/8792. Available from: http://www.intechopen.com/books/biomimetics-learning-from-nature/lipid-based-biomimetics-in-drug-and-vaccine-delivery

[4] Carmona-Ribeiro AM, Barbassa B, Melo LD. Antimicrobial Biomimetics. In Cavrak M, editor., Biomimetic Based Applications, Olajinica: InTech: 2011. pp. 227-284. doi:10.5772/14400. Available from: http://www.intechopen.com/books/biomimetic-based-applications/antimicrobial-biomimetics

[5] Dai C, Li W, Cui Y, Sun Y, Wu W, Xu Z, Liu Y, Yang Z, Wu V. The effect of functional groups on the sphere-to-wormlike micellar transition in quaternary ammonium surfactant solutions, Colloids Surf A: Physicochem Eng Aspects. 2016;**500**:32-39. doi:10.1016/j.colsurfa.2016.04.024. (http://www.sciencedirect.com/science/article/pii/S0927775716302564)

[6] Carmona-Ribeiro AM, Vieira DB, Lincopan N. Cationic surfactants and lipids as anti-infective agents. Anti-Infect Agents Med Chem 2006;**5**:33-54. doi:10.2174/187152106774755572

[7] Vieira DB, Carmona-Ribeiro AM. Cationic lipids and surfactants as antifungal agents: mode of action. J Antimicrob Chemother 2006;**58**:760-767. doi:10.1093/jac/dkl312

[8] De Melo Carrasco LD, Sampaio JL, Carmona-Ribeiro AM. Supramolecular cationic assemblies against multidrug-resistant microorganisms: activity and mechanism of action. Int J Mol Sci. 2015;**16**(3):6337-6352. doi:10.3390/ijms16036337

[9] Carrasco LD, Bertolucci R Jr, Ribeiro RT, Sampaio JL, Carmona-Ribeiro AM. Cationic nanostructures against foodborne pathogens. Front Microbiol. 2016;**7**:1804. doi:10.3389/fmicb.2016.01804

[10] Zhang C, Cui F, Zeng GM, Jiang M, Yang ZZ, Yu ZG, Zhu MY, Shen LQ. Quaternary ammonium compounds (QACs): a review on occurrence, fate and toxicity in the environment. Sci Total Environ. 2015;**518-519**:352-362. doi:10.1016/j.scitotenv.2015.03.007

[11] Garcia MT, Kaczerewska O, Ribosa I, Brycki B, Materna P, Drgas M. Biodegradability and aquatic toxicity of quaternary ammonium-based gemini surfactants: effect of the spacer on their ecological properties. Chemosphere. 2016;**154**:155-60. doi:10.1016/j.chemosphere.2016.03.109

[12] Brycki B, Waligórska M, Szulc A. The biodegradation of monomeric and dimeric alkylammonium surfactants. J Hazard Mater. 2014;**280**:797-815. doi:10.1016/j.jhazmat.2014.08.021

[13] Minbiole KPC, Jennings MC, Ator LE, Black JW, Grenier MC, LaDow JE, Caran KL, Seifert K, Wuest WM. From antimicrobial activity to mechanism of resistance: the

multifaceted role of simple quaternary ammonium compounds in bacterial eradication. Tetrahedron. 2016;72:3559-3566. doi:10.1016/j.tet.2016.01.014

[14] Israelachvili JN, Mitchell DJ, Ninham BW. Theory of self-assembly of lipid bilayers and vesicles. Biochim Biophys Acta. 1977;**470**(2):185-201. doi:10.1016/0005-2736(77)90099-2

[15] Israelachvili JN. Intermolecular and Surface Forces. 3rd ed. San Diego: Academic Press; 2011. 674 p. doi:10.1016/B978-0-12-375182-9.10025-9

[16] Marra J. Effects of counterion specificity on the interactions between quaternary ammonium surfactants in monolayers and bilayers. J Phys Chem. 1986;**90**(10):2145-2150. doi:10.1021/j100401a031

[17] Guldbrand L, Jonsson B, Wennerstrom H, Linse P. Electrical double layer forces. A Monte Carlo study. J Chem Phys. 1984;**80**:2221-2228. doi:10.1063/1.446912

[18] Pashley RM, McGuiggan PM, Ninham BW, Brady J, Evans DF. Direct measurements of surface forces between bilayers of double-chained quaternary ammonium acetate and bromide surfactants. J Phys Chem. 1986;**90**(8):1637-1642. doi:10.1021/j100399a037

[19] Nascimento DB, Rapuano R, Lessa MM, Carmona-Ribeiro AM. Counterion effects on properties of cationic vesicles. Langmuir 1998;**14**(26):7387-7391. doi:10.1021/la980845c

[20] Rapuano R, Carmona-Ribeiro AM. Physical adsorption of bilayer membranes on silica. J Colloid Interface Sci. 1997;**193**(1):104-111. doi:10.1006/jcis.1997.5060

[21] Rapuano R, Carmona-Ribeiro AM. Supported bilayers on silica. J Colloid Interface Sci. 2000;**226**:299-307. doi:10.1006/jcis.2000.6824

[22] Moura SP, Carmona-Ribeiro AM. Cationic bilayer fragments on silica at low ionic strength: competitive adsorption and colloid stability. Langmuir. 2003;**19**(17):6664-6667. doi:10.1021/la034334o

[23] Moura SP, Carmona-Ribeiro AM. Adsorption behavior of DODAB/DPPC vesicles on silica. J Colloid Interface Sci. 2007;**313**(2):519-26. doi:10.1016/j.jcis.2007.04.061

[24] Pereira EM, Petri DF, Carmona-Ribeiro AM. Adsorption of cationic lipid bilayer onto flat silicon wafers: effect of ion nature and concentration. J Phys Chem B. 2006;**110**(20):10070-10074. doi: 10.1021/jp060737w

[25] Carmona-Ribeiro AM, Midmore BR. Synthetic bilayer adsorption onto polystyrene microspheres. Langmuir. 1992;**8**(3):801-806. doi:10.1021/la00039a013

[26] Tsuruta LR, Lessa MM, Carmona-Ribeiro AM. Interactions between dioctadecyldimethylammonium chloride or bromide bilayers in water. Langmuir. 1995;**11**(8):2938-2943. doi:10.1021/la00008a016

[27] Tsuruta LR, Lessa MM, Carmona-Ribeiro AM. Effect of particle size on colloid stability of bilayer-covered polystyrene microspheres. J Colloid Interface Sci. 1995;**175**(2):470-475. doi:10.1006/jcis.1995.1477

[28] Carmona-Ribeiro AM, De Moraes Lessa M. Interactions between bilayer membranes and látex. Colloids Surf. A: Physicochem. Eng. Aspects. 1999;**153**(1-3):355-361. doi:10.1016/S0927-7757(98)00532-9

[29] Pereira EMA, Vieira DB, Carmona-Ribeiro AM. Cationic bilayers on polymeric particles: effect of low NaCl concentration on surface coverage. J Phys Chem B. 2004;**108**(31):11490-11495. doi:10.1021/jp048555u

[30] Lincopan N, Espíndola NM, Vaz AJ, Carmona-Ribeiro AM. Cationic supported lipid bilayers for antigen presentation. Int J Pharm. 2007;**340**(1-2):216-22. doi:10.1016/j.ijpharm.2007.03.014

[31] Rosa H, Petri DF, Carmona-Ribeiro AM. Interactions between bacteriophage DNA and cationic biomimetic particles. J Phys Chem B. 2008;**112**(51):16422-16430. doi:10.1021/jp806992f

[32] Pereira EMA, Petri DF, Carmona-Ribeiro AM. Synthetic vesicles at hydrophobic surfaces. J Phys Chem B 2002;**106**(34):8762-8767. doi:10.1021/jp020735l

[33] Pereira EM, Kosaka PM, Rosa H, Vieira DB, Kawano Y, Petri DF, Carmona-Ribeiro AM. Hybrid materials from intermolecular associations between cationic lipid and polymers. J Phys Chem B. 2008;**112**(31):9301-9310. doi:10.1021/jp801297t

[34] Melo LD, Palombo RR, Petri DF, Bruns M, Pereira EM, Carmona-Ribeiro AM. Structure-activity relationship for quaternary ammonium compounds hybridized with poly(methyl methacrylate). ACS Appl Mater Interfaces. 2011;**3**(6):1933-1939. doi:10.1021/am200150t

[35] Vieira DB, Pacheco LF, Carmona-Ribeiro AM. Assembly of a model hydrophobic drug into cationic bilayer fragments. J Colloid Interface Sci. 2006;**293**(1):240-247. doi:10.1016/j.jcis.2005.06.046

[36] Vieira DB, Carmona-Ribeiro AM. Cationic nanoparticles for delivery of amphotericin B: preparation, characterization and activity in vitro. J Nanobiotechnol. 2008;6:6. doi:10.1186/1477-3155-6-6

[37] Pacheco LF, Carmona-Ribeiro AM. Effects of synthetic lipids on solubilization and colloid stability of hydrophobic drugs. J Colloid Interface Sci. 2003;**258**(1):146-154. doi:10.1016/S0021-9797(02)00103-0

[38] Lincopan N, Mamizuka EM, Carmona-Ribeiro AM. In vivo activity of a novel amphotericin B formulation with synthetic cationic bilayer fragments. J Antimicrob Chemother. 2003;**52**(3):412-418. doi:10.1093/jac/dkg383

[39] Carmona-Ribeiro AM. Bilayer-forming synthetic lipids: drugs or carriers? Curr Med Chem. 2003;**10**(22):2425-1446. doi:10.2174/0929867033456611

[40] Lincopan N, Mamizuka EM, Carmona-Ribeiro AM. Low nephrotoxicity of an effective amphotericin B formulation with cationic bilayer fragments. J Antimicrob Chemother. 2005;**55**(5):727-734. doi:10.1093/jac/dki064

[41] Lincopan N, Carmona-Ribeiro AM. Lipid-covered drug particles: combined action of dioctadecyldimethylammonium bromide and amphotericin B or miconazole. J Antimicrob Chemother. 2006;**58**(1):66-75. doi:10.1093/jac/dkl153

[42] Lincopan N, Santana MR, Faquim-Mauro E, da Costa MH, Carmona-Ribeiro AM. Silica-based cationic bilayers as immunoadjuvants. BMC Biotechnol. 2009;**9**:5. doi:10.1186/1472-6750-9-5

[43] Carmona-Ribeiro AM. Lipid bilayer fragments and disks in drug delivery. Curr Med Chem. 2006;**13**(12):1359-70. doi:10.2174/092986706776872925

[44] Barbassa L, Mamizuka EM, Carmona-Ribeiro AM. Supramolecular assemblies of rifampicin and cationic bilayers: preparation, characterization and micobactericidal activity. BMC Biotechnol. 2011;**11**:40. doi:10.1186/1472-6750-11-40

[45] Lima EG, Gomes LR, Carmona-Ribeiro AM. Stable indomethacin dispersions in water from drug, ethanol, cationic lipid and carboxymethyl-cellulose. Pharm Nanotechnol. 2016;**4**(2):126-135. doi:10.2174/2211738504666160304195436. (Available from http://www.eurekaselect.com/node/140208/article)

[46] Kikuchi IS, Carmona-Ribeiro AM. Interactions between DNA and synthetic cationic liposomes. J Phys Chem B. 2000;**104**(13):2829-2835. doi:10.1021/jp9935891

[47] Kikuchi, IS, Viviani W, Carmona-Ribeiro AM. Nucleotide insertion in cationic bilayers. J Phys Chem A. 1999;**103**:8050-8055. doi:10.1021/jp9911090

[48] Nantes IL, Correia FM, Faljoni-Alario A, Kawanami AE, Ishiki HM, Amaral AT, Carmona-Ribeiro AM. Nucleotide conformational change induced by cationic bilayers. Arch Biochem Biophys. 2003;**416**(1):25-30. doi:10.1016/S0003-9861(03)00280-7

[49] Rozenfeld JH, Oliveira TR, Lamy MT, Carmona-Ribeiro AM. Interaction of cationic bilayer fragments with a model oligonucleotide. Biochim Biophys Acta. 2011;**1808**(3):649-655. doi:10.1016/j.bbamem.2010.11.036

[50] Rozenfeld JH, Silva SR, Ranéia PA, Faquim-Mauro E, Carmona-Ribeiro AM. Stable assemblies of cationic bilayer fragments and CpG oligonucleotide with enhanced immunoadjuvant activity in vivo. J Control Release. 2012;**160**(2):367-373. doi:10.1016/j.jconrel.2011.10.017

[51] Carvalho LA, Carmona-Ribeiro AM. Interactions between cationic vesicles and serum proteins Langmuir. 1998;**14**(21):6077-6081. doi:10.1021/la980345j

[52] Tsuruta LR, Quintilio W, Costa MH, Carmona-Ribeiro AM. Interactions between cationic liposomes and an antigenic protein: the physical chemistry of the immunoadjuvant action. J Lipid Res. 1997;**38**(10):2003-2011.

[53] Lincopan N, Carmona-Ribeiro AM. Protein assembly onto cationic supported bilayers. J Nanosci Nanotechnol. 2009;**9**(6):3578-3586. doi:10.1166/jnn.2008.003

[54] Lincopan N, Espíndola NM, Vaz AJ, da Costa MH, Faquim-Mauro E, Carmona-Ribeiro AM. Novel immunoadjuvants based on cationic lipid: preparation, characterization and activity in vivo. Vaccine. 2009;**27**(42):5760-71. doi:10.1016/j.vaccine.2009.07.066

[55] Carvalho CA, Olivares-Ortega C, Soto-Arriaza MA, Carmona-Ribeiro AM. Interaction of gramicidin with DPPC/DODAB bilayer fragments. Biochim Biophys Acta. 2012;**1818**(12):3064-3071. doi:10.1016/j.bbamem.2012.08.008

[56] Ragioto DA, Carrasco LD, Carmona-Ribeiro AM. Novel gramicidin formulations in cationic lipid as broad-spectrum microbicidal agents. Int J Nanomed. 2014;**9**:3183-3192. doi:10.2147/IJN.S65289

[57] Carmona-Ribeiro AM, de Melo Carrasco LD. Novel formulations for antimicrobial peptides. Int J Mol Sci. 2014;**15**(10):18040-18083. doi:10.3390/ijms151018040

[58] Correia FM, Petri DF, Carmona-Ribeiro AM. Colloid stability of lipid/polyelectrolyte decorated latex. Langmuir. 2004;**20**(22):9535-9540. doi:10.1021/la048938j

[59] Melo LD, Mamizuka EM, Carmona-Ribeiro AM. Antimicrobial particles from cationic lipid and polyelectrolytes. Langmuir. 2010;**26**(14):12300-12306. doi:10.1021/la101500s

[60] Leão-Silva AC, Naves AF, Pereira EM, Petri DF, Carmona-Ribeiro AM. Assembly of horseradish peroxidase within supported cationic bilayers. Biotechnol Prog. 2011;**27**(5):1433-41. doi:10.1002/btpr.640

[61] Naves AF, Palombo RR, Carrasco LD, Carmona-Ribeiro AM. Antimicrobial particles from emulsion polymerization of methyl methacrylate in the presence of quaternary ammonium surfactants. Langmuir. 2013;**29**(31):9677-9684. doi:10.1021/la401527j

[62] Carmona-Ribeiro AM. Preparation and characterization of biomimetic nanoparticles for drug delivery. Methods Mol Biol. 2012;**906**:283-294. doi:10.1007/978-1-61779-953-2_22

[63] Carmona-Ribeiro AM. Biomimetic particles in drug and vaccine delivery. J Liposome Res. 2007;**17**(3-4):165-72.

[64] Carmona-Ribeiro AM. Biomimetic nanoparticles: preparation, characterization and biomedical applications. Int J Nanomed. 2010;**5**:249-59.

[65] Carmona-Ribeiro AM. Biomimetic systems in nanomedicine. In: Torchilin, V. (Ed,) Handbook of Nanobiomedical Research. Singapore: World Scientific; 2014. Available from: http://www.worldscientific.com/doi/abs/10.1142/97898 14520652_0063.

[66] Carmona-Ribeiro AM. Chapter 4—Interactions between bilayer vesicles, biomolecules and interfaces. In Handbook of Surfaces and Interfaces of Materials, edited by Hari Singh Nalwa, Burlington: Academic Press, 2001, pp. 129-165. doi:10.1016/B978-01251 3910-6/50061-X

[67] Mamizuka EM, Carmona-Ribeiro AM. Cationic liposomes as antimicrobial agents. In: Antonio Méndez Vilas. (Org.). Communicating Current Research and Educational Topics and Trends in Applied Microbiology. 1ed. Badajoz: Formatex, 2007, v. 2, pp. 636-647. ISBN-13: 978-84-611-9423-0. Available from http://www.formatex.org/microbio/pdf/pages636-647.pdf

[68] Carmona-Ribeiro AM, Yoshida LS, Chaimovich H. Salt effects on the stability of dioctadecyldimethylammonium chloride and sodium dihexadecyl phosphate vesicles. J Phys Chem. 1985;**89**(13):2928-2933. doi:10.1021/j100259a045

[69] Carmona-Ribeiro AM, Chaimovich H. Salt-induced aggregation and fusion of dioctadecyldimethylammonium chloride and sodium dihexadecylphosphate vesicles. Biophys J. 1986;**50**(4):621-628. doi:10.1016/S0006-3495(86)83501-9

[70] Claesson PM, Carmona-Ribeiro AM, Kurihara K. Dihexadecyl phosphate monolayers: intralayer and interlayer interactions. J Phys Chem. 1989;**93**(2):917-922. doi:10.1021/j100339a071

[71] Carmona-Ribeiro AM. Synthetic amphiphile vesicles. Chem Soc Rev. 1992;**21**:209-214. doi:10.1039/CS9922100209

[72] Carmona-Ribeiro AM. Interactions between charged spheric vesicles. J Phys Chem. 1993;**97**(45):11843-11846. doi:10.1021/j100147a043

[73] Jamróz D, Kepczynski M, Nowakowska M. Molecular structure of the dioctadecyldimethylammonium bromide (DODAB) bilayer. Langmuir. 2010;**26**(19):15076-15079. doi:10.1021/la102324p

[74] Feitosa E, Adati RD, Hansson P, Malmsten M. Thermal and structural behavior of dioctadecyldimethylammonium bromide dispersions studied by differential scanning calorimetry and X-ray scattering. Langowski J, ed. PLoS One. 2012;**7**(9):e44702. doi:10.1371/journal.pone.0044702.

[75] Kepczynski M, Bednar J, Kuźmicz D, Wydro P, Nowakowska M. Spontaneous formation of densely stacked multilamellar vesicles in dioctadecyldimethylammonium bromide/oleosiloxane mixtures. Langmuir. 2010;**26**(3):1551-1556. doi:10.1021/la904094e

[76] Kepczynski M, Lewandowska J, Witkowska K, Kędracka-Krok S, Mistrikova V, Bednar J, Wydro P, Nowakowska M. Bilayer structures in dioctadecyldimethylammonium bromide/oleic acid dispersions. Chem Phys Lipids. 2011;**164**(5):359-367. doi:10.1016/j.chemphyslip.2011.04.007

[77] Andersson M, Hammarstroem L, Edwards K. Effect of bilayer phase transitions on vesicle structure and its influence on the kinetics of viologen reduction. J Phys Chem. 1995;**99**:14531-14538. **doi:**10.1021/j100039a047

[78] Carmona-Ribeiro AM, Chaimovich H. Preparation and characterization of large dioctadecyldimethylammonium chloride liposomes and comparison with small sonicated vesicles. Biomembranes. 1983;**733**(1):172-179. doi:10.1016/0005-2736(83)90103-7

[79] Chou TH, Liang CH, Lee YC, Yeh LH. Effects of lipid composition on physicochemical characteristics and cytotoxicity of vesicles composed of cationic and anionic dialkyl lipids. Phys Chem Chem Phys. 2014;**16**(4):1545-53. doi:10.1039/c3cp54176b

[80] Kuo AT, Chang CH. Elucidating the effects of cholesterol on the molecular packing of double-chained cationic lipid langmuir monolayers by infrared reflection-absorption spectroscopy. J Oleo Sci. 2015;**64**(4):455-465. doi:10.5650/jos.ess14266

[81] Sobral CN, Soto MA, Carmona-Ribeiro AM. Characterization of DODAB/DPPC vesicles. Chem Phys Lipids. 2008;**152**(1):38-45. doi:10.1016/j.chemphyslip.2007.12.004

[82] Wu FG, Wu RG, Sun HY, Zheng YZ, Yu ZW. Demixing and crystallization of DODAB in DPPC-DODAB binary mixtures. Phys Chem Chem Phys. 2014;**16**(29):15307-15318. doi:10.1039/c4cp01707b

[83] Lawrence MJ. Surfactant systems: their use in drug delivery. Chem Soc Rev. 1994;**23**:417-424. doi:10.1039/CS9942300417

[84] Maswal M, Pandith AH, Islam N, Dar AA. Co-solubilization of the hydrophobic drugs carbamazepine and nifedipine in aqueous nonionic surfactant media. J Solut. Chem. 2013;**42**:1374-1392. doi:10.1007/s10953-013-0036-4

[85] Kumar A, Kaur G, Kansal SK, Chaudhary GR, Mehta SK. Enhanced solubilization of curcumin in mixed surfactant vesicles. Food Chem. 2016;**199**:660-666. doi:10.1016/j.foodchem.2015.12.077

[86] Pereira GG, Detoni CB, Balducci AG, Rondelli V, Colombo P,Guterres SS, Sonvico F. Hyaluronate nanoparticles included in polymer films for the prolonged release of vitamin E for the management of skin wounds. Eur J Pharm Sci. 2016;**83**:203-211. doi:10.1016/j.ejps.2016.01.002

[87] Kaminski GA, Sierakowski MR, Pontarolo R, Santos LA, de Freitas RA. Layer-by-layer polysaccharide-coated liposomes for sustained delivery of epidermal growth factor. Carbohydr Polym. 2016;**140**:129-135. doi:10.1016/j.carbpol.2015.12.014

[88] Silva JP, Oliveira IM, Oliveira AC, Lúcio M, Gomes AC, Coutinho PJ, Oliveira ME. Structural dynamics and physicochemical properties of pDNA/DODAB:MO lipoplexes:effect of pH and anionic lipids in inverted non-lamellar phases versus lamellar phases. Biochim Biophys Acta. 2014;**1838**(10):2555-2567. doi:10.1016/j.bbamem.2014.06.014

[89] Silva JP, Oliveira AC, Lúcio M, Gomes AC, Coutinho PJ, Oliveira ME. Tunable pDNA/DODAB:MO lipoplexes: the effect of incubation temperature on pDNA/DODAB:MO lipoplexes structure and transfection efficiency. Colloids Surf B: Biointerfaces. 2014;**121**:371-379. doi:10.1016/j.colsurfb.2014.06.019

[90] Oliveira AC, Raemdonck K, Martens T, Rombouts K, Simón-Vázquez R, Botelho C, Lopes I, Lúcio M, González-Fernández Á, Real Oliveira ME, Gomes AC, Braeckmans K. Stealth monoolein-based nanocarriers for delivery of siRNA to cancer cells. Acta Biomater. 2015;**25**:216-29. doi:10.1016/j.actbio.2015.07.032

[91] Oliveira AC, Martens TF, Raemdonck K, Adati RD, Feitosa E, Botelho C, Gomes AC, Braeckmans K, Real Oliveira ME. Dioctadecyldimethylammonium:monoolein nanocarriers for efficient in vitro gene silencing. ACS Appl Mater Interfaces. 2014;**6**(9):6977-6989. doi:10.1021/am500793y

[92] Carneiro C, Correia A, Lima T, Vilanova M, Pais C, Gomes AC, Real Oliveira ME, Sampaio P. Protective effect of antigen delivery using monoolein-based liposomes in experimental hematogenously disseminated candidiasis. Acta Biomater. 2016;**39**:133-145. doi:10.1016/j.actbio.2016.05.001

[93] Aps LR, Tavares MB, Rozenfeld JH, Lamy MT, Ferreira LC, Diniz MO. Bacterial spores as particulate carriers for gene gun delivery of plasmid DNA. J Biotechnol. 2016;**228**:58-66. doi:10.1016/j.jbiotec.2016.04.027

[94] Salas C, Genzer J, Lucia LA, Hubbe MA, Rojas OJ. Water-wettable polypropylene fibers by facile surface treatment based on soy proteins. ACS Appl Mater Interfaces. 2013;**5**(14):6541-6548. doi:10.1021/am401065t

[95] Chen X, Wang Y, Tong J, Xia S, Zhou Y, Wu K Electrochemical sensing platform for tetrabromobisphenol A at pM level based on the synergetic enhancement effects of graphene and dioctadecyldimethylammonium bromide. Anal Chim Acta. 2016;**935**:90-96. doi:10.1016/j.aca.2016.06.052

[96] de Brito AK, Nordi CS, Caseli L. Algal polysaccharides as matrices for the immobilization of urease in lipid ultrathin films studied with tensiometry and vibrational spectroscopy: physical-chemical properties and implications in the enzyme activity. Colloids Surf B Biointerfaces. 2015;**135**:639-645.

[97] Yonamine Y, Cervantes-Salguero K, Minami K, Kawamata I, Nakanishi W, Hill JP, Murata S, Ariga K. Supramolecular 1-D polymerization of DNA origami through a dynamic process at the 2-dimensionally confined air-water interface. Phys Chem Chem Phys. 2016;**18**(18):12576-81. doi:10.1039/c6cp01586g

[98] Swiatkowska A, Kosman J, Juskowiak B. FRET study of G-quadruplex forming fluorescent oligonucleotide probes at the lipid monolayer interface. Spectrochim Acta A Mol Biomol Spectrosc. 2016;**152**:614-21. doi:10.1016/j.saa.2015.01.102

[99] Jennings MC, Minbiole KPC, Wuest WM. Quaternary ammonium compounds: an antimicrobial mainstay and platform for innovation to address bacterial resistance. ACS Infect Dis. 2015:**1**:288–303. doi:10.1021/acsinfecdis.5b00047

[100] Carmona-Ribeiro AM. Chapter Thirteen—Nanomaterials based on lipids for vaccine development, In Micro and Nano Technologies, edited by Mariusz Skwarczynski and Istvan Toth, William Andrew Publishing, 2017, pp. 241-257, Micro and Nanotechnology in Vaccine Development, ISBN 9780323399814, doi:10.1016/B978-0-323-39981-4.00013-0. (http://www.sciencedirect.com/science/article/pii/B9780323399814000130)

[101] Holten-Andersen L, Doherty TM, Korsholm KS, Andersen P. Combination of the cationic surfactant dimethyl dioctadecyl ammonium bromide and synthetic mycobacterial cord factor as an efficient adjuvant for tuberculosis subunit vaccines. Infect Immun. 2004;**72**(3):1608-1617.

[102] Agger EM. Novel adjuvant formulations for delivery of anti-tuberculosis vaccine candidates. Adv Drug Deliv Rev. 2016;**102**:73-82. doi:10.1016/j.addr.2015.11.012

[103] Carmona-Ribeiro AM, Ortis F, Schumacher RI, Armelin MCS. Interactions between cationic vesicles and cultured mammalian cells. Langmuir. 1997;**13**:2215-2218. doi:10.1021/la960759h

[104] Vieira DB, Carmona-Ribeiro AM. Synthetic bilayer fragments for solubilization of amphotericin B. J Colloid Interface Sci. 2001;**244**:427-431. doi:10.1006/jcis.2001.7975

[105] Cocquyt J, Olsson U, Olofsson G, Van der Meeren P. Temperature quenched DODAB dispersions: fluid and solid state coexistence and complex formation with oppositely charged surfactante. Langmuir. 2004;**20**(10):3906-3912. doi:10.1021/la036080c

[106] Abbas AK, Lichtman AH, Pillai S. Cellular and Molecular Immunology. 6th ed. Philadelphia: Saunders Elsevier; 2006. 566 p.

[107] Davidsen J, Rosenkrands I, Christensen D, Vangala A, Kirby D, Perrie Y, Agger EM, Andersen P. Characterization of cationic liposomes based on dimethyldioctadecylammonium and synthetic cord factor from M. tuberculosis (trehalose 6,6'-dibehenate)—a novel adjuvant inducing both strong CMI and antibody responses. Biochim Biophys Acta. 2005;**1718**(1-2):22-31. doi:10.1016/j.bbamem.2005.10.011

[108] Smith Korsholm K, Agger EM, Foged C, Christensen D, Dietrich J, Andersen C S, Geisler C, Andersen P. The adjuvant mechanism of cationic dimethyldioctadecylammonium liposomes. Immunology. 2007;**121**:216-226. doi:10.1111/j.1365-2567.2007.02560.x

[109] Fifis T, Gamvrellis A, Crimeen-Irwin B, Pietersz GA, Li J, Mottram PL, McKenzie IF, Plebanski M. Size-dependent immunogenicity: therapeutic and protective properties of nano-vaccines against tumors. J Immunol. 2004;**173**(5):3148-3154. doi:10.4049/jimmunol.173.5.3148

[110] Foged C, Brodin B, Frokjaer S, Sundblad A. Particle size and surface charge affect particle uptake by human dendritic cells in an in vitro model. Int J Pharm. 2005;**298**(2):315-322. doi:10.1016/j.ijpharm.2005.03.035

[111] Xiang SD, Scholzen A, Minigo G, David C, Apostolopoulos V, Mottram PL, Plebanski M. Pathogen recognition and development of particulate vaccines: does size matter? Methods. 2006;**40**(1):1-9. doi:10.1016/j.ymeth.2006.05.016

[112] Vollmer J, Krieg AM. Immunotherapeutic applications of CpG oligodeoxynucleotide TLR9 agonists. Adv Drug Deliv Rev. 2009;**61**:195-204.

[113] Manolova V, Flace A, Bauer M, Schwarz K, Saudan P, Bachmann MF. Nanoparticles target distinct dendritic cell populations according to their size. Eur J Immunol. 2008;38:1404-1413.

[114] Katz D, Lehrer S, Galan O, Lachmi B, Cohen S, Inbar I, Samina I, Peleg B, Heller D, Yadin H, Chai D, Freeman E, Schupper H, Fuchs P. Unique immunomodulating properties of dimethyl dioctadecyl ammonium bromide (DDA) in experimental viral vaccines. Adv Exp Med Biol. 1996;**397**:115-125.

[115] Perrie Y, Kastner E, Kaur R, Wilkinson A, Ingham AJ. A case-study investigating the physicochemical characteristics that dictate the function of a liposomal adjuvant. Hum Vaccin Immunother. 2013;**9**(6):1374-1381. doi:10.4161/hv.24694

[116] Carmona-Ribeiro AM (2014). Cationic nanostructures for vaccines. In Ht Duc (editor) Immune Response Activation, InTech, doi:10.5772/57543. Available from: http://www.intechopen.com/books/immune-response-activation/cationic-nanostructures-for-vaccines

[117] Hafner AM, Corthésy B, Merkle HP. Particulate formulations for the delivery of poly(I:C) as vaccine adjuvant. Adv Drug Deliv Rev. 2013;**65**(10):1386-1399. doi:10.1016/j.addr.2013.05.013

[118] Zhang Y, Wang Z, Gemeinhart RA. Progress in microRNA delivery. J Control Release. 2013;**172**(3):962-974. doi:10.1016/j.jconrel.2013.09.015

Chapter 6

Saponin-Based, Biological-Active Surfactants from Plants

Dorota Kregiel, Joanna Berlowska,
Izabela Witonska, Hubert Antolak,
Charalampos Proestos, Mirko Babic,
Ljiljana Babic and Bolin Zhang

Additional information is available at the end of the chapter

http://dx.doi.org/10.5772/68062

Abstract

Plants have the ability to synthesize almost unlimited number of substances. In many cases, these chemicals serve in plant defense mechanisms against microorganisms, insects, and herbivores. Generally, any part of the plant may contain the various active ingredients. Among the plant, active compounds are saponins, which are traditionally used as natural detergents. The name 'saponin' comes from the Latin word 'sapo,' which means 'soap' as saponins show the unique properties of foaming and emulsifying agents. Steroidal and triterpenoid saponins can be used in many industrial applications, from the preparation of steroid hormones in the pharmaceutical industry to utilization as food additives that exploit their non-ionic surfactant properties. Saponins also exhibit different biological activities. This chapter has been prepared by participants of the Marie Sklodowska-Curie Action—Research and Innovation Staff Exchange (RISE) in the framework of the proposal 'ECOSAPONIN.' Interactions between the participants, including chemists, physicists, technologists, microbiologists and botanists from four countries, will contribute to the development of collaborative ties and further promote research and development in the area of saponins in Europe and China. Although this chapter cannot provide a comprehensive account of the state of knowledge regarding plant saponins, we hope that it will help make saponins the focus of ongoing international cooperation.

Keywords: plants, saponins, surfactants, emulsifiers, biological activity

1. Introduction

An increasing trend in the food, pharmaceutical, and cosmetic industry is the utilization of natural plant extracts or plant-derived compounds, as an alternative to the application of chemical or synthetic antimicrobials to combat spoilage microflora and pathogens [1, 2]. Furthermore, the nontoxic nature of chemicals in plants, positive healthy properties, consumer perception and acceptance of their use has been well demonstrated [3, 4].

There are estimated 250,000–500,000 species of plants on Earth. A relatively small percentage (1–10%) of these is consumed as food by both humans and animal species. It is possible that a greater number are used for medicinal purposes. People on all continents have long applied poultices and imbibed infusions of hundreds, if not thousands, of indigenous plants. Currently, antimicrobial plant extracts are of especial interest to chemists and microbiologists due to growing public awareness of the negative effects of the over-use of antibiotics and disinfectants [5].

Plants have the ability to synthesize an almost limitless array of substances. In many cases, these chemicals serve as plant-defense mechanisms against predation by microorganisms, insects, and herbivores. Some, such as terpenoids, give plants their flavors; others—quinones and tannins are responsible for plant pigmentation. Any part of the plant may contain active components. For instance, roots of ginseng plants contain active saponins and essential oils, while eucalyptus leaves are harvested for their essential oils and tannins. Some trees contain useful substances in their bark, leaves, and shoots [6]. Some of the same herbs and spices used by humans to season food can yield useful medicinal compounds. Among different compounds derived from plants, saponins deserve a special mention. These chemicals may be considered as a part of plants' defense systems. They have been included in a large group of protective molecules found in plants named 'phytoanticipins' or 'phytoprotectants' [7].

The physiochemical and biological properties of saponins have led to a number of traditional and industrial applications. They have traditionally been used as natural detergents. The combination of a hydrophobic aglycone backbone and hydrophilic sugar molecules confers foaming and emulsifying properties of saponins [8]. The name 'saponin' is derived from the Latin word 'sapo,' meaning soap, as a soapy lather forms when plants containing saponins are agitated in water. They also exhibit a variety of biological activities. Plant-derived triterpenoid and steroidal saponins have been used in the production of steroid hormones in the pharmaceutical industry, as food additives, fire extinguishers and in other industrial applications. Other interesting biological applications include their use in anti-inflammatory, hypocholesterolemic and immune-stimulating remedies [9, 10].

2. Molecular characteristics

Saponins are a class of substances with a rigid skeleton of at least four hydrocarbon rings to which sugars in groups of one or two are attached (usually not more than 10 units). Traditionally, they are subdivided into triterpenoid and steroid glycosides. Steroidal saponins are mainly compounds containing 27 carbon atoms forming the core structures: spirostan (16β,22:22α,26-diepoxy-cholestan) and furostan (16β,22-epoxycholestan) [11–13] (**Figures 1** and **2**).

http://dx.doi.org/10.5772/68062

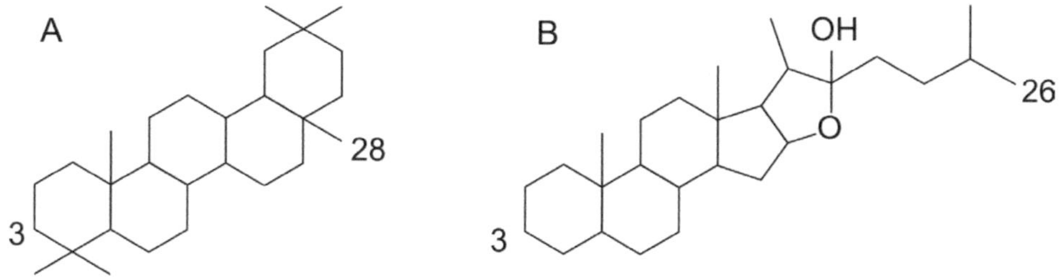

Figure 1. Structures of (A) triterpenoid and (B) steroidal saponins [8].

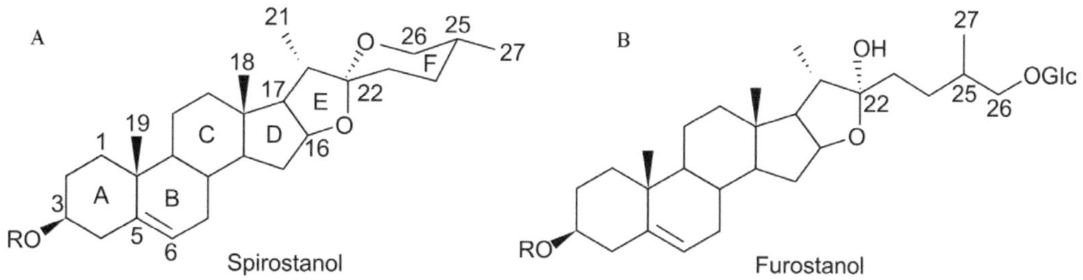

Figure 2. Structures of (A) spirostanol and (B) furostanol saponins [14].

There are 11 main classes of saponins: dammaranes, tirucallanes, lupanes, hopanes, oleananes, taraxasteranes, ursanes, cycloartanes, lanostanes, cucurbitanes, and steroids. The oleanane skeleton is the most common, present in most orders of the Plant Kingdom [15, 16].

Saponins with the carbohydrate or oligosaccharide groups attached at the C-3 position are monodesmosidic, while saponins with carbohydrates attached at both the C-3 and C-26 or C-28 positions are bidesmosidic. The variety of a glycones, carbohydrates, and different attachment positions result in numerous types of saponins. The carbohydrate chains of saponins usually include: D-glucose, D-galactose, L-rhamnose, L-arabinose, D-xylose, D-apiose, D-fucose, and D-glucuronic acid. The steroidal saponins usually show furostanol or spirostanol form. Additionally, both steroidal and triterpene saponins may contain other functional groups: –OH, –COOH, $–CH_3$ that give them additional diversity [17].

The chemical structure of saponins may be transformed during storage or processing. The linkages between the sugar chain and the aglycones as well as between the sugar residues can undergo hydrolysis during acid or base treatment, hydrothermolysis or enzymatic/microbial transformations, resulting in the formation of aglycones, prosapogenins (partially hydrolyzed saponins), and sugar residues [17]. Therefore, the selection of methods appropriate to storage of plant material is a key part of each efficient technology [18–20].

3. Plant sources

The presence of saponins has been reported in more than 100 families of plants and in a few marine sources such as star fish and sea cucumber. Triterpene saponins are present in many taxonomic plant groups. In particular, they can be found in parts of dicotyledonous plants (*Dicotyledones*) such as the seeds of *Hippocastani*, roots and flowers of *Primulae*, leaves of *Hedrae*, roots of *Ginseng*, bark of *Quillaja*, roots of *Glycyrrbizae*, roots of *Senegae*, leaves of *Polygalae Amarae*, roots of *Saponariae*, seeds of *Glycine max* and leaves of *Herniariae*. Legumes such as soybeans, beans and peas are rich sources of triterpenoid saponins. Steroidal saponins are typically found in members of the *Agavaceae, Alliaceae, Asparagaceae, Dioscoreaceae, Liliaceae, Amaryllidaceae, Bromeliaceae, Palmae* and *Scrophulariaceae* families and accumulate in abundance in crop plants such as yams, alliums, asparagus, fenugreek, yucca and ginseng. Diosgenin, the steroidal aglycone obtained by hydrolysis of dioscin, a saponin abundant in the tubers of *Dioscorea villosa* (wild yam), is the precursor for commercial synthesis of steroids such as cortisone, progesterone and pregnenolone. Steroidal glycoalkaloids are commonly found in members of the *Solanaceae* family including tomato, potato, aubergines and capsicum [8]. Cereals and grasses are generally deficient in saponins, with some notable exceptions, such as the *Avena* species (oats) which accumulates both triterpenoid and steroidal saponins. The phylogenetic tree with plant subclasses from which saponins have been isolated and characterized is presented in **Figure 3**.

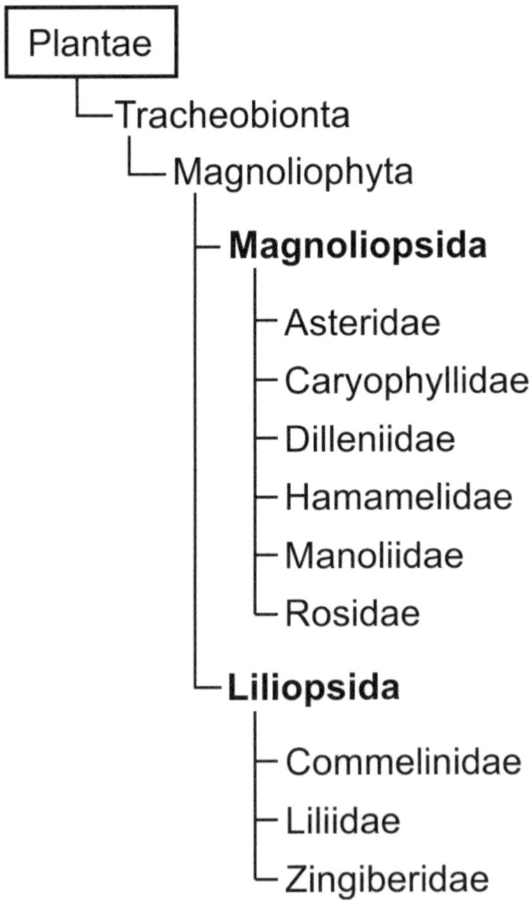

Figure 3. The phylogenetic tree with plant subclasses [16].

Some studies have suggested that variations in saponin distribution and composition in plants may be a reflection of varying needs for plant protection. In some plants, for example, *Phytolacca dodecandra* (gopo berry) and *Dioscorea pseudojaponica* (yam), maximal saponin accumulation has been noted during fruit and tuber development and has been suggested to protect reproductive organs. However, it was documented that in several plant species, the production of saponins is induced in response to biotic (herbivory and pathogen attack) and abiotic (humidity, nutrient starvation, light, temperature) stresses [8].

The main sources of saponins in human diet are legumes, mainly broad beans, kidney beans and lentils. Saponins are also present in *Allium* species (onion, garlic), asparagus, oats, spinach, sugarbeet, tea and yam. Nevertheless, the main plant sources of saponins used in medicine and industrial applications are soap bark tree (*Quillaja saponaria*), Mojave yucca (*Yucca schidigera*), licorice (*Glycyrrhiza* species), ginseng (*Panax* species), fenugreek (*Trigonellafoenum-graceum*), alfalfa (*Medicago sativa*), horse chestnut (*Aesculus hippocastanum*), soapwort (*Saponaria officinaux*), gypsophila genus (*Gypsophila paniculata*) and sarsaparilla (*Smilax* species).

Some of the better-known botanicals rich in saponins are presented in **Table 1**.

Plant		**Saponin content [%]**
Latin name	**Common name**	
Aesculis hipocastanum	Horse-chestnut	3
Avena sativa	Oat	0.1–0.13
Beta vulgaris	Sugar beet (leaves)	5.8
Chenopodium quinoa	Quinoa	0.14–2.3
Cicer arietinum	Chickpea	0.23
Crocus savitus	Saffron crocus	1.2–3.4
Glycine max	Soybean	0.22–0.49
Glycyrrhiza glabbra	Licorice (root)	22.2–32.3
Hedera helix	Ivy	5
Medicago sativa	Alfalfa	0.14–1.71
Panax ginseng	Chinese ginseng	2–3
Panax quinquefolius	American ginseng	1.42–5.58
Pisum sativum	Green pea	0.18–4.2
Polygala spp.	Milkwort	8–10
Primula spp.	Primula	5–10
Quillaja saponaria	Quillaja bark	9–10
Saponaria officinalis	Soapwort	2–5
Smilax officinalis	Sarsaparilla	1.8–2.4
Trigonellafoenum-graecum	Fenugreek	4–6
Yucca schidigera	Yucca	10

Table 1. The better-known plants—sources of saponins [21, 22].

In Northern Europe, the main sources of saponins are: *Saponaria officinalis, Calendula officinalis, Salvia, Digitalis, Verbascum, Solanum* species, sugar beet, oats, etc.

Calendula officinalis (*Asteraceae*) is well-known medicinal plant in Poland. It is also popular in gardens as a decorative annual species. Traditionally, it has been used topically for many eruptive skin diseases and abrasions, as well as for gastric and menstrual discomfort, as a plant with antiseptic, mild diaphoretic and antispasmodic properties. Calendula contains significant amounts of oleananesaponins, which form two distinct series of related compounds, called 'glucosides' and 'glucuronides' according to the structure of the respective precursor. Extracts from marigold flowers are still used in ointments, cosmetic creams and hair-shampoos [15].

In sugar beet leaves, saponins have been reported at level of 5%, and in roots 0.1–0.3%. However, during raw beet processing, these saponins are mostly concentrated in the waste products. For example, the concentration of saponins in sugar beet pulp water reaches 1.2% [23]. Similar concentrations of saponins have been detected in the filtration residues and molasses. In Polish research laboratories, several triterpene-based saponin structures have been isolated and characterized [24]. Given the scale of worldwide sugar production from sugar beet, this plant can be considered as an industrial source of saponins [25]. Sugar beet as a high economic value crop will have a prosperous perspective of application in the food, bioenergy, and pharmacy industries [26].

In Southern Europe, the region around the Mediterranean Sea is rich in grapes. Saponin glycosides in red wine are known as heart protective, due to their LDL cholesterol-lowering and HDL cholesterol-increasing effects. The saponins in red wine also help prevent clumping of red blood cells. Many of plant species rich in saponins are used traditionally in Greece for making herbal teas, as flavorings and seasonings and have been tested for various pharmacological activities [27]. Mediterranean thyme (*Thymus capitatus*) is a common plant in the Mediterranean region, growing in arid rocky places and flowering between May and August. It is commonly used as a medicinal and culinary herb, owing to its strong and agreeable odor, mainly attributed to its essential oil. Other constituents include saponins and organic acids. Thyme has several medicinal uses including antiseptic, expectorant, antispasmodic and anthelminthic properties. Greek agave plants contain saponins and fructans. Many other representative species of the Mediterranean flora including *Melissa officinalis* (balm), *Origanum vulgare* (wild marjoram), *Origanum dictamnus* (dittany of Crete or hop marjoram), *Hyssopus officinalis* (hyssop), *Dioscorea villosa* (wild yam), *Viola tricolor* (wild violet, wild pansy, heartsease, Johnny jump-ups), *Salvia officinalis* (sage), *S. officinalis* (common soapwort), *Tribulus terrestris* (tribulus) contain saponins with antioxidant and anti-inflammatory properties and can boost the human immune system [28, 29]. The genus *Ruscus* (*Asparagaceae* family) is native to the Mediterranean, Southern and Western Europe. The underground parts of *Ruscus* plants are a source of steroidal saponins. *Ruscus* extracts were extensively used, especially in Germany and France, for the treatment of chronic venous insufficiency, varicose veins, hemorrhoids, and orthostatic hypotension [30].

China is rich in various plant sources of saponins, which are often unknown in Europe. *Mussaenda pubescens (Rubiaceae), Bupleurum chinense, Clinopodium chinense* var. *parviflorum* and

http://dx.doi.org/10.5772/68062

Clematis chinensis Osbeck (*Ranunculaceae*) and *Yucca elephantipes* are Chinese folk medicine plants used as diuretics, antiphlogistics, diaphoretics and antipyretic agents and have also been used to detoxify mushroom poisons and terminate early pregnancy. *Yucca* (*Agavaceae*) plants are native to China. The leaf extract of *Y. elephantipes* with saponins has been reported to have antiviral activity against tobacco mosaic virus and to exhibit antifungal activity against the pathogenic yeasts *Candida albicans* and *Cryptococcus neoformans* [31].

The interesting plant in China is *Caragana*, also known as peashrub, a member of *Fabaceae*. More than 80 *Caragana* species were recorded, and several of them have a long history of use in traditional Chinese medicine, for example, in the treatment of cervical and breast cancer. Seeds of this legume represented an interesting source of triterpenoid saponins of the soyasaponin B type [32].

The *Glycyrrhiza* genus (*Leguminosae*family) consists of about 30 species and is widely distributed all over the world. In China, three species *G. uralensis, G. glabra* and *G. inflata* are officially used as licorice and recorded in Chinese Pharmacopoeia. Biological studies showed that licorice has a variety of biological effects, such as antioxidant, antiviral, anti-cancer, antidepressant, anti-inflammatory, anti-carcinogenesis, hepatoprotective and neuroprotective bioactivities [33, 34].

The important source of natural medicines is *Panax* genus. Three valuable *Panax* species *P. ginseng, P. quinquefolius*, and *P. notoginseng* are of great interest to medicine and food industry, and they are widely used in healthcare products, foods and food additives. To the end of 2012, at least 289 saponins were reported from eleven different *Panax* species [35]. Most of them are glycosides of triterpenoid aglycones [36]. Ginseng has been used as a herbal medicine in China for thousands of years due to its wide pharmacological properties, such as anticancer, antidiabetic, antifatigue, anti-ageing, hepatoprotective and neuroprotective [37]. It was also documented that *P. notoginseng* saponins suppress radiation-induced osteoporosis by regulating bone formation and resorption [38].

Calamus leptospadix grows as a non-climbing palm in the Sub-Himalayan region. Extract of *C. leptospadix* was characterized by Borah and co-workers, and they documented presence of a triterpenoid saponin with antimicrobial properties against both *Escherichia coli* and *Candida albicans* [39].

Stauntonia brachyanthera is an evergreen shrub belonging to the family of *Lardizabalaceae,* mainly distributed in the southwest of China. This plant is traditionally used to treat various diseases. Its fruit, zhuyaozi, is very popular in the southwest of China because of its fresh taste and abundant nutrients. The chemical study on this fruit resulted in the isolation of triterpenoid saponins. This research provided useful clues for the fruit of *S. brachyanthera* as a new resource of food for hepatoprotection [40].

Camellia oleifera, originated in China, is an important source of edible oil obtained from its seeds. This plant has been used as a natural detergent, and its extract rich in saponins is commercially utilized as a foam-stabilizing and emulsifying agent. The percentage of crude saponins extract that was obtained from the defatted seed meal of *C. oleifera* was 8.34% [41].

Plant saponins show region-specific character. It was found that variety of soybean from China is richer in saponins than those from Japan, Canada or United States [42]. *Tribulus terrestris* samples collected in Bulgaria, Greece, Serbia, Macedonia, Turkey, Georgia, Iran, Vietnam and India were analyzed by LC-ESI/MS/MS, and the results revealed distinct differences in the saponin profiles depending on region of sample collection, plant part studied and stage of plant development. The samples from Bulgaria, Turkey, Greece, Serbia, Macedonia, Georgia and Iran exhibited similar features but the Vietnamese and Indian samples exhibit totally different chemical profile. The obtained results suggested the existence of one chemotype common to the East South European and West Asian regions [43]. Studies conducted by Montero and co-workers showed that several licorice (*Glycyrrhiza glabra*) samples collected at different locations were characterized by specific metabolite profiles. Therefore, it was concluded that obtained 2D-chromatograms from the different licorice samples can be used as typical patterns that could potentially be related to geographical location and authentication of plant source [44].

To obtain saponins from plant material different extraction methods may be used, using solvents as water, methanol, ethanol or hydroalcoholic mixtures in Soxhlet extractors or in orbital shakers. In addition, other solvents such as glycerol and aqueous or alcoholic surfactants solutions were also reported. Novel procedures use lower amounts of solvent but additional physical/chemical treatment: multi-stage extraction, pressure, microwaves, ultrasounds or supercritical fluid extraction. These methods can led to an increase in the process efficiency. However, it should be considered that under harsher conditions (higher temperature and pressure), saponins can be hydrolyzed and degraded, so rather mild processes should be used [45–49].

4. Natural surfactants and emulsifiers

Saponins, due to the presence of a lipid-soluble aglycone and water-soluble sugar chain, show amphiphilic nature. In this way, foam formation (with liquid-gaseous phases), an emulgator effect (with liquid-liquid phases) and dispersion abilities (with liquid-solid phases) are achieved. Saponins with one sugar chain have the best foaming characteristics. The compounds with two or three sugar chains show decreasing of foaming ability. Some saponins without foaming character have also been observed [17].

In aqueous solution, saponin molecules align themselves vertically on the surface with their hydrophobic ends oriented away from the water. This has the effect of reducing the surface tension of the water, causing it to foam. In aqueous solutions, surfactants form micelles above a critical concentration called critical micelle concentration (CMC). Below this concentration, molecules remain unassociated. Micelles have a lipophilic center, and this creation of a fat-loving compartment explains why detergents can dissolve grease and oils (**Figure 4**).

The size and structure of micelles are dependent on the type of saponin. For example, saponins from *S. officinalis* and soya bean form small micelles consisting of only two molecules, while the aggregates of *Quillaya saponaria* saponin consist of 50 molecules. It was documented that the properties and the aggregation number (number of monomers) of micelles forming

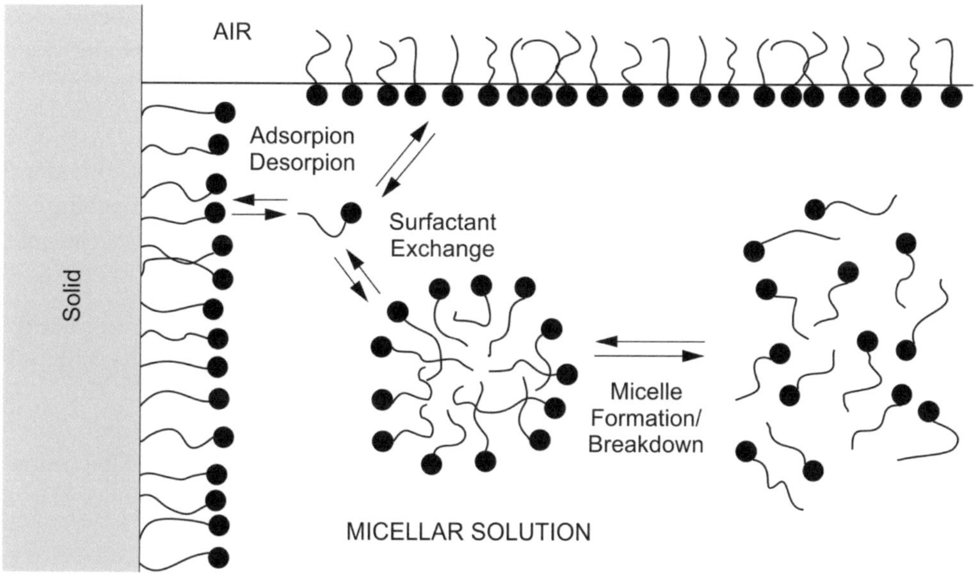

Figure 4. Micelle formation [50].

by *Quillaya* saponins are affected by temperature, salt concentration, and pH level. For saponins from *Q. saponaria,* CMC is equal from 0.5 to 0.8 g/l at temperature 25°C and decreases with increasing salt dose [17]. The micelle shapes depend on the saponin molecule. For example, micelles formed by *Saponaria* and *Quillaya saponins* are elongated or even filamentous, while those formed by saponins of *G. max* are rather circular. Probably, the reason for these differences is the chemical structure of aglycone.

The presence of carboxylic acid in the saponin molecule may strongly influence the surface activity. Additionally, the location of this acid in the molecule is particularly important. For example, *G. max* saponin contains -COOH group in its hydrophilic part. The carboxylic group dissociates in aqua phase and forms free carboxyl anion, responsible for increasing the solubility of saponin in water environment. In contrast, saponins of *Sapindus mukorossi* (Chinese washnut) also contain the carboxylic groups but they attach to the hydrophobic aglycone. In consequence of this mechanism, the dissociation level of –COOH groups is very low. Saponins can also form mixed 'sandwich-like' or 'pile of coins-like' micelles with bile acids. These are much larger than the micelles of saponins alone, and they differ depending on the structure of the aglycone. In the presence of bile acids, saponins from *S. officinalis* and *Q. saponaria* form filamentous structures, while *G. max* saponins have an open structure. The ability of saponins to form large stable micelles with bile acids gives important implications for dietary mechanisms. Saponins in food and feed increase fecal excretion of bile acids. Additionally, the incorporation of cholesterol into saponin micelles increases their size, CMC, viscosity, and the aggregation level resulting in the solubility enhancement of cholesterol. The micelles formed are too large for the digestive tract to absorb. This mechanism leads to decreasing of the plasma cholesterol concentration. Saponin*Q. saponaria* was found to solubilize cholesterol significantly better than linear hydrocarbon chain surfactants [51].

Interactions between saponin and membrane-bound cholesterol lead pore formation and increasing of membrane permeabilizing properties. This specific effect of saponins depends on the combination of various factors: the membrane composition, the type of saponin, and—especially—the nature of aglycone [52].

Saponins also affect the permeability of intestinal cells by forming complexes with sterols in mucosal cell membranes. This leads to increase in intestinal mucosal cells permeability. Thus, this facilitates the uptake of substances to which the gut would normally be impermeable, for example, milk alergen α-lactoglobulin [17].

Quillaja saponins also had a solubilizing effect on some toxic polycyclic aromatic hydrocarbons, which increases linearly with saponin concentration at values higher than CMC. A similar linear correlation has been observed between the concentration of the saponins from *Sapindus mukorossi* and aqueous solubility of hexachlorobenzene and naphthalene [21]. Saponins also enhance solubility of Yellow OB, and progesterone [8] Purified saponins and saponin mixtures resulted in both enhancements and reductions in water solubility of quercetin, digitoxin, rutin and aesculin [53].

Emulsifiers play two key roles in the creation of successful emulsion-based products. They: (i) facilitate the initial formation of fine lipid droplets during homogenization and (ii) enhance the stability of the lipid droplets once they have been formed. Oil-in-water emulsions may be formed using either high- or low-energy approaches. High-energy approaches utilize mechanical devices (homogenizers): high shear mixers, colloid mills, high-pressure valve homogenizers, microfluidizers, and sonicators. Low-energy homogenization relies on the spontaneous formation of emulsions when the composition or environment of certain emulsifier-oil-water mixtures is changed in a particular way. *Quillaja* saponin is a natural effective emulsifier to form and stabilize oil/water emulsions with very small oil beads ($d < 200$ nm). They are stable in wide range of environmental parameters (pH, ionic strength, temperature). This fact makes saponins of *Q. saponaria* suitable for wide application in food products [54].

Quillaja saponins currently find commercial scale in food industry as emulsifiers with milk and egg proteins, for example, β-lactoglobulin, β-casein or egg lysozyme by electrostatic and hydrophobic interactions as well as by specific sugar binding sites [55].

5. Biological activity

Due to their amphiphilic nature, saponins show a wide range of biological activities. Various crude isolates, extracts, and saponin containing plants were utilized in the investigation of biological activity in the earlier studies; however, progress in the isolation/purification and characterization techniques has enabled the investigation of the bioactivity of well characterized [56, 57]. Saponins have been shown to swell and rupture erythrocytes causing a release of hemoglobin. The effect of saponin on erythrocyte death or hemolysis may limit the therapeutic use of the substances. On the other hand, saponins have been proposed for the treatment of a variety of diseases, including diabetes, obesity and osteoporosis [58]. Pharmacological effects of saponins include stimulation of immune responses. Their efficacy against cancer has been attributed to their ability to inhibit cell proliferation, to counteract angiogenesis and to stimulate apoptosis [59–61].

The toxicity of saponins to insects (insecticidal activity), parasite worms (anthelmintic activity), molluscs (molluscicidal), and fish (piscidal activity), and their antifungal, antiviral, and antibacterial activity is well documented. Toxicity of saponins to warm blooded animals is dependent on the source, composition, and concentration of these compounds. The results of in vivo studies with rats, mice, and rabbits implied that saponins are not absorbed in the alimentary channel but hydrolyzed enzymatically to sapogenins [21].

The action of saponins, by enhancing the immune response to antigens, has been documented since 1940s. *Quillaja* saponins are exclusively used in the production of saponin adjuvants, and this immune function was also reported for soya, quinoa, gypsophila and *Saponaria* saponins [62]. Due to the structural complexity and toxicity of plant saponins, their use in human vaccines is limited, but the progress in new processing and purification techniques with maintaining of immunological adjuvant activity is important to create saponins as new generation vaccines [63].

Several mechanisms have been proposed to explain the hypocholesterolaemic activity of saponins. Possible mechanisms may involve the capacity of saponins to: (i) form insoluble complexes with cholesterol, (ii) affect micelle formation, (iii) interfere with bile acid metabolism, (iv) inhibit lipase activity, or (v) regulate cholesterol homeostasis via monitoring the expression of the key regulatory genes of proteins or enzymes related to cholesterol metabolism [58, 64]. Cholesterol-lowering activity of saponins has been demonstrated in both animal and human trials. Animal diet containing purified saponins or concentrated saponin extracts containing, for example, digitonin (saponin from *Digitalis purpurea*), saikosaponin (saponins from *Bupleurumfalcatum* and related plants) and saponins from *Saponaria*, soya, chick pea, *Yucca*, alfalfa, fenugreek, *Quillaja*, *Gypsohila*, and garlic resulted in reductions of cholesterol concentrations [21].

Anticancer activity has been reported for soya saponins, ginsenosides, saikosaponin, diosgenin and glycyrrhizic acid. In particular, the potential of soybean saponins as anticarcinogens has been studied in recent years. Anticancer activities of saponin containing plants such as ginseng and licorice were also investigated [65].

The study of the relationship between chemical structure of aglycones and colon anticancer activity of soybean saponins revealed that the soya sapogenols were more bioactive than the glycosidic saponins. Other aglycones with anticancer activity include dammaranesapogenins from ginseng, betulinic acid, and oleanolic acid. These two last compounds were also reported to possess anti-viral, anti-inflammatory, hepatoprotective, anti-ulcer, antibacterial, hypoglycaemic, anti-fertility, and anticariogenic activities. However, the conversion of saponins to their aglycones may also result in the loss of activity. For example, the hydrolysis of saponins by ruminal bacteria results in the loss of antiprotozoal activity. Similarly, the deacylation of *Quillaja* saponins decreases their adjuvant activity [66].

6. Antimicrobial activity

The antimicrobial effects of saponins extracted from plants have been studied in *Solanum,* oats, seeds of *Capsicum annuum,* alfalfa, garlic, *Yucca, Quillaja,* etc. The saponin extracts were tested against numerous Gram-positive and Gram-negative bacteria, yeasts and molds. However, the results were varied due to the high diversity of plant saponins [67]. For example, saponins from *Yucca* exhibit antimicrobial activity against Gram-positive cells but do not act on Gram negative bacteria. However, *S. officinalis* extracts showed antibacterial action against Gram negative, avian pathogenic *Escherichia coli* (APEC) strains [68, 69]. In general, the antibacterial activity of saponins is often weak, whereas significant antifungal activity has been observed. The primary mode of action of saponins toward fungi involves pore formation and loss of membrane integrity. The mechanism of action is an analogous to hemolytic activity of saponins. It was proposed a model of action for avenacin—triterpene saponin of oats. The first step involves the insertion of the aglycone fragments into the membrane and then their binding to sterols [70]. The following stage conducts to the interaction of sugar residues and formation of sterol-saponin complexes. This phenomenon leads to the rearrangement of membrane lipids, formation of pores and-finally—lysis of cells [71, 72]. Yeast studies on *Quillaja* saponins conducted in Poland found that saponin treatment lead to increased cell membrane permeability in different yeast strains, and therefore, it was concluded that *Quillaja* saponins facilitate the process of obtaining yeast salt-free lysates [73]. It is interesting that *Yucca* and *Quillaja* saponins increased growth of bacterial *Escherichia coli* cells up to a certain concentration, and thereafter decreased growth [74]. Arabski and co-workers demonstrated that saponin *Q. saponaria* at dose of 12 μg/mL enhanced the six *E. coli* strains growth [75]. Naturally, cholesterol-free Gram-negative bacteria cell-wall outer membranes are around 90% covered with lipopolysaccharide (LPS). Therefore, it was concluded that saponin may interact with the lipid A part of LPS and thereby increase the permeability of the bacterial cell wall. Sublethally injured or weakened cells may become more susceptible to the action of conventional disinfectants, even at reduced concentrations. It was suggested that lipid A-saponin complexes could promote antibiotic (colistin, ampicillin) or disinfectant action toward inherently resistant microbial cells [75]. The similar results were obtained by Alberice and co-workers [76]. They documented that application of saponin extract in the food industry would be economically viable and sustainable. The results indicated that saponin alone can be used by the industry as a bactericide to reduce the risk of juice spoilage by Gram-positive cells *Alicyclobacillus acidoterrestris.*

7. Commercial applications

Y. schidigera and *Q. saponaria* are the two major commercial sources of saponins added to cosmetics as well as food products as emulsifiers and long-lasting foaming agents [17].

Y. schidigera is a native plant from southwestern United States and Mexico. Native Americans used it to make soap. The trunk of the plant is mechanically shredded, and yucca juice is produced by mechanical squeezing in a press. The obtained juice is concentrated by evaporation. *Y. schidigera* syrup (concentrated juice; *Yucca* extract), and dried and finely powdered logs (*Yucca* powder) are of particular interest to cosmetic, pharmaceutical and beverage industries as well as animal nutrition [77]. These products possess foaming features that are of particular interest in cosmetic, soft drinks (root beer), food and feed industries [78].

In the United States, *Yucca* is listed in The Code of Federal Regulation [79]. In Japan, *Yucca* extract (extract of whole plant of *Yucca arborescens* or *Y. schidigera*) is listed in the List of Existing Food Additives [80]. Because steroidal saponins in *Yucca* exhibit antifungal activities, *Yucca* extract has been added to food as a 'shelf life extender' in the Japanese market. Yucca powder water extracts can be successfully used in confectionery/food industries for improving both product quality and shelf stability. Sucharzewska and co-workers documented that *Yucca* extract contains two groups of beneficial substances. One group is formed by steroidal saponins, which may improve product quality (porosity, density, and hardness), and the second one is created by antioxidants that are able to reduce fat oxidation and extend food quality during shelf-life time [81]. It is also worth to note that *Yucca* extracts may be used as natural, non-toxic deodorizers. The studies conducted in Poland show that combined treatment with microbial preparations and *Yucca* extract can significantly reduce the concentration of odorants in poultry manure [82]. Natural saponin extracts, namely those that may be obtained by steam treating the pulp of *Yucca* with water, in combination with proteins exhibit a synergistic effect, eliminating odors from the breath and oral cavity of humans, as well as from other environments [83].

Tenon and co-workers used HPLC/ELSD technique for *Yucca* steroidal saponin quantification. This method is effective for routine industrial analyses for saponin fingerprints and capable of distinguishing saponin profiles from taxonomically distant species [78].

The second saponin source of commercial value is *Q. saponaria*. The term 'quillaia' refers to the dried inner bark of the tree, which is a large evergreen with shiny, leathery leaves and a thick bark, native to China and several South American countries, principally Bolivia, Chile, and Peru [84, 85]. The bark of this tree was used as shampoo in for hundreds of years. *Quillaja* extracts contain over 100 triterpenoid saponins. The basic structure of them is the hydrophobic triterpenoid quillaic acid known as sapogenin, and the hydrophilic sugar moieties are attached at two positions: di- or trisaccharide at C3 and oligosaccharide at C28 [85]. Young plants usually exhibit less heterogeneous saponins profiles than those obtained from mature extracts [87].

A large amount of *Quillaja* saponin is mainly utilized as a surfactant. It is also used in beverages, food ingredients, shampoos, liquid detergents, toothpastes and extinguishers as an emulsifier and long-lasting foaming agent. Additionally, a saponin mixture possessing immune-adjuvant properties was given a pharmaceutical application, as a suspension stabilizer [88].

The beneficial effects of extracts from *Yucca* and *Quillaja* are well documented. The extracts from these plants may influence microbial fermentation. Inhibition of gut microbes, particularly *Streptococcus bovis, Butyrivibrio fibrisolvens, Escherichia coli* and rumen protozoa has been reported [74]. Extracts of *Y. schidigera* and *Q. saponaria* have been used as 'food grade' saponins. This term is widely used by manufacturers, and it is defined as any grade or preparation of saponin which is approved for use in food and beverages under the United States Food and Drug Administration (FDA).

According to the Codex Alimentarius Commission, extracts from *Q. saponaria* may be used as a foaming agent in 'water-based flavored drinks', including 'sport' or 'electrolyte' drinks and particulate drinks (GSFA category 14.1.4, 500 mg/kg maximum use level). In soft drinks, unpurified *Quillaja* extracts are used at dose up to 200 mg/kg. However, in syrups intended for dispensable frozen beverages (FCBs) or frozen lemonades, *Quillaja* extracts may be up to 500 mg/kg on dry solid basis [87].

Although *Quillaja* and *Yucca* saponins are not considered Generally Recognized As Safe (GRAS) by FDA, they have been assigned as GRAS by Flavor and Extract Manufacturers' Association of the United States (FEMA) with FEMA number 2973 [21, 87].

Quillaja extracts are classified as type 1 and type 2 based on their saponin content, 20–26% and 75–90%. *Quillaja* extract, type 2, is used in Japan as an emulsifier for preparations containing lipophilic colors or flavors that are added to soft drinks, fermented vegetables, and dressing [87]. Other saponins used food additives include enzymatically modified soybean saponin, Pfaffia and Yucca extracts, and tea seed saponins [80].

In the European Union, *Quillaja* extract is classified as the foaming agent for use in water-based, flavored non-alcoholic drinks and labeled as E999 (200 mg/l calculated as anhydrous extract) [87].

The physiochemical properties of saponins can also be utilized in food processing applications, thus, while complex formation of saponins with cholesterol has been used for the removal of cholesterol from dairy products such as butter oil [89–91]. It was documented that the natural food-grade surfactant isolated from the bark of the *Q. saponaria* Q-Naturale® may be able to replace synthetic surfactants in food and beverages [92]. The interaction of saponins with cell membranes has been considered for the selective precipitation of fat globule membranes from cheese whey. In this application, saponins are used to increase the hydrophobicity of the fat membrane to facilitate flocculation and precipitation of the formed complexes.

As a natural surfactant, *Q saponaria* saponins demonstrated good performance in manufacturing orange oil nanoemulsions. This fact may permit the manufacture of good quality orange oil-based nanoemulsions in beverage and alcohol-free mouthwash applications [93, 94]. *Quillaja* saponins show a high surface activity and functionality to solubilize a lutein ester extract for its incorporation in food matrices [86]. Additionally, it was documented that the mixtures containing *Quillaja* saponins and lecithins were rather unaffected upon heating from 25 to 75°C. Therefore, these results provide important insights into selecting surfactants to be used in specific food applications, for example, whether the food will be heat treated or not. This type of structure modulation through different environmental conditions and heating may also be useful for structure design in pharmaceutical applications [95].

Dried roots of licorice represent an important agricultural product. The name 'glycyrrhiza' originates from the Greek words 'glykosrhiza,' which mean 'sweet root.' Licorice is used as a sweetener and a flavor enhancer for foods in China and other countries. It is approved by Food and Drug Administration USA as a food additive, regarded with the 'GRAS' label and registered as CFR 184.1408 [33].

Saponins can be used to enhance both the effectiveness of cleaning/disinfection processes. They are considered natural detergents and are used as additives in washing powders, and additives for liquid/powder cleaning. The addition of a small amount of a saponin to an aqueous environment provides a product that is an effective water clarifier and solid surface cleanser. These compositions may be used to clean metals, metal-plated surfaces, ceramics, wood, glass, etc. The use of natural plant products as detergents could provide cheaper, safer and more consumer-acceptable alternatives to synthetic compounds.

8. Conclusion

Saponins are diverse compounds traditionally used as natural detergents. Their physico-chemical and biological properties are wide exploited in food, cosmetics and pharmaceuticals. Information on the composition (qualitative and quantitative), properties of the saponins present in the raw material, and the effects of processing on their composition and properties are key elements of successful process design.

Author details

Dorota Kregiel[1]*, Joanna Berlowska[1], Izabela Witonska[1], Hubert Antolak[1], Charalampos Proestos[2], Mirko Babic[3], Ljiljana Babic[3] and Bolin Zhang[4]

*Address all correspondence to: dorota.kregiel@p.lodz.pl

1 Lodz University of Technology, Poland

2 National and Kapodistrian University of Athens, Greece

3 University of Novi Sad, Serbia

4 Beijing Forestry University, China

References

[1] Killeen GF, Madigan CA, Connolly CR, Walsh GA, Clark C, Hynes MJ, Timmins BF, James P, Headon DR, Power RF. Antimicrobial saponins of *Yucca schidigera* and the implications of their in vitro properties for their in vitro impact. Journal of Agricultural and Food Chemistry. 1998;46:3178-3186. doi:10.1021/jf970928j

[2] Selim SA, Adam ME, Hassan SM,AlbalawiAR. Chemical composition, antimicrobial and antibiofilm activity of the essential oil and methanol extract of the Mediterranean cypress (*Cupressussempervirens*L.). BMC Complementary and Alternative Medicine. 2014;14:179. doi:10.1186/1472-6882-14-179

[3] Proestos C, Rashed K, Roidaki A, Sinanoglou VJ. Antioxidant capacity and antimicrobial activity of selected aromatic Egyptian plants. Promising raw materials for "superfoods" and dietary supplements. Agro Food Industry HI-TECH. 2016;**27(4)**:35-38.

[4] RoidakiA, Kollia E, Panagopoulou E, Chiou A, VarzakasT, MarkakiP, Proestos C. Super foods and super herbs: antioxidant and antifungal activity. Current Research in Nutrition and Food Science. 2016;**4(SI.2)**:138-145. doi:10.12944/crnfsj.4.special-issue-october.19

[5] Goud MJP, Komraiah A, Rao KN, Ragan A, Raju VS, Charya MAS. Antibacterial activity of some folklore medicinal plants from South India. African Journal of Traditional, Complementary and Alternative Medicines. 2008;**5(4)**:421-426. doi:10.4314/ajtcam.v5i4.446

[6] Cowan MM. Plant products as antimicrobial agents. Clinical Microbiology Reviews. 1999;**12(4)**:564-582.

[7] Francis G, Kerem Z, Makkar HPS, Becker K. The biological action of saponins in animal systems: a review. British Journal of Nutrition. 2002;**88**:587-605. doi:10.1079/BJN2002725

[8] Moses T, Papadopoulou KK, Osbourn A. Metabolic and functional diversity of saponins, biosynthetic intermediates and semi-synthetic derivatives. Critical Reviews in Biochemistry and Molecular Biology. 2014;**49(6)**:439-462. doi:10.3109/10409238.2014.95362

[9] Cheeke PR, Piacente S, Oleszek W. Anti-inflammatory and anti-arthritic effects of *Yucca schidigera*: a review. Journal of Inflammation. 2006;**3**:6 doi:10.1186/1476-9255-3-6

[10] Liu J, Li Y, Shi H, Wang T, Wu X, Sun X, Yu L. Components characterization of total tetraploid jiaogulan (*Gynostemma pentaphyllum*) saponin and its cholesterol-lowering properties. Journal of Functional Foods. 2016;**23**:542-555. doi:10.1016/j.jff.2016.03.013

[11] Sparg SG, Light ME, van Staden J. Biological activities and distribution of plant saponins. Journal of Ethnopharmacology. 2004;**94**:219-243. doi:10.1016/j.jep.2004.05.016

[12] Thakur M, Melzig MF, Fuchs H, Weng A. Chemistry and pharmacology of saponins: Special focus on cytotoxic properties. Botanics Targets Therapy. 2011;**1**:19-29. doi:10.2147/BTAT.S17261

[13] Saxena M, Saxena J, Nema R, Singh D, Gupta A. Phytochemistry of medicinal plants. Journal of Pharmacognosy and Phytochemistry. 2013;**1(6)**:168-182.

[14] Challinor VL,De Voss JJ. Open-chain steroidal glycosides, a diverse class of plant saponins. Natural Product Reports.2013;**30(3)**:429-454. doi:10.1039/c3np20105h

[15] Szakiel A, Ruszkowski D, Janiszowska W. Saponins in *Calendula officinalis* L.—structure, biosynthesis, transport and biological activity. Phytochemistry Reviews. 2005;**4**:151-158. doi:10.1007/s11101-005-4053-9

[16] Vincken JP,Heng L,de Groot A,Gruppen H. Saponins, classification and occurrence in the plant kingdom. Phytochemistry. 2007;**68(3)**:275-297. doi:10.1016/j.phytochem.2006.10.008

[17] Oleszek W, Hamed A. Saponin-based surfactants. In: Kjellin M, Johansson I editors. Surfactants from Renewable Sources Resources. John Wiley & Sons Ltd. Chichester, UK; 2010. pp. 239-249. doi:10.1002/9780470686607.ch12

[18] Mitrevski V, Lutovska M, Mijakovski V, Pavkov I, Babić M, Radojčin M. Adsorption isotherms of pear at several temperatures. Thermal Science. 2015;**19(3)**:1119-1129. doi:10.2298/TSCI140519082M

[19] Pavkov I, Babić L, Babić M, Radojčin M, Stamenković Z. Mathematical modelling of convective drying nectarine halves (*Pyrus persica* L.). Savremena Poljoprivredna Tehnika. 2013;**39(2)**:103-112.

[20] Hossain MB, Brunton NP, Rai DK. Effect of drying methods on the steroidal alkaloid content of potato peels, shoots and berries. Molecules. 2016;**21(4)**:403. doi:10.3390/molecules21040403

[21] Güçlü-Ustündağ O, Mazza G. Saponins: properties, applications and processing. Critical Reviews in Food Science and Nutrition. 2007;**47(3)**:231-258. doi:10.1080/10408390600698197

[22] Mir MA, Parihar K, Tabasum U, Kumari E. Estimation of alkaloid, saponin and flavonoid content in various extracts of *Crocus sativa*. Journal of Medicinal Plants Studies. 2016;**4(5)**:171-174.

[23] Brezhneva TA, Nikolaevskii VA, Selemenev VF, Slivkin AI, Muad AA, Khind T, Safonova EF. Isolation of saponins from sugar beet roots and preliminary characterization of their adaptogen properties. Pharmaceutical Chemistry Journal. 2001;**35(3)**:159-161. doi:10.1023/A:1010462013789

[24] Mikołajczyk-Bator K, Błaszczyk A, Czyżniejewski M, Kachlicki P. Identification of saponins from sugar beet (*Beta vulgaris*) by low and high-resolution HPLC-MS/MS. Journal of Chromatography B. 2016;**1029-1030**:36-47. doi:10.1016/j.jchromb.2016.06.038

[25] Binczarski M, Witonska I, Berlowska J, Dziugan P, Piotrowski J. Sweetening juices by beet juice. Agro Przemysł. 2013;**1**:66-69 (In Polish)

[26] Zhang Y, Nan J, Yu B. OMICS technologies and applications in sugar beet. Frontiers in Plant Science. 2016;**7**:900. doi:10.3389/fpls.2016.00900

[27] Couladis M, Tzakou O, Verykokidou E, Harvala C. Screening of some Greek aromatic plants for antioxidant activity. Phytotherapy Research. 2003;**17**:194-195. doi:10.1002/ptr.1261

[28] Kwak WJ, Han CK, Chang HW, Kim HP, Kang SS, Son KH. Loniceroside C, an anti-inflammatory saponin from *Lonicera japonica*. Chemical and Pharmaceutical Bulletin (Tokyo). 2003;**51(3)**:333-335. doi:10.1248/cpb.51.333

[29] Skotti E, Anastasaki E, Kanellou G, Polissiou M, Tarantilis PA. Total phenolic content, antioxidant activity and toxicity of aqueous extracts from selected Greek medicinal and aromatic plants. Industrial Crops and Products. 2014;**53**:46-54. doi:10.1016/j.indcrop.2013.12.013

[30] Masullo M, Pizza C, Piacente S. *Ruscus* genus: a rich source of bioactive steroidal saponins. Planta Medica. 2016;**82**:1513-1524. doi:10.1055/s-0042-119728

[31] Zhang Y, Zhang YJ, Jacob MR, Li XC,Yang CR. Steroidal saponins from the stem of *Yucca elephantipes*. Phytochemistry.2008;**69(1)**:264-270. doi:10.1016/j.phytochem.2007.06.015

[32] Taylor WG, Sutherland DH, Richards KW, Zhang H. Oleanane triterpenoid saponins of *Caraganaarborescens* and their quantitative determination. Industrial Crops and Products. 2015;**77**:74-80. doi:10.1016/j.indcrop.2015.08.034

[33] Tao W, Duan J, Zhao R, Li X, Yan H, Li J, Guo S, Yang N, Tang Y. Comparison of three officinal Chinese pharmacopoeia species of *Glycyrrhiza* based on separation and quantification of triterpene saponins and chemometrics analysis. Food Chemistry. 2013;**141(3)**:1681-1689. doi:10.1016/j.foodchem.2013.05.073

[34] Quirós-Sauceda AE, Ovando-Martínez M, Velderrain-Rodríguez GR, González-Aguilar GA, Ayala-Zavala JF. Licorice (*Glycyrrhiza glabra* Linn.) oils. In: Preedy V. editor. Essential Oils in Food Preservation, Flavor and Safety. Elsevier Inc. Amsterdam, Netherlands; 2016. pp. 523-530. doi:10.1016/B978-0-12-416641-7.00060-2

[35] Yang W, Hu Y, Wu W, Ye M, Guo D. Saponins in the genus *Panax* L. (*Araliaceae*): a systematic review of their chemical diversity. Phytochemistry. 2014;**106**:7-24.doi:10.1016/j.phytochem.2014.07.012

[36] Shin B-K, Kwon SW, Park JH. Chemical diversity of ginseng saponins from *Panax ginseng*. Journal of Ginseng Research. 2015;**39(4)**:287-298. doi:10.1016/j.jgr.2014.12.005

[37] Li K-K, Gong X-J. A review on the medicinal potential of *Panax ginseng* saponins in diabetes mellitus. RSC Advances. 2015;**5**:47353-47366. doi:10.1039/c5ra05864c

[38] Wenxi D, Shufang D, Xiaoling Y, Liming Y. *Panaxnoto ginseng* saponins suppress radiation-induced osteoporosis by regulating bone formation and resorption. Phytomedicine.2015:**22(9)**:813-819. doi:10.1016/j.phymed.2015.05.056

[39] Borah B,Phukon P, Hazarika MP, Ahmed R, Sarmah DK, Wann SB, Das A, Bhau BS. *Calamus leptospadix* Griff. a high saponin yielding plant with antimicrobial property. Industrial Crops and Products. 2016;**82**:127-132. doi:10.1016/j.indcrop.2015.11.075

[40] Meng D-L, Xu L-H, Chen C, Yan D, Fang Z-Z, Cao Y-F. A new resource of hepatic protectant, nor-oleanane triterpenoid saponins from the fruit of *Stauntonia brachyanthera.* Journal of Functional Foods. 2015;**16**:28-39. doi:10.1016/j.jff.2015.04.003

[41] Chen Y-F, Yang C-S, Chang M-S, Ciou Y-P, Huang Y-C. Foam properties and detergent abilities of the saponins from *Camellia oleifera.* International Journal of Molecular Sciences. 2010;**11**:4417-4425. doi:10.3390/ijms11114417

[42] Yoshiki Y, Kudou S, Okubo K Relationship between chemical structures and biological activities of triterpenoid saponins from soybean. Bioscience, Biotechnology, and Biochemistry. 1998;**62(12)**: 2291-2299. doi:10.1271/bbb.62.2291

[43] Dinchev D, Janda B, Evstatieva L, Oleszek W, Aslani MR, Kostova I. Distribution of steroidal saponins in *Tribulus terrestris* from different geographical regions. Phytochemistry. 2008;**69**:176-186. doi:10.1016/j.phytochem.2007.07.003

[44] Montero L, Ibáñez E, Russo M, di Sanzo R, Rastrelli L, Piccinelli AL, Celano R, Cifuentes A, Herrero M. Metabolite profiling of licorice (*Glycyrrhiza glabra*) from different locations using comprehensive two-dimensional liquid chromatography coupled to diode array and tandem mass spectrometry detection. Analytica Chimica Acta. 2016;**913**:145-159. doi:10.1016/j.aca.2016.01.040

[45] Doughari JH. Phytochemicals: extraction methods, basic structures and mode of action as potential chemotherapeutic agents. In: Rao V editor. Phytochemicals—A Global Perspective of Their Role in Nutrition and Health. InTech. Rijeka, Croatia; 2012. pp. 1-32. doi:10.5772/1387

[46] Ribeiro BD, Alviano DS, Barreto DW, Coelho MA. Functional properties of saponins from sisal (*Agave sisalana*) and juá (*Ziziphusjoazeiro*): Critical micellar concentration, antioxidant and antimicrobial activities. Colloids and Surfaces A: Physicochemical and Engineering Aspects. 2013;**436**:736-743. doi:10.1016/j.colsurfa.2013.08.007

[47] Bitencourt RG, Queiroga CL, Junior IM, Cabral FA. Fractionated extraction of saponins from Brazilian ginseng by sequential process using supercritical CO_2, ethanol and water. Journal of Supercritical Fluids. 2014;**92**:272-281. doi:10.1016/j.supflu.2014.06.009

[48] Cheok CY, Salman HAK, Sulaiman R. Extraction and quantification of saponins: A review. Food Research International. 2014;**59**:16-40. doi:10.1016/j.foodres.2014.01.057

[49] Moghimipour E, Handali S. Saponin: Properties, methods of evaluation and applications. Annual Research & Review in Biology. 2015;**5(3)**:207-220. doi:10.9734/ARRB/2015/11674

[50] Zana R. Dynamics of Surfactant Self-Assemblies: Micelles, Microemulsions, Vesicles and Lyotropic Phases. CRC Press. London, UK; 2005. 515 p. doi:10.1081/DIS-200067928

[51] Mitra S, Duncan SR. Cholesterol solubilization in aqueous micellar solutions of *Quillaja* saponin, bile salts, or nonionic surfactants. Journal of Agricultural and Food Chemistry. 2001;**49(1)**:384-394. doi:10.1021/jf000568r

[52] Bachran C, Bachran S, Sutherland M, Bachran D, Fuchs H. Preclinical studies of saponins for tumor therapy. In: Rahman A, Choudhar MI, Perry G, editors. Recent Advances in Medicinal Chemistry. Vol. 1. Bentham Science Publishers Ltd. Amsterdam, Netherlands; 2014. pp. 272-302. doi:10.1016/B978-0-12-803961-8.50009-9

[53] Walthelm U, Dittrich K, Gelbrich G, Schöpke T. Effects of saponins on the water solubility of different model compounds. Planta Medica. 2001;**67(1)**:49-54. doi:10.1055/s-2001-10876

[54] McClements DJ, Gumus CE. Natural emulsifiers—Biosurfactants, phospholipids, biopolymers, and colloidal particles: Molecular and physicochemical basis of functional performance. Advances in Colloid and Interface Science. 2016;**234**:3-26. doi:10.1016/j.cis.2016.03.002

[55] Kezwon A, Wojciechowski K. Interaction of *Quillaja* bark saponins with food-relevant proteins. Advances in Colloid and Interface Science. 2014;**209**:185-195. doi:10.1016/j.cis.2014.04.005

[56] Oda K, Matsuda H, Murakami T, Katayama S, Ohgitani T, Yoshikawa M. Adjuvant and haemolytic activities of 47 saponins derived from medicinal and food plants. Biological Chemistry. 2000;**381(1)**:67-74. doi:10.1515/BC.2000.009

[57] Gurfinkel DM, Rao AV. Soya saponins: the relationship between chemical structure and colon anticarcinogenic activity. Nutrition and Cancer. 2003;**47(1)**:24-33. doi:10.1207/s15327914nc4701_3

[58] Marrelli M, Conforti F, Araniti F, Statti GA. Effects of saponins on lipid metabolism: A review of potential health benefits in the treatment of obesity. Molecules. 2016;**21**:1404. doi:10.3390/molecules21101404

[59] Bissinger R, Modicano P, Alzoubi K, Honisch S, Faggio C, Abed M, Lang F. Effect of saponin on erythrocytes. International Journal of Hematology. 2014;**100(1)**:51-59. doi:10.1007/s12185-014-1605

[60] Du J-R, Long F-Y, Chen C. Research progress on natural triterpenoid saponins in the chemoprevention and chemotherapy of cancer. In: Bathaie SZ, Tamanoi F editors. The Enzymes, Vol 36. Natural Products and Cancer Signaling: Isoprenoids, Polyphenols and Flavonoids. 2014; Elsevier Inc. London, UK; pp. 95-130. doi:10.1016/B978-0-12-802215-3.00006-9

[61] Elekofehinti OO. Saponins: Anti-diabetic principles from medicinal plants—A review. Pathophysiology. 2015;**22**:95-103. doi:10.1016/j.pathophys.2015.02.001

[62] Vinay TN, Park CS, Kim HY, Jung SJ. Toxicity and dose determination of Quillaja saponin, aluminum hydroxide and squalene in olive flounder (*Paralichthysolivaceus*). Veterinary Immunology and Immunopathology. 2014;**158**:73-85. doi:10.1016/j.vetimm.2013.03.007

[63] Netala VR, Ghosh SB, Bobbu P, Anitha D, Tartte V. Triterpenoid saponins: a review on biosynthesis, applications and mechanism of their action. International Journal of Pharmacy and Pharmaceutical Sciences. 2015;**7(1)**:24-28.

[64] Zhao D. Challenges associated with elucidating the mechanisms of the hypocholesterolaemic activity of saponins. Journal of Functional Foods.2016;**23**:52-65. doi:10.1016/j.jff.2016.02.023

[65] Nasri H, Baradaran A, Shirzad H, Rafieian-Kopaei M. New concepts in nutraceuticals as alternative for pharmaceuticals. International Journal of Preventive Medicine. 2014;**5(12)**:1487-1499.

[66] Marciani DJ, Ptak RG, Voss TG, Reynolds RC, Pathak AK, Chamblin TL, Scholl DR, May RD. Degradation of *Quillaja saponaria* Molina saponins: loss of the protective effects

of a herpes simplex virus 1 subunit vaccine. International Immunopharmacology. 2002;**2(12)**:1703-11. doi:10.1016/S1567-5769(02)00192-3

[67] Lanzotti V, Barile E, Antignani V, Bonanomi G, Scala F. Antifungal saponins from bulbs of garlic, *Allium sativum* L. var. Voghiera. Phytochemistry. 2012;**78**:126-134. doi:10.1016/j.phytochem.2012.03.009

[68] Guil-Guerrero JL, Ramos L, Moreno C, Zúñiga-Paredes JC, Carlosama-Yepez M, Ruales P. Antimicrobial activity of plant-food by-products: A review focusing on the tropics. Livestock Science. 2016;**189**:32-49.

[69] Nabinejad A. Antibacterial effects of *Saponaria officinalis* extracts against avian pathogenic *Escherichia coli* (APEC). African Journal of Agricultural Research. 2013;**8(18)**:2068-2071. doi:10.5897/AJAR11.1390

[70] Korchowiec B, Gorczyca M, Wojszko K, Janikowska M, Henry M, Rogalska E. Impact of two different saponins on the organization of model lipid membranes. Biochimica et Biophysica Acta (BBA)—Biomembranes. 2015;**1848**:1963-1973. doi:10.1016/j.bbamem.2015.06.007

[71] Lacaille-Dubois MA, Wagner H. A review of the biological and pharmacological activities of saponins. Phytomedicine. 1996;**2(4)**:363-386. doi:10.1016/S0944-7113(96)80081-X

[72] Coleman JJ, Okoli I, Tegos GP, Holson EB, Wagner FF, Hamblin MR, Mylonakis E. Characterization of plant-derived saponin natural products against *Candida albicans*. ACS Chemical Biology. 2010;**5(3)**:321-332.doi:10.1021/cb900243b

[73] Berlowska J, Dudkiewicz M, Kregiel D, Czyzowska A, Witonska I. Cell lysis induced by membrane-damaging detergent saponins from *Quillaja saponaria*. Enzyme and Microbial Technology. 2015;**75-76**:44-48. doi:10.1016/j.enzmictec.2015.04.007

[74] Sen S, Makkar HPS, Muetzel S, Becker K. Effect of *Quillaja saponaria* saponins and *Yucca schidigera* plant extract on growth of *Escherichia coli*. Letters in Applied Microbiology. 1998;**27**:35-38.

[75] Arabski M, Węgierek-Ciuk A, Czerwonka G, Lankoff A, Kaca W. Effects of saponins against clinical *E. coli* strains and eukaryotic cell line. Journal of Biomedicine and Biotechnology. 2012; Article ID 286216, doi:10.1155/2012/286216

[76] Alberice JV, Funes-Huacca ME, Guterres SB, Carrilho E. Inactivation of *Alicyclobacillus acidoterrestris* in orange juice by saponin extracts combined with heat-treatment. International Journal of Food Microbiology. 2012;**159**:130-135. doi:10.1016/j.ijfoodmicro.2012.08.004

[77] Piacente S, Pizza C, Oleszek W. Saponins and phenolics of *Yucca schidigera* Roezl: Chemistry and bioactivity. Phytochemistry Reviews. 2005;**4**:177-190. doi:10.1007/s11101-005-1234-5

[78] Tenon M, Feuillère N, Roller M, Birtic S. Rapid, cost-effective and accurate quantification of *Yucca schidigera* Roezl. steroidal saponins using HPLC-ELSD method. Food Chemistry. 2017;**221**:1245-1252. doi:10.1016/j.foodchem.2016.11.033

[79] U.S. FDA Code of Federal Regulation, 21, part 172, subpart F—Flavoring Agents and Related Substances, §172.510 [Internet]. 2016. Available from: https://www.accessdata.fda.gov/scripts/cdrh/cfdocs/cfcfr/CFRSearch.cfm?fr=172.510 [Accessed: 2017-01-17]

[80] JETRO Specifications and Standards for Foods, Food Additives (Abstract) 2010 [Internet] 2011. Available from: https://www.jetro.go.jp/ext_images/en/reports/regulations/pdf/foodext2010e.pdf [Accessed: 2017-01-17]

[81] Sucharzewska D, Stochmal A, Oleszek W. The effect of *Yucca schidigera* extract on the physical structure and on the oxidative stability of sugar-candy foam products. Lebensmittel-Wissenschaft&Technologie. 2003;**36**:347-351. doi:10.1016/S0023-6438(03)00016-1

[82] Matusiak K, Oleksy M, Borowski S, Nowak A, Korczyński M, Dobrzański Z, Gutarowska B. The use of *Yucca schidigera* and microbial preparation for poultry manure deodorization and hygienization. J Environ Manage. 2016;**170**:50-59. doi:10.1016/j.jenvman.2016.01.007

[83] Böttger S, Hofmann K, Melzig MF. Saponins can perturb biologic membranes and reduce the surface tension of aqueous solutions: a correlation? Bioorganic and Medicinal Chemistry. 2012;**20(9)**: 2822-2828. doi:10.1016/j.bmc.2012.03.032

[84] Guo S, Kenne L, Lundgren LN, Rönnberg B, Sundquist BG. Triterpenoid saponins from *Quillaja saponaria*. Phytochemistry.1998;**48(1)**:175-180.

[85] WHO Food Additives Series: 48 Safety Evaluation of CertainFood Additives and Contaminants [Internet]. 2002 http://www.inchem.org/documents/jecfa/jecmono/v48je01.htm [Accessed: 2017-01-17]

[86] Tippel J, Lehmann M, von Klitzing R, Drusch S. Interfacial properties of *Quillaja* saponins and its use for micellisation of lutein esters. Food Chemistry. 2016;**212**:35-42. doi:10.1016/j.foodchem.2016.05.130

[87] FAO. 61st JECFA. *Quillaia* extracts [Internet]. 2004 http://www.fao.org/fileadmin/templates/agns/pdf/jecfa/cta/61/QUILLAIA.pdf [Accessed 2017-01-17]

[88] Patent US8808692B2 Compositions comprising immunoreactive reagents and saponins, and methods of use thereof [Internet] 2014. Available from: https://docs.google.com/viewer?url=patentimages.storage.googleapis.com/pdfs/US8808692.pdf [Accessed 2017-01-17]

[89] Patent US 5326579 A Process to remove cholesterol from dairy products [Internet]. 1994. Available from: https://docs.google.com/viewer?url=patentimages.storage.googleapis.com/pdfs/US5326579.pdf [Accessed 2017-01-17]

[90] Patent US 5370890 A Aqueous process to remove cholesterol from food products [Internet] Available from: https://docs.google.com/viewer?url=patentimages.storage.googleapis.com/pdfs/US5370890.pdf [Accessed 2017-01-17]

[91] Chang EJ, Oh HI, Kwak HS. Optimization of cholesterol removal conditions from homogenized milk by treatment with saponin. Asian-Australasian Journal of Animal Sciences. 2001;**14(6)**:844-849. doi:10.5713/ajas.2001.844

[92] Yang Y, Leser ME, Sher AA, McClements DJ. Formation and stability of emulsions using a natural small molecule surfactant: *Quillaja* saponin (Q-Naturale®). Food Hydrocolloids. 2013;**30**:589-596. doi:10.1016/j.foodhyd.2012.08.008

[93] Zhang J, Bing L, Reineccius GA. Formation, optical property and stability of orange oil nanoemulsions stabilized by *Quallija*saponins. LWT—Food Science and Technology. 2015;**64**:1063-1070. doi:10.1016/j.lwt.2015.07.034

[94] Zhang J,Bing L,Reineccius GA. Comparison of modified starch and *Quillaja* saponins in the formation and stabilization of flavor nanoemulsions. Food Chemistry. 2016;**192**:53-59. doi:10.1016/j.foodchem.2015.06.078

[95] Reichert CL, Salminen H, Leuenberger BH, Weiss J. Influence of heat on miscibility of *Quillaja* saponins in mixtures with a co-surfactant. Food Research International. 2016;**88**:16-23. doi:10.1016/j.foodres.2016.03.034

Chapter 7

Amino Acid-Based Surfactants for Biomedical Applications

Lídia Pinheiro and Célia Faustino

Additional information is available at the end of the chapter

http://dx.doi.org/10.5772/67977

Abstract

The growing demand for surfactants worldwide has a profound impact on the environment and public health. The quest for environmentally friendly "green" surfactants has driven research toward bio-based surfactants from renewable sources with improved performances and low toxicity. Amino acid-based surfactants (AAS) are a promising class of biocompatible and biodegradable surfactants for biomedical applications due to their improved safety profiles that meet the requirements of both physiological and ecological compatibility. Natural amino acids are chiral compounds and important raw materials for production of AAS. The amino acid pool allows the synthesis of multifunctional surfactants with chiral properties that can be tailored for specific technological and/or biomedical applications. The nature of the amino acid residue, the chirality, and the ability for hydrogen bond formation strongly influences the surface active properties and self-assembly behavior of AAS. This review summarizes recent developments in AAS structure-property relationships providing valuable information for modulation of the surface active and biological properties of AAS to meet specific biomedical applications. The interaction of AAS with biointerfaces and biological molecules is also addressed concerning cellular toxicity and potential therapeutic applications of AAS as antimicrobial agents, drug delivery vehicles, and a promising alternative to viral vectors in gene therapy.

Keywords: amino acid, surfactant, micelles, drug delivery, gene delivery

1. Introduction

Surfactants are surface active molecules characterized by a polar headgroup linked to a long hydrocarbon chain. According to the nature of their headgroup, surfactants are classified as nonionic, anionic, cationic, or zwitterionic [1]. The amphiphilic nature of surfactants is the basis of their characteristic properties, such as the ability to adsorb at interfaces, lower the surface tension of water or spontaneously self-assembly in aqueous systems forming micelles once the critical micelle concentration (CMC) is reached.

Surfactants are widely used as wetting agents, detergents, emulsifiers, and softeners in a wide range of industries [1]. The type of application is determined by the balance between the hydrophilic and hydrophobic regions in the surfactant molecule. Petrochemical-based surfactants represent the large majority of surfactants in the market today [2, 3]. The huge consumption of surfactants worldwide calls for sustainable surfactant production from natural renewable sources in order to reduce the impact on the environment, answers consumers' demand, and keeps up with increasing regulatory pressure. Thus, there is an urgent need for the development of novel environmentally friendly surfactants with low toxicity and improved performances based on natural building blocks that can be produced by clean and sustainable technologies.

Natural amino acids are potential building blocks for surfactant synthesis due to their availability, biocompatibility, and multifunctionality [4–6]. Protein hydrolysates from waste proteins are ideal raw materials for the production of amino acid-based surfactants thus contributing to the valorization of secondary products, to the reduction of pollution load, and to the sustainable development of the bioindustry [7, 8].

Amino acid surfactants (AAS) are biocompatible and biodegradable surfactants obtained by condensation of natural amino acids with fatty acids (or their derivatives) of oleochemical source [5, 9, 10]. Hydrolysis of triglycerides from animal fat or vegetable oils furnishes a wide variety of saturated and unsaturated fatty acids with straight hydrocarbon chains and an even number of carbon atoms due to their biosynthetic route [3, 11, 12]. The use of non-edible waste cooking oils is also a viable alternative, and further contributes to reduce the environmental burden [13].

Moreover, AAS can be produced in large scale by green chemistry approaches, including enzyme-catalyzed synthesis using immobilized lipases and proteases, although chemical processes still prevail due to high yields and low production costs [14–19]. Being specialty surfactants, AAS still represent a low market share, but the trend toward green surfactants entirely produced from natural renewable sources by environmentally friendly technologies can change this scenario [3].

2. Chemical structure and classification

The presence of an amino acid as the polar headgroup characterizes AAS. The 20 standard amino acids used as the building blocks of proteins are the natural choice as raw materials for the production of AAS. The proteinogenic amino acids are L-α-amino acids (except gycline, which is achiral) classified according to the nature of their side-chain. Due to the presence of both amino and carboxylic acid groups, amino acids are amphoteric compounds and exist as zwitterions at physiological pH.

http://dx.doi.org/10.5772/67977

The wide diversity of AAS, with different structures and properties, is only possible due to the nature and multifunctionality of the amino acid residue. According to the site of introduction of the hydrophobic chain and to the charge of the amino acid side-chain, anionic, cationic, or zwitterionic AAS can be obtained.

The hydrophobic chain can be introduced through acyl, ester, amide, or alkyl linkage [5, 9]. Due to the stability of the alkyl bond compared to biodegradable amide and ester bonds, it will not be considered further. Thus, introduction of the hydrophobic chain at the amino group by acylation with a fatty acid leads to anionic *N*-acyl AAS, while condensation of the carboxyl group of the amino acid with fatty alcohols or fatty amines produces cationic alkyl ester and alkyl amide AAS, respectively (**Figure 1**).

Moreover, according to the number of hydrophobic chains introduced, single-chain, double-chain, or dimeric (gemini) AAS can be obtained. Gemini AAS are made of two amino acid headgroups and two hydrophobic chains per molecule joined by a spacer chain at or near the

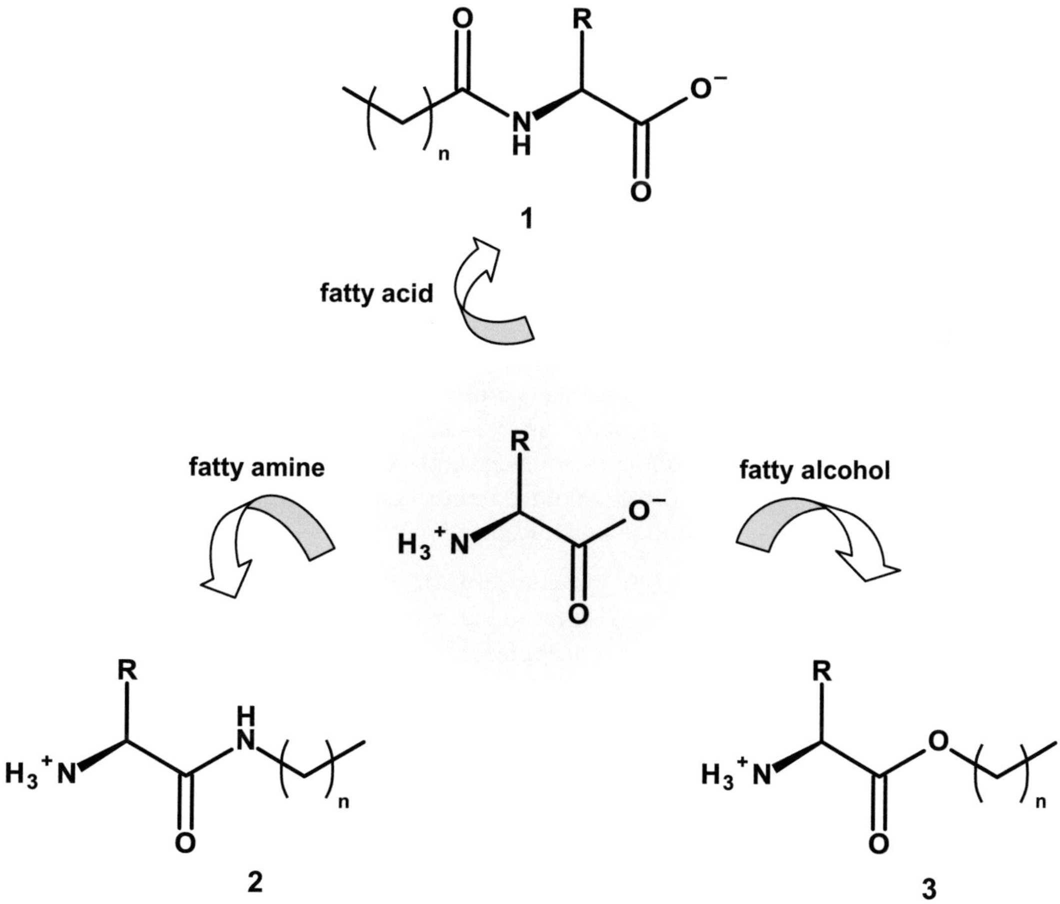

Figure 1. Types of amino acid-based surfactants: acyl (**1**), ester (**2**), and amide (**3**) bond derivatives.

headgroups. Gemini AAS show better performance compared to their monomeric counterparts, such as lower CMC, higher efficiency in surface and interfacial tension reduction, and higher solubilization capacity [10, 20]. Cystine, the dimer of cysteine where the thiol groups have been oxidized to form a disulfide bond, is a potential raw material for the synthesis of gemini AAS.

3. Properties and self-assembly behavior

3.1. Single-chain surfactants

The nature of the amino acid residue determines the main differences on the adsorption, aggregation, and biological properties among the AAS. The self-assembly of surfactants in aqueous media has been extensively studied and some correlations between AAS structure and surfactant properties have been established. Thus, increasing the length of the hydrophobic chain is usually accompanied by a decrease in the CMC, similarly to conventional surfactants, where the hydrophobic interaction is the driving force for the self-assembly process. For the AAS, non-covalent interactions arising from the side-chains of the amino acid residues provide additional contributions, and effective solvation of the headgroups is also a relevant factor influencing self-assembly [6, 10, 21].

The CMC is an important parameter for the biomedical application of AAS, since interactions with biointerfaces and cellular components largely differ in the presence of micelles or monomeric surfactant molecules. Although the CMC often decreases with increasing hydrophobicity of the amino acid, specific intra- and/or intermolecular interactions between the amino acid residues may alter this trend [22, 23].

The presence of aromatic or bulky substituents and the presence of hydrogen bond donor/acceptor groups can strongly influence molecular packing at interfaces and micelle stabilization. Intermolecular hydrogen bonding interactions between headgroups can occur for AAS with amino acid residues with hydroxyl, amido, amino, and carboxylic groups in their side-chains, contributing to their adsorption and micellization properties in solution [6, 10, 21].

The *N*-acyl phenylalanine AAS usually show lower CMC than the other *N*-acyl amino acids with the same acyl-chain length due to intramicellar π-π interactions between the aromatic rings of the amino acid side-chains that contribute for micelle stabilization. These interactions also occur in the solid state, contributing to the higher Kraft temperature of phenylalanine AAS. On the other hand, the complex self-assembly behavior of proline AAS is associated with van der Waals repulsions between the bulky side-chains while the incapability to form hydrogen bonds is reflected in high surface area as a result of less compact molecular packing at the air/water interface [10, 21, 23, 24].

The CMC values found for *N*-decanoyl leucine, methionine, proline, and serine followed a trend related to the hydrophilicity of the amino acid headgroup, with the less hydrophilic surfactant showing the lower CMC [24]. DLS measurements showed that the leucine and methionine derivatives formed spherical micelles with sizes around 3–5 nm, while the serine and proline surfactants formed larger supramolecular aggregates (11–14 nm), probably elongated rod-like micelles, due to the presence of the hydroxyl group on the serine AAS and conformational rigidity in the proline AAS. The steric hindrance of the pyrrolidine moiety in the proline AAS as well as the hydroxyl-containing side-chain in the serine derivative that prevent simple insertion of the hydrophobic chain inside red cell membrane were also responsible for their lower hemolytic activity [24].

The hydroxyl in the phenol ring of tyrosine dodecyl ester hydrochloride has a strong influence on the conformation of the molecule, providing more compact structures at the interface and contributing to lower interfacial area relatively to the more hydrophobic phenylalanine derivative, as demonstrated from molecular modeling studies [25].

For the surfactants with amide bonds, such as the *N*-acyl AAS, the capacity of the amide bond to participate in intra- and intermolecular hydrogen bonding can strongly influence the surfactant properties both at interfaces and in solution. The role of the amide bond for the self-assembly of AAS in solution, and their adsorption at the air/water interface and at solid surfaces, was studied by different techniques using sodium *N*-lauroyl glycinate and sodium *N*-lauroylsarcosinate as model surfactants [26]. The former can form intermolecular hydrogen bonds via the amide group but not the latter, due to the methylation of the amide nitrogen. The amide bond was found to contribute to the hydrophilicity of both surfactants, with slightly lower CMC being observed for the sarcosinate derivative due to the hydrophobicity of the additional methyl group. The ability of the glycinate surfactant to form intermolecular hydrogen bonds led to tighter packing at the air/water interface and at hydrophobic surfaces. The higher ionization degree found for the sarcosinate surfactant was also in accordance with a less tight packing of the surfactant in micelles in aqueous solution.

Besides micelles, AAS can form other supramolecular structures in aqueous medium due to the chiral nature of the amino acid residue, which may induce molecular packing into self-assembled tubules, rods, helical and twisted ribbons, or fibers. Moreover, the chiral nature of AAS allows them to interact selectively with enantiomers of chiral solutes, which has important practical implications. Chiral AAS are routinely employed as column-packing material in chiral chromatography for the separation of racemic mixtures and as enantiodiscriminating NMR solvents. The chiral discrimination of AAS has been reported in the solid state, in liquid crystalline phases, in Langmuir monolayers, and even in micelles [6, 10].

Racemic mixtures of the sodium salts of *N*-lauroyl phenylalanine and *N*-lauroyl valine, as well as *N*-acyl glutamic acid disodium salts, exhibited higher CMC than that of the pure enantiomers. The effect was attributed to the differences in the conformation of the amino acid moiety of surfactants at the micelle surface. Racemic *N*-stearoylserine also showed slightly higher CMC compared to the pure L- or D- enantiomers and strong variations in the circular dichroism spectra of the enantiomerically pure micelles suggested formation of a repetitive arrangement of the polar headgroups at the micellar surface stabilized by intermolecular hydrogen bonds between the amide groups [6, 10]. The preference for homochiral (L-L or D-D) over heterochiral (D-L) interaction depends on the stability of the hydrogen bond formed between the amino acid residues in combination with stereochemical effects.

The influence of chirality on the micellar properties of AAS seems to be dependent on structural effects related to the amino acid residue, such as size and stereochemical hindrance, since no differences have been found between the CMC of the racemic mixtures and that of enantiomerically pure enantiomers for the sodium salts of *N*-palmitoyl-phenylalanine, leucine, threonine, methionine, and proline, and potassium salts of *N*-lauroylalanine or cationic surfactants derived from *N*-lauroylarginine [23]. Supramolecular chirality has been successfully employed in asymmetric organic synthesis [27].

The AAS developed at the industrial scale are mainly anionic *N*-acyl AAS due to their mild properties, low toxicity, biodegradability, and facile synthesis by the Schotten-Baumman process involving condensation of the amino acid with a fatty acyl chloride (from a fatty acid) under aqueous alkaline conditions. Moreover, many *N*-acyl AAS have emulsifying and antimicrobial properties with potential value as additives for formulations in the food, pharmaceutical, and cosmetic industries.

Although most AAS reported in the literature are based on compounds made from condensation of an amino acid with pure fatty acids, many commercial AAS are produced from amino acid and/or fatty acid mixtures obtained from protein hydrolysates or triglyceride hydrolysis, respectively, due to cost advantage [28]. Often AAS with mixed fatty acid chains have superior performance when compared to pure compounds because of synergistic interactions. Cocoyl glycinate exhibited considerably lower CMC than the sodium salts of lauroyl glycinate, lauroyl sarcosinate, and *N*-stearoyl amino acids, which may be related to the presence of acyl glycinates of different acyl chain lenghts [19].

The sodium salts of *N*-acyl phenylalanines and *N*-acyl isoleucines prepared from fatty acid mixtures obtained from coconut, palm, palm kernel, jathropa, karanja, *Sterculia foetida*, castor, and high oleic sunflower oils exhibited superior surface active properties like surface tension, CMC, calcium tolerance, wetting power, foaming, and emulsion stability compared to reference surfactant sodium lauryl sulphate (SLS) [12, 29]. Except for the *N*-acylphenylalaninate from coconut oil, all the other *N*-acyl phenylalanines showed promising cytotoxicity against human cancer cell lines [12].

The *N*-acyl AAS from aspartic and glutamic acid are mild surfactants widely used in cosmetics and personal care formulations due to their low toxicity and mildness to the skin and eyes. The disodium salts of *N*-lauroyl aspartate and *N*-lauroyl glutamate show CMC values of 73 and 74 mmol/L, respectively, much higher than that of sodium *N*-lauroyl glycinate (14 mmol/L) due to the presence of the additional carboxylate group [30]. The CMC and the micellar ionization degree were also influenced by the choice of monovalent counterion, increasing in the order $Li^+ < Na^+ < K^+$ which is the opposite trend usually found for anionic surfactants in solution. These results were interpreted according to the hard and soft acid-base concept, i.e., the hard carboxylate headgroup binds stronger to the harder lithium cation, in agreement with the Hofmeister series [31].

Significant differences were found in the ability of the dicarboxylate surfactants to chelate calcium [30]. Lauroyl aspartate formed an intramolecular complex with calcium ions while lauroyl glutamate formed an intermolecular complex which resulted in higher calcium tolerance,

http://dx.doi.org/10.5772/67977

tight packing at the air/water interface, and very low surface tension values (below 30 mN/m). On the other hand, the ability to form intramolecular chelates favored adsorption at calcium-containing surfaces with lauroyl aspartate adsorbing strongly on hydroxyapatite while lauroyl glutamate showed weak adsorption [30].

The degree of neutralization also influences the properties of the dicarboxylate AAS. According to TEM studies, a gel network structure was formed both in the bulk phase and in the foam films of sodium lauroyl glutamate at a certain temperature and pH range [32]. Variation of gel strength in foam film with changes in pH and temperature influenced foam stability. Formation of a weak gel in foam film favored foam stability while formation of hard gel had the opposite effect. The highest foam stability, as well as minimum surface tension, were found at pH 7 when no gel was formed in solution suggesting a more dense arrangement of the surfactant molecules at the air/water interface of the foam film as described by molecular dynamics simulation.

N^{α},N^{ε}-Dioctyl lysinate salts with different counterions (Li^+, Na^+, K^+, Lys^+, and $Tris^+$) have been studied both in the dry state and in aqueous solution. The CMC was found to be nearly independent of the counterion, which had a strong influence in hemolytic activity. Surfactants interact with erythrocyte membranes in a biphasic way by protecting against hypotonic hemolysis at low concentrations but inducing hemolysis at higher concentrations [33]. All the compounds protected erythrocytes against hypotonic hemolysis, with HC_{50} values (concentration of surfactant that induces 50% hemolysis in isotonic medium) in the range 260.7–560.0 μg/mL that increased in the order $Na^+ < K^+ < Lys^+ < Li^+ < Tris^+$. The maximum protective concentration of each surfactant was close to its HC_{50} value and below the CMC value while the antihemolytic potency was around 35% except for the potassium salt which showed a value of 76% [33].

The structure of the dioctyl lysinate salts in the dry state depended on the size of the counterion. Large organic counterions favored lamellar arrangements while small inorganic counterions favored bicontinuous cubic structures [34]. The influence of acyl chain length on the dry state structure of diacyl lysine surfactants showed that long alkyl chains favored a lamellar structure while medium length chains produced cubic bicontinuous structures. On the other hand, short chains promoted formation of reverse hexagonal structure similar to that of Aerosol-OT in the dry state, a behavior that was attributed to lack of flexibility of the chain to adopt a packed conformation.

AAS can spontaneously form vesicles with other amphiphiles in aqueous media. Catanionic mixtures of sodium N^{α},N^{ε}-dilauroyl lysinate and dodecyltrimethylammonium bromide (DTAB) exhibited several single and multiphase regions [35]. Addition of increasing concentrations of the lysinate surfactant to pure DTAB solution leads to mixed micelle formation and micellar growth until a given mixing ration at which vesicles assemble and coexist with small micelles. In the DTAB-rich system, stable unilamellar vesicles were observed with an average size in the order of 30–40 nm according to self-diffusion measurements and cryo-TEM imaging.

On the other hand, the pure lysinate surfactant crystallized into micrometer-sized tubules upon cooling from an isotropic micellar solution that induce gelation of the system [36].

Hydrophobic interactions and hydrogen-bonding between the polar headgroups contributed to the stability and overall rigidity of the tubules. The chiral center in the amino acid headgroup was held responsible for tubular self-assembly; however, electrostatic interactions also played a role in the process since tubules were not formed at low pH when the surfactant exists mainly in the neutral form. The phase behavior of the lysine AAS showed the phase sequence micellar → hexagonal → lamellar → hydrated crystals which is expected for single-chain surfactants, suggesting that the double-chain lysinate AAS adopts an overall cone-shaped configuration instead of a cylindrical one [36].

Long chain cationic AAS from arginine are usually biodegradable surfactants showing antimicrobial activity and low toxicity. Several cationic single-chain arginine AAS have been studied, including N^{α}-acyl arginine methyl ester hydrochloride, arginine *N*-alkyl amide dihydrochloride, and arginine *O*-alkyl ester dihydrochloride, obtained by the synthetic pathways shown in **Figure 1**. For all the surfactants studied, increasing the hydrophobic chain length was accompanied by a decrease in the CMC, as expected. The CMC values and the surface tension at the CMC were lower than the ones found for commercial quaternary ammonium surfactants with the same alkyl chain length, where the cationic charge is closer to the α-carbon of the hydrophobic chain than in the arginine surfactants. For the same alkyl chain length, CMC was lower and molecular surface area at the air/water interface was higher for the dicationic surfactants compared to the monovalent compound which indicates less tight packing at the interface due to an increase in the inter- and intramolecular electrostatic repulsions among the headgroups [5].

Considering the type of linkage between the arginine headgroups and the hydrophobic chain, the ester bond resulted in surfactants with higher biodegradation rates when compared to the ones with amide bonds. Based on the hemolytic activity measured by HC_{50}, both the monocationic esters and the dicationic amides were classified as non-hemolyzing agents (HC_{50} < 1000 μg/mL compared to 4–15 μg/mL for commercial cationic surfactants) and nonirritating to the eyes according to the *in vivo* eye irritation Draize test [14].

The phase behavior of N^{α}-acyl arginine methyl ester hydrochloride showed the classical phase progression hexagonal → cubic → lamellar liquid crystal typical of single-chain surfactants, like DTAB, with increasing concentration in the micellar solution phase. Reversed vesicles in the lecithin- N^{α}-lauroyl arginine methyl ester-squalene-water system have also been reported [5].

Amino acid-glyceride conjugates with a glycerolipid structure have also been studied. The 1-monoacyl glyceroarginine AAS formed micelles with CMC values in the range 0.2–6 mmol/L that decreased with increasing acyl chain length [5]. The 1,2-diacyl glyceroarginine derivatives formed lamellar liquid crystals and their dispersions at 0.1% in water led to spontaneously self-assembly into stable multilamellar vesicles. The diacyl glyceroarginine AAS were also found to stabilize both water-in-oil (w/o) and oil-in-water (o/w) droplets, forming multiple emulsions that constitute potential alternatives to diglycerides and lecithins with additional antimicrobial properties [5, 14].

3.2. Gemini surfactants

Gemini AAS surfactants usually show better performance, such as lower CMC, lower surface tension, lower Kraft temperature, and higher solubilization power, when compared to their

http://dx.doi.org/10.5772/67977

monomeric counterparts. The solubilization power of micelles has pharmaceutical relevance since micellar solubilization of hydrophobic drugs improves their water-solubility and stability against chemical and/or enzymatic degradation, thus enhancing drug bioavailability. Gemini AAS also show a rich polymorphic phase behavior and a variety of self-assembled aggregates has been observed, such as micelles, bilayers, and vesicles, depending on the nature of the amino acid polar headgroup and on the lengths of the hydrophobic tail and the spacer chain [5, 10].

The properties of anionic gemini AAS were first reported by Tsubone and coworkers who studied the sodium salts of *N*-acyl AAS derived from aspartic acid [37]. Very low CMC values, in the micromolar range, were observed for these geminis as well as an inversion of the tendency of the CMC to decrease with increasing acyl chain length for surfactants with acyl chains longer than 14 carbons accompanied by an increase in surface tension values, due to the formation of small size soluble aggregates (dimmers) below the CMC. The premicellar aggregates were devoided of surface activity but decreased the concentration of free monomers, thus reducing the surface activity. Other unusual findings were the absence of a break in the conductivity-surfactant concentration plots for the aspartate geminis and an increase in the pH of the solution with surfactant concentration in the CMC neighborhood. These findings suggested protonation of the carboxylate anion with simultaneous release of Na^+ during micellization. Hydrogen bond formation between the carboxyl and the amide groups, leading to an increase in the size of the headgroup, was responsible for the inhibition of micellization. The phenomenon had already been observed for monomeric AAS of the kind and results from the characteristic surfactant structure containing both *N*-acyl amide and carboxylate groups [38]. Moreover, the skin irritation potential of the gemini surfactant was lower than that of the monomeric counterpart or the nontoxic sodium sarcosinate surfactant as determined from human response to *in vivo* closed patch tests.

Several gemini surfactants formerly derived from cysteine have been synthesized since the nucleophilic thiol group of cysteine can be readily oxidized to cystine by the formation of a disulfide bond. Cystine is a potential building block for gemini surfactants for biomedical applications, including controlled-release drug delivery systems. Depending on the hydrophobic chain length, either micelles or vesicles can be formed, and the disulfide bond can be easily cleaved by endogenous reduction agents, such as glutathione, thus regenerating the free thiol group and liberating the encapsulated bioactive agents. Sodium *N,N'*-didecanoyl- and *N,N'*-dilauroyl cystine has been prepared and the ease of reduction of the disulfide bonds of the gemini surfactants was used to control the surface properties and aggregation behavior of these switchable surfactants [39]. Reduction of the gemini surfactants with dithiothreitol led to vesicle disruption while oxidation of the corresponding monomers to gemini surfactants regenerated the vesicles.

Faustino and coworkers synthesized anionic *N,N'*-dicarbamoyl gemini AAS from cystine by condensing the disodium salt of the dimeric amino acid with octylisocyanate. Their behavior in aqueous media at physiological pH and interaction with biomolecules of pharmaceutical relevance was characterized by conductivity, surface tension, and fluorescence quenching methods with pyrene as probe. The gemini AAS were found to interact with bile acids, membrane phospholipids, oligosaccharides, and serum albumin protein [10, 40–42].

The gemini surfactant was less efficient in surface tension reduction than its monomeric counterpart, which was attributed to the film formation by the hydrocarbon chains of the former at the air/water interface so that they cannot adsorb effectively at the interface, an unusual phenomenon also found for other gemini surfactants from cystine with the same alkyl chain lenght [43]. Chirality was found to influence the surface active properties of the gemini AAS and their interaction with chiral biomacromolecules but not their micellar properties in solution since similar CMC values and Gibbs energy of micellization were obtained for gemini AAS derived from L- and D-cystine, and their racemate [40, 41]. On the other hand, a less favorable packing at the air/water interface for the racemic mixture compared to the pure enantiomers was suggested by the higher surface tension and the higher Gibbs energy of adsorption observed for the former [40].

The surfactant cystine dioctyl ester dihydrochloride showed remarkable surface activity with a CMC value of 14.2 μmol/L, which was about one order of magnitude lower than that reported for other cystine surfactants with the same alkyl chain length [44]. Results from Langmuir film balance experiments showed that the cationic gemini forms stable viscoelastic films at the interface with molecular modeling studies pointing to a tilted orientation of the surfactant at the interface. SEM studies suggest that the gemini surfactant forms elongated micelles in aqueous solution.

Cationic gemini surfactants derived from arginine by linking two long chain N^{α}-acyl-L-arginine residues through amide bonds to a diamine spacer of variable chain length, bis(Args), have been studied. Unlike their monomeric counterparts, aqueous solutions of the geminis exhibited unconventional aggregation behavior and two distinct CMC values were obtained from surface tension (CMC_1) and conductivity measurements (CMC_2). The higher CMC_2 values were consistent with the formation of regular micelles while the very low CMC_1 values could be attributed to non-globular small-size aggregates or to big lamellar-type aggregates according to the length of the spacer chain [5].

Recently, cationic gemini surfactants derived from lysine intended for biomedical applications were synthesized from N^{α}-lauroyl lysine or N^{ε}-lauroyl lysine by linking the monomers through amide bonds to 1,6-hexanediamine or spermidine as spacers. Their CMC values were similar to the ones obtained by conductivity measurements for cationic arginine geminis of the same alkyl chain and spacer lengths, and about one order of magnitude lower than the ones for their similar cationic monomeric counterparts, the N^{α}- and N^{ε}-lauroyl lysine methyl esters, respectively. For both type of lysine geminis, the position of the cationic charge, located either at the α-amino or at the side-chain ε-amino group of the lysine residue, did not significantly affect the CMC which was dependent only on the hydrophobic character of the surfactants [45].

4. Biomedical applications

4.1. Antimicrobial agents

Development of new antimicrobial agents is mandatory, on the face of the fast growth of drug-resistant bacteria and fungi [46]. Cationic amino acid-based surfactants, which mimic natural antimicrobial peptides, can be seen as promising alternative antibacterial and antifungal agents

http://dx.doi.org/10.5772/67977

when compared to the currently used antibiotic compounds [14, 46]. Antibacterial monocatenary, dicatenary and gemini surfactants, with a creative molecular design that goes through new modes of action and diverse targets makes the difference to the existing conventional antibiotics.

These cationic antimicrobial AAS show an optimal association between the cationic charge and the hydrophobic moiety, which is the key to its activity against bacteria, fungi, and yeast [46]. Conventional antibiotics usually target specific enzymes or DNA. However, cationic AAS interacts with cellular membranes, leading to depolarization, lysis, and cell death, probably resulting from an advantageous incorporation into the hydrophobic lipid bilayer, which hinder bacterial resistance [46–48].

Cationic AAS represent promising alternatives to the typical cationic surfactants derived from quaternary ammonium salts that had been long-time utilized as biocidal agents (being part of antiseptics, dressing, catheters, and sutures). The hemolytic activity and cytotoxicity of the latter does not make them convenient for biomedical applications. Unlike cationic AAS, they are not easily biodegradable, hence toxic to aquatic organisms. The antimicrobial activity of cationic AAS depends on their structures and size (namely, the amino acid residue and the chain length as key parameters), the molecule hydrophilic/lipophilic balance, and the cationic charge density [46].

The arginine amino acid is an optimum raw material to prepare cationic surfactants with significant antiseptic and biocidal properties, due to the presence of a guanidine side chain [14, 46, 49]. Pinazo and coworkers [14, 46, 50] developed different synthetic routes (chemical, enzymatic, or a combination of both) to prepare a broad range of single chain and gemini arginine-based surfactants. The obtained minimum inhibitory concentration (MIC) were lower than the corresponding CMC values, which suggests that the species interacting with the bacterial membrane are monomers instead of aggregates. The synthesized compounds, with a broad range of action, exhibited good inhibitory activity against Gram-positive bacteria, Gram-negative bacteria, and yeast.

In the single-chain arginine-based surfactants, the alkyl chain length influenced the antimicrobial efficiency. For long chain N^{α}-acyl arginine methyl or ethyl ester, the maximum activity of surfactants was obtained for an alkyl chain length of 12 carbons [14, 46]. For the arginine-*N*-alkyl amide surfactants the variation of MIC with the alkyl chain was less noticeable than in the former compounds (whose structural difference from the amide surfactants lies in the number of positive charges per molecule). In contradiction with the preceding compounds, arginine-*O*-alkyl ester dihydrochlorides showed a very pronounced reduction in the antimicrobial activity, as the ester bond (instead of an amide bond) linking the hydrophobic group with the polar head can be easily hydrolyzed by bacteria.

On the other hand, conjugates of arginine with 1-monoacyl- and 1,2-diacyl-glycerolipids which can be considered analogs of partial glycerides and phospholipids, combine the physicochemical properties of glycerol derivatives and polar arginine-based surfactants. These attributes confers them some advantages over common phospholipids, namely, endowing the amphiphilic structure with antimicrobial activity due to the cationic features related to arginine [46].

Cationic lysine-based surfactants, in which the hydrophobic part is connected to the carboxylic lysine group through an ester or amide bond, have also been studied [46, 50, 51]. Several lysine-derived surfactants such as a lauroyl amide, a guanidinylated lauroyl amide, and a polyol-modified carbohydrate-templated lauroyl amide of lysine were tested using a group of clinically relevant isolates of Gram-positive (including MRSA and MRSE) and Gram-negative bacteria [51]. The results revealed that the substitution of lysine by carbohydrate-templated lysine analogs improved the hydrophobicity of the polar group and reduced the antibacterial potency of the corresponding cationic lipid. However, enhancement of the antibacterial activity was observed by guanidinylation of the two lysine amino groups.

Other studies concerning the effects of the position of the cationic charge on the biological properties of these compounds have been performed with cationic N^{ε}-acyl lysine methyl ester, N^{α}-lauroyl lisine methyl and ethyl ester hydrochloride analogs, N^{α}-lauroyl-N^{ε}-trimethyl lysine derivatives, and N^{ε}-miristoyl-N^{α}-trimethyl lysine methyl ester surfactants [46, 52]. The N^{α}-lauroyl lysine methyl and ethyl ester hydrochlorides presented MIC values in the same range as those of arginine analogs, explaining the wide spectrum of antimicrobial activity against Gram-positive and Gram-negative bacteria of cationic AAS. The results obtained showed that the stereochemistry of these surfactants did not influence their antimicrobial behavior. The maintenance of bioactivity regardless of optical purity is an advantageous issue, considering the difficult task in isolation of pure diastereoisomers [46].

N^{α}-Lauroyl-N^{ε}-trimethyl lysine derivatives displayed similar antimicrobial activity, in spite of the introduction of three methyl groups into the amino group of these molecules, probably to their identical cationic charge density. However, considering the pK_a values of the N^{ε}-acyl lysine methyl ester derivatives, their cationic charge density was expected to be lower than the N^{α}-lauroyl-N^{ε}-trimethyl lysine surfactants. Transferring the alkyl chain from the α-amino to the ε-amino group of lysine, considerably weakened the antimicrobial activity of N^{ε}-acyl lysine methyl ester derivatives, displaying no inhibitory effects upon Gram-negative bacteria. Furthermore, fixing the cationic charge in the N^{ε}-miristoyl-N^{α}-trimethyl lysine methyl ester surfactants, led to an enhanced antimicrobial activity.

Thereby for amino acid-based cationic surfactants in which the cationic charge is found on a protonated amine group, the antimicrobial activity is influenced by the pK_a of the latter [46]. Good inhibitory effects can be observed for surfactants with pK_a values higher than 9, whereas for those with pK_a values lower than 7, the antimicrobial efficacy is less pronounced, especially against Gram-negative microorganisms [46]. Generally, antimicrobial surfactants have frequently less impact on Gram-negative bacteria, because the lipopolysaccharide-packed outer envelopes hamper the access of amphiphilic compounds.

Cationic surfactants based on L-tryptophan and on L-tyrosine have been proved to be excellent gelators, revealing notable bactericidal properties [46]. Considering the cationic surfactants based on L-tryptophan with chloride as the counterion, optimum inhibitory features (against Gram-positive and particularly against Gram-negative with MICs of 0.5–5.0 μg/mL) were found for alkyl chain lengths of 10–14 carbons. However, molecules with alkyl chain length of 15 and 17 carbons still showed activity against Gram-positive bacteria, but not inhibiting the growth of Gram-negative bacteria. Exchanging the chloride counter-ion for the more hydrophobic organic carboxylates increased the activity against Gram-positive bacteria and fungi, improving also the biocompatibility towards eukaryotic cells [46].

Hydrogelator surfactants based on L-tryptophan and on L-tyrosine were used as templates for *in situ* synthesis of silver nanoparticles in order to increase the antimicrobial power. Since pure compounds only disturbed Gram-positive bacteria, the supramolecular assemblies of silver nanoparticles allowed the development of soft nanocomposites showing a wider bioactivity range for both Gram-positive and Gram-negative bacteria [53].

Antimicrobial activity for cationic surfactants from phenylalanine and tyrosine, in which the alkyl chain is linked to the carboxylic group of the amino through an ester bond, has been reported [54]. The antimicrobial activity was high for Gram-positive bacteria and low for Gram-negative bacteria. Antibacterial properties were affected by the alkyl chain length (increasing with this descriptor) and by both electrostatic and hydrophobic interactions between surfactants and the bacterial membranes.

The enhanced antimicrobial activity of cationic gemini AAS when compared with their monomeric counterparts can possibly be explained by their low CMC values, good solubility, the presence of two positively charged headgroups, and two hydrophobic chains per molecule [14, 20, 46, 47].

The antimicrobial behavior of gemini surfactants is influenced by a number of factors such as the length of the spacer chain, the length of the alkyl chain, the site where the cationic charge is positioned, and the net cationic charge of the molecules. According to the majority of the studies reported [46], the antimicrobial activity of the gemini surfactants is generally higher than of the corresponding monocatenary compounds, on account of their structural and functional characteristics. Growth inhibition of a comprehensive array of microorganisms (including Gram-positive and Gram-negative bacteria) was observed for a cationic gemini surfactant prepared by condensation of *N*-lauroyl glycine betaine with cystine dimethyl ester hydrochloride, with MICs ranging from 0.125 to 16 μg/mL [46].

Cationic gemini surfactant from arginine (consisting of N^{α}-acyl arginine linked by amide bonds to a polymethylene spacer chain) and from lysine (consisting of N^{α}-acyl- N^{ε}-acyl lysine with a hexamethylene or a spermidine spacer linked by amide bonds to the carboxyl groups) also displayed antimicrobial activity against a wide range of Gram-positive and Gram-negative bacteria [45, 46]. When the acyl chain was kept constant, the antimicrobial activity was shown to decrease with long spacer chains: the longer the spacers, the greater the ability to form viscous solutions enclosing large aggregates. Since big aggregates hardly interact with erythrocyte membranes, gemini surfactants with long spacers are much less hemolytic than their single-chain counterparts [55].

Concerning the alkyl chain length (whose influence is similar in monocatenary homologs) gemini surfactants with 10–12 carbon tails showed the best performance in terms of antimicrobial behavior.

Regarding the location of the cationic charge, arginine-based gemini surfactants usually display a higher antimicrobial efficiency when compared with the monomeric compounds.

The net cationic charge appreciably affected the antimicrobial activity of several gemini compounds, increasing with the pK_a value of the molecules, and modulating their capacity to disrupt the bacterial membrane, a similar pattern to that shown by their monomeric counterparts [20, 45, 46].

Antimicrobial gemini surfactants from arginine and lysine have shown lower hemolytic activity than their single chain homologs [45, 46], which seem to depend on the alkyl chain length, and also on the spacer length and cationic charge density (for the same alkyl chain length). Aggregate size in solution (which depends on the molecular architecture of the surfactants) is another feature influencing the hemolytic activity; big micellar aggregates in aqueous medium hinder the interaction with biological membranes [46, 55]. Gemini surfactants with short spacer chains are more hemolytic than their single chain homolog, whereas gemini surfactants with long spacers are much less hemolytic than their single chain counterpart [55].

Catanionic mixtures of oppositely charged surfactants have shown to improve physicochemical and biological properties, when compared to the individual components. Within this framework, mixtures of lichenysin (an anionic biosurfactant) and two amino acid-based gemini cationic surfactants (N^{α},N^{ω}-bis(N^{α}-lauroyl lysine) α,ω-hexylendiamide and N^{α},N^{ω}-bis(N^{α}-lauroyl α,ω-propylendiamide) were explored [56]. Lichenysin is a cyclic lipopeptide produced by *Bacillus licheniformis* similarly to surfactin, with a high surface tension activity in water (29 mN/m) and a very low CMC (15 mg/L), but without its antimicrobial activity.

The antimicrobial activity of the surfactant binary systems were evaluated *in vitro* against a wide range of Gram-negative and Gram-positive bacteria (including MRSA strains) and *Candida albicans*. A significant bacterial growth inhibition was observed for the 8:2 lichenysin-N^{α},N^{ω}-bis(N^{α}-lauroyl α,ω-propylendiamide) mixture, showing a clearly synergistic effect. The differences between the two cationic AAS mixtures with lichenysin, in terms of the synergistic effect, can be ascribed to the different pK_a values of the surfactants. These results suggested a "hybrid surfactant" formation which could produce a more powerful hydrophobic interaction with the lipid bilayer, with additional stronger electrostatic interaction due to the presence of the guanidine group, present in the mixture and acting similar to a cationic surfactant [56].

Moreover, the antimicrobial properties of cationic gemini AAS may be enhanced by the use of cosurfactants, reinforcing the potential for biomedical applications of amino acid-derived surfactants.

On the other hand, as the negative charge density at the cell membrane in fungi is lower than in bacteria, the majority of cationic AAS do not exhibit antifungal properties. However, alanine-based Gemini surfactants were effective in preventing the formation of mycoses on mucous membranes of patients with suppressed immunity caused by different strains of *Candida albicans* with deletions of gene-encoded multidrug resistance transporters [57]. The obtained good results suggested their use as surface-coating agents against fungal colonization.

Cationic amino acid-based surfactants have also been proved to have antiviral activity [14]. Acyl amino acid derivatives produced inhibition on influenza neuraminidase. On the other hand, several N^{α}-palmitoyl amino acids, which have been incorporated into model membranes, seemed to influence the transition temperature between the bilayer to hexagonal phase, a property linked to antiviral activity against the Cantell strain of the Sendai virus (parainfluenza type 1) [14].

4.2. Drug delivery

To enable controlled or responsive self-assembly systems with special characteristics, new functional surfactants or mixtures of different types of surfactants are constantly being developed and formulated.

Nanocarriers have gained recent widespread interest due to their targeted drug delivery, hence positively impacting on the systemic side effects often seen by avoiding other organs, having also the advantage of protecting the drugs from degradation and increasing drug solubility. AAS can be promising novel biomaterials in drug delivery systems, given their biocompatible properties and low cytotoxicity.

Cells usually take up drug carriers through endocytosis that limits the internalized active compounds to vesicles (endosomes). Surface properties, such as hydrophobicity and surface charge, have a major impact on cellular uptake of particulate drug delivery systems, therefore the incorporation of charged surfactants into these carriers might improve targeting to specific cells. In addition, surfactants with pH-responsive membrane-disruptive activity may further destabilize endosomal compartments [58].

The membrane-disruptive activity of N^{ε}-myristoyl lysine methyl ester and N^{ε}-palmitoyl lysine methyl ester surfactants was evaluated using erythrocytes as a model of an endosomal membrane [58]. Due to the positive charge on the α-amino group of lysine, both surfactants showed pH-responsive hemolytic activity. The overall hemolysis results suggested that both surfactants might achieve maximum membrane lytic activity in the late endosomes, and the hemolytic kinetics demonstrated their ability to disrupt endosomal membranes before vesicular evolution from endosomes to lysosomes. These outcomes identify these lysine surfactants as potential bioactive excipients in drug delivery systems [58].

In the same context, a series of chitosan–tripolyphosphate nanoparticles for intracellular drug delivery were designed using two pH-sensitive cationic AAS from the family of N^{α},N^{ε}-dioctanoyl lysine as bioactive compounds [59]. The results showed that by inserting the lysine-based amphiphiles into chitosan nanoparticles, pH-sensitive membranolytic and potentially endosomolytic nanocarriers were developed, which, therefore, demonstrated ideal viability for intracellular drug delivery. The enhanced kinetics of the hemolytic activity supported the ability of these functional nanodevices to disrupt endosomal membranes before vesicular evolution from endosomes to lysosomes, where many drugs may suffer degradation.

On the whole, the results suggested the possible potential of these pH responsive nanocarriers to promote an improved delivery of bioactive compounds to the intracellular compartments, although further *in vitro* and *in vivo* studies are needed to substantiate this hypothesis [59].

Cationic vesicular systems prepared from biocompatible diacylglyceroarginine surfactants can be eventually used as carriers in controlled drug release formulations [46, 60]. These vesicles were able to encapsulate drugs such as ciprofloxacin, with percentage of encapsulated drug depending on both the physicochemical properties of the carrier and the nature of the drug. Antimicrobial activity of empty and ciprofloxacin-loaded vesicles (against some Gram-positive and Gram-negative bacteria) was noticeable, with drug-loaded vesicles showing similar or higher bioactivity than the free drug solution. Additionally, the encapsulated ciprofloxacin preserved its antimicrobial activity. Adding dipalmitoyl phosphatidylcholine as a membrane additive diminished the antimicrobial power of the cationic vesicles without drug, but improved the antimicrobial activity of vesicles loaded with ciprofloxacin. The dual pharmacological functions (related to the nature of the encapsulated drug and related to the intrinsic antibacterial properties of the surfactant-based carriers) turned these formulations into innovative potential candidates for drug delivery.

A synergistic formulation, combining a natural antimicrobial cationic surfactant from lysine (N^{ε}-myristoyl lysine methyl ester) with the sodium salt of hyaluronic acid was developed in order to be used as a coating for viscose fabric in wound healing and textile medicine [61]. The amount of amine groups deposited on the viscose fabric surface is a key factor when aiming for an antimicrobial functionalization of textiles. The interaction studies proved that the lysine-based surfactant and the biopolymer formed a complex bearing a slightly positive charge at neutral pH, and the viscose samples thus treated showed very pronounced antimicrobial properties when tested against several Gram-positive and Gram-negative bacteria, and some pathogenic fungi.

Novel AAS derived from bis(carboxymethyl) lysine with saturated and polyunsaturated fatty acyl chains of variable chain length and unsaturation degree attached at the ε-amino group were developed to improve solubilization of a water insoluble anticancer drug [62]. Their cytoxicity was evaluated *in vitro* by the MTT and LDH assays on endothelial cells. The arachidonoyl and pentacosanoyl derivatives were less cytotoxic than polysorbate 80 used as the model solubilizer. The alkyl chain length and the unsaturation degree strongly influenced toxicity. The saturated surfactants showed similar hemolytic activity, due to their low CMC values and the linear configuration of their hydrophobic chain. The arachidonoyl and 10,12-pentacosadinoyl derivatives were less hemolytic than polysorbate 80.

The nonadecanoyl, pentacosanoyl, and 10,12-pentacosadinoyl derivatives were found to increase drug solubility from <0.15 μg/mL up to 7 mg/mL, with 46% (w/w) drug loading, which was attributed to their linear and flexible hydrophobic chain configuration, in accordance with the molecular modeling studies. The potential use of these surfactants as solubilizers is dependent on the selection of the hydrophobic moiety based on the compromise between the strength of the hydrophobic interaction with the drug, leading to improved solubility, and the affinity for the cell membrane leading to toxicity [62].

As invasive fungal infections are a major cause of concern in immunocompromised patients, AAS can be considered interesting substitutes for the solubilization of amphotericin B (AmB), a hydrophobic polyene antibiotic used in the therapy of systemic fungal infections. Due to its poor water solubility, AmB is commercialized as a colloidal suspension using sodium deoxycholate as the solubilizing agent. However, severe toxic effects of this formulation are associated with AmB aggregation and thus more suitable delivery vehicles are required. The anionic *N,N′*-bis(octylcarbamoyl) gemini AAS derived from cystine was found to form micelles at a lower CMC than the bile salt under physiological mimetic conditions, being also a better solubilizing agent for AmB [63]. The increased solubility of the drug in the gemini micellar solutions is due to the dimeric structure of the surfactant which contributes to a higher hydrophobicity and thus to a higher molar fraction of gemini surfactant in the micellar form as a result of a lower CMC. The gemini micelles solubilized AmB in a monomeric form, contributing to a less toxic formulation, although AmB efficacy was slightly reduced as indicated by MIC values.

Equimolar mixtures of the same anionic gemini AAS with bile salts sodium cholate and sodium deoxycholoate were evaluated as potential delivery agents for AmB [48]. Results showed that mixed micellar systems improved the solubilization of AmB (in its monomeric and less toxic form), and exhibited *in vitro* antifungal activity against *Candida albicans* comparable to that of commercial formulation. The potential safety profile of the gemini AAS and the possibility of reductive cleavage of the disulfide bond to control drug release from gemini AAS micelles, turn these formulations (either pure or in combination with other amphiphiles promoting synergistic interactions) into novel and promising drug delivery systems for the solubilization of amphiphilic drugs sparingly water-soluble other than AmB, contributing to increase the therapeutic window of the drug [48, 63].

4.3. Transfection vectors

DNA compaction in livings cells is a critical process, being important to regulate cellular activities through gene transcription, cellular proliferation, and differentiation. The ability to compact DNA protects it against enzymatic degradation and releases it after reaching the desired compartment in the target cell are crucial requirements for the design of efficient vectors for gene delivery.

Gene transfer vectors commonly used are mostly based on viruses which arises considerable related biosafety concerns associated with carcinogenicity, immunogenicity, and broad tropism. Hence, non-viral vectors such as cationic liposomes offer a nonimmunogenic and safe method for systemic gene delivery but are, in general, less efficient than viral vectors [14, 20, 46, 64].

DNA molecules are known to interact with single or double chain cationic surfactants, as well as with cationic gemini surfactants. Hydrophobic and electrostatic interactions occur between DNA and surfactants: the cationic group promotes the displacement of sodium cations nearby the nucleic acid, whereas hydrophobic interactions take place between the alkyl tails of surfactants; these two cooperative mechanisms promote the formation of complexes between DNA and surfactants with potential application in gene therapy [64].

AAS conjugated with biogenic (poly)amines, such as spermidine and spermine, with the amino acid residue acting as a linker between the hydrophobic chain(s) and the hydrophilic headgroup, have been developed as synthetic alternatives to viral vectors. The best known example of this class is the dioctadecylamidoglycylspermine (DOGS), an efficient transfection agent with glycine as spacer [20].

Cationic gemini AAS are able to bind DNA with several advantages compared to classic monovalent surfactants: lower cellular toxicity, lower CMC, and higher tendency to self-assemble [65]. Addition of a helper lipid (dioleyl phosphatidylethanolamine (DOPE)), induces polymorphic phase behavior, with the appearance of inverted micellar and cubic structures, leading to an increased transfection efficiency.

Gemini surfactants, *N,N*-bis(dimethylalkyl)-α,ω-alkanediammonium halide derivatives, which are known to be flexible vectors for non-viral gene delivery, have been modified at the spacer chain by introduction of an amino acid (glycine or lysine) to improve transfection efficiency. Gene delivery efficiency was evaluated in epithelial cells for topical cutaneous and mucosal applications, in the presence of DOPE. The superior performance of these spacer-substituted gemini surfactants in transfecting epithelial cells might be attributed to their better flexibility and biocompatibility conferred by the amino acid residue, when compared to the surfactants possessing unsubstituted spacers. These results demonstrate the feasibility of using amino acid-substituted gemini surfactants as gene carriers for the treatment of diseases affecting epithelial tissue [65].

Transfection efficiency has also been attempted *in vitro* in several cell lines, including human hepatocarcinoma and human breast adenocarcinoma, using cationic AAS bearing serine, alanine, and β-alanine headgroups [66]. The transfection efficacy was more significant for cationic AAS with alanine and β-alanine headgroups than with their serine headgroup counterparts, presumably due to the enhanced sensitivity of DNA associated with the hydroxyl-containing serine headgroup [66]. Gene transfer efficacies of cationic amphiphiles can be significantly modulated by minor structural variations in both the polar headgroup and hydrophobic tail regions of the surfactant.

Cationic liposomal formulations composed of a mixture of dioleoyl trimethylammoniumpropane (DOTAP) and cholesterol (Chol), and a pH-sensitive formulation comprising DOTAP, Chol, DOPE, and cholesteryl hemisuccinate (CHEMS) were developed as a new gene delivery system to plasmid DNA precondensed with arginine *N*-lauroyl amide dihydrochloride, along with the incorporation of blood protein transferrin (Tf) [67].

The transfection efficiency of these systems was directly related with the presence of the nontoxic arginine surfactant and the lipidic composition. Better transfection profiles were found for the complexes based on the pH-sensitive liposomal formulation. The pair DOPE:CHEMS is believed to act synergistically with the arginine surfactant and Tf, contributing to the escape of DNA complexes from the endosome, therefore improving the transfection, when compared to complexes composed of DOTAP:Chol liposomes.

4.4. Interactions with biomolecules and biointerfaces

Interactions of AAS with biological systems are relevant for potential biotechnological and biomedical applications. In the biological domain, the interaction between proteins and surfactants is of great importance since it can clarify the action of surfactants as denaturants and solubilizing agents for proteins [20]. Proteins are known to cooperatively bind many surfactants forming a protein-surfactant complex where the hydrophobic moieties of the surfactant cause protein unfolding by interacting with the non-polar amino acid residues [40].

The different interactions between glutamic acid-based gemini surfactants and hemoglobin, when compared with their corresponding single-chain homolog, were reported in the literature [20]. The weaker denaturing ability of gemini surfactants concerning hemoglobin can be assigned to their large size, and the extension of denaturation decreased when spacer length increased.

http://dx.doi.org/10.5772/67977

Interactions between anionic cystine-based gemini surfactants with the globular protein BSA were investigated, proving to be influenced by temperature and pH [40]. The gemini surfactant stereochemistry also affected the interactions, since the association of enatiomerically pure compounds with BSA is favored when compared to the racemic mixture, and pure L-stereochemistry is preferred over D-stereochemistry. On the other hand, the cationic gemini surfactant cystine dilauroyl amide dihydrochloride, with longer dodecyl tails but a cationic headgroup, showed a weaker interaction with BSA, with an association constant of 1.68×10^4 L/mol according to fluorescence quenching data [63].

The interest in enhanced properties of amphiphilic compounds in the biomedical areas has led to the development of mixed surfactants systems formulations [10]. Mixed micelles comprising bile salts solubilize significant biological compounds, such as cholesterol and fatty acids, representing promising drug delivery systems. The mixed micelle formation in basic solutions between an anionic gemini AAS derived from cystine and bile salts sodium cholate and sodium deoxycholate has been studied [10]. Micellization in these mixtures was found to depend not only on the hydrophobic effect, which aims at minimizing the hydrophobic surface, but also on hydrogen bonding ability, which is determined by some structural factors, including the number, position, and orientation of the hydroxyl groups of bile salts and packing geometry restrictions.

Studies have also been performed with mixed systems comprising the anionic gemini AAS and phospholipids. Synergism was found for surfactant mixtures between the gemini and diheptanoyl phosphatidylcholine (DHPC, a micelle-forming lipid) as well as for surfactant mixtures with dimyristoyl phosphatidylcholine (DMPC, a vesicle-forming lipid). These findings were due to the reduction of electrostatic repulsions between the anionic headgroups of the surfactant as a result of intercalation of the zwitterionic phospholipids in the mixed micelles [42]. Structural effects can also be involved since the short disulfide spacer draws the two hydrophobic chains of the gemini molecule close together thus increasing alkyl chain density and also the charge density of the headgroups, leading to strong intermolecular interactions with other amphiphiles in solution.

Biocompatible thermoresponsive gels, produced by mixtures of the ethyl(hydroxyethyl)cellulose polysaccharide and both monomeric and gemini arginine surfactants bis(Args) have been described [68]. When compared with the monomeric surfactant, the gemini compound showed superior gel formation capacity (needed to induce a sol-gel transition), generating thermoresponsive gels at concentrations 1000-fold lower. The cytotoxicity of the polymer-surfactant systems, evaluated through *in vitro* experiments on a human epithelial cervical carcinoma cell line, was significantly compensated by their superior efficiency at low concentrations. This fact seems particularly interesting for applications requiring temperature-induced thickening useful in pharmaceutical and biomedical areas.

Aqueous dispersions of pure gemini surfactants from arginine or mixtures of bis(Args) with phospholipids lead to stable cationic colloidal systems with a promising use as drug delivery systems [20]. While single chain surfactants and gemini with short spacer chains promoted solutions with micellar aggregates, gemini with long spacers gave rise to large aggregates promoting viscous solutions or gels [55]. As big aggregates do not interact so easily with biological membranes, gemini surfactants with long spacers show lower cytotoxicity, thus antimicrobial and hemolytic activities are strongly affected by aggregates size.

5. Conclusions

AAS possess good surface active and emulsifying properties, low toxicity, and high biodegradability, which are attractive properties for applications in food, personal care products, and pharmaceutics. AAS are based on naturally occurring renewable sources having a strong influence in their environmental impact, and their preparation is economically feasible. Moreover, their wide structural diversity and different physicochemical and biological properties can be tailored to meet a specific application by appropriate choice of the amino acid residue and linkage of the hydrophobic chain.

The outcome of numerous studies about the cationic AAS, particularly gemini compounds, placed them as ideal candidates for biomedical applications that require positively charged amphiphiles, since they show promising biological properties, namely, antimicrobial and DNA transfecting. Given their distinctive physicochemical and biological properties, new possible pharmaceutical devices based on cationic AAS may be considered a viable alternative to the classical formulations, showing good stability, low hemolytic effects, and also a natural antimicrobial activity, which is not provided by conventional ones.

However, since many *in vitro* tests used to measure the toxicity of AAS are inconclusive, and that these formulations are intended for human use, more *in vivo* tests should continue to be conducted. The correct choice and combination of cell lines and bioassays in toxicity studies for a safe and reliable screen of AAS with potential interest in pharmaceutical industry is thus critical.

Notwithstanding, future perspectives point to the preparation of a larger library of compounds for better robustness in biomedical applications aiming at rationally designing and developing more effective therapeutic agents and delivery systems based on AAS.

Author details

Lídia Pinheiro* and Célia Faustino

*Address all correspondence to: lpinheiro@ff.ulisboa.pt

Research Institute for Medicines (iMed.ULisboa), Faculty of Pharmacy, Universidade de Lisboa, Lisbon, Portugal

References

[1] Rosen MJ. Surfactants and Interfacial Phenomena. 3rd ed. New Jersey: Wiley Interscience; 2004. 444 p. DOI: 10.1002/0471670561.

[2] Morán C, Pérez L, Pons R, Pinazo A, Infante MR. Aminoacids, lactic acid and ascorbic acid as raw materials for biocompatible surfactants. In: Kjellin M, Johansson I, editors. Surfactants from Renewable Resources. Chichester: Wiley; 2010. p. 85-107. DOI: 10.1002/9780470686607.ch5.

[3] Foley P, Kermanshahi A, Beach ES, Zimmerman JB. Derivation and synthesisof renewable surfactants. Chem. Soc. Rev. 2012;**41**:1499-1518. DOI: 10.1039/C1CS15217C.

[4] Scott E, Peter F, Sanders J. Biomass in the manufacture of industrial products—the use of proteins and amino acids. Appl. Microbiol. Biotechnol. 2007;**75**:751-762. DOI: 10.1007/s00253-007-0932-x.

[5] Pinazo A, Pons R, Pérez L, Infante MR. Amino acids as raw material for biocompatible surfactants. Ind. Eng. Chem. Res. 2011;**50**:4805-4817. DOI: 10.1021/ie1014348.

[6] Bordes R, Holmberg K. Amino acid-based surfactants—do they deserve more attention?. Adv. Colloid Interface Sci. 2015;**222**:79-91. DOI: 10.1016/j.cis.2014.10.013.

[7] Wu M-H, Wan L-Z, Zhang Y-Q. A novel sodium N-fatty acyl amino acid surfactant using silkworm pupae as stock material. Sci. Rep. 2014;**4**:4428. DOI: 10.1038/srep04428.

[8] Bautista ME, Pérez L, García MT, Cuadros S, Marsal A. Valorization of tannery wastes: lipoamino acid surfactant mixtures from the protein fraction of process wastewater. Chem. Eng. J. 2015;**262**:399-408. DOI: 10.1016/j.cej.2014.10.004.

[9] Infante MR, Pérez L, Pinazo A, Clapés P, Morán MC, Angelet M, Garcia MT, Vinardell MP. Amino acid-based surfactants. C. R. Chim. 2004;**7**:583-592. DOI: 10.1016/j.crci.2004.02.009.

[10] Faustino CMC, Serafim CS, Ferreira IN, Branco MA, Calado ART, Garcia-Rio L. Mixed micelle formation between an amino acid-based anionic gemini surfactant and bile salts. Ind. Eng. Chem. Res. 2014;**53**:10112-10118. DOI: 10.1021/ie5003735.

[11] Hassanzadeh M, Kambarani M, Tayebi L, Yazdian F. Kinetic study of sodium cocoyl sarcosinate synthesis and factors affecting the reaction on bench and pilot scales. J. Surf. Deter. 2012;**15**:551-556. DOI: 10.1007/s11743-012-1333-4.

[12] Sreenu M, Prasad RBN, Sujitha P, Kumar CG. Synthesis and surface-active properties of sodium *N*-acylphenylalanines and their cytotoxicity. Ind. Chem. Eng. Res. 2015;**54**:2090-2098. DOI: 10.1021/ie503764v.

[13] Zhang Q-Q, Cai B-X, Xu W-J, Gang H-Z, Liu J-F, Yang S-Z, Mu B-Z. The rebirth of waste cooking oil to novel bio-based surfactants. Sci. Rep. 2015;**5**:9971. DOI: 10.1038/srep09971.

[14] Morán MC, Pinazo A, Pérez L, Pinazo A, Clapés P, Angelet M, Garcia MT, Vinardell MP, Infante MR. "Green" amino acid-based surfactants. Green Chem. 2004;**6**:233-240. DOI: 10.1039/b400293h.

[15] Villeneuve P. Lipases in lipophilization reactions. Biotechnol. Adv. 2007;**25**:515-536. DOI: 10.1016/j.biotechadv.2007.06.001.

[16] Morcelle SR, Liggieri CS, Bruno MA, Priolo N, Clapés P. Screening of plant peptidases for the synthesis of arginine-based surfactants. J. Mol. Catal. B: Enzymatic. 2009;**57**:177-182. DOI: 10.1016/j.molcatb.2008.08.013.

[17] Reznik GO, Vishwanath P, Pynn MA, Sitnik JM, Todd JJ, Wu J, Jiang Y, Keenan BG, Castle AB, Haskell RF, Smith TF, Somasundaran P, Jarrell KA. Use of sustainable chemistry to

produce an acyl amino acid surfactant. Appl. Microbiol. Biotechnol. 2010;**86**:1387-1397. DOI: 10.1007/s00253-009-2431-8.

[18] Franssen MCR, Steunenberg P, Scott EL, Zuilhofac H, Sanders JPM. Immobilised enzymes in biorenewables production. Chem. Soc. Rev. 2013;**42**:6491-6533. DOI: 10.1039/c3cs00004d.

[19] Zhang G, Xu B, Han F, Zhou Y, Liu H, Li Y, Cui L, Tan T, Wang N. Green synthesis, composition analysis and surface active properties of sodium cocoyl glycinate. Am. J. Anal. Chem. 2013;**4**:445-450. DOI: 10.4236/ajac.2013.49056.

[20] Pérez L, Pinazo A, Pons R, Infante MR. Gemini surfactants from natural amino acids. Adv. Colloid Interface Sci. 2014;**205**:134-155. DOI: 10.1016/j.cis.2013.10.020.

[21] Haldar S, Maji SK. Role of non-covalent interactions in the molecular organization of *N*-n-hexadecanoyl amino acid amphiphiles with hydrophobic C_α-side chains in Tris buffer (pH 9.3). Colloids Surf A Physicochem. Eng. Aspects. 2013;**420**:10-21. DOI: 10.1016/j.colsurfa.2012.12.012.

[22] Ohta A, Toda K, Morimoto Y, Asakawa T, Miyagishi S. Effect of the side chain of *N*-acyl amino acid surfactants on micelle formation: an isothermal titration calorimetry study. Colloids Surf A Physicochem. Eng. Aspects. 2008;**317**:316-322. DOI: 10.1016/j.colsurfa.2007.10.028.

[23] Gerova M, Rodrigues F, Lamère J-F, Dobrev A, Fery-Forgues S. Self-assembly properties of some chiral *N*-palmitoyl amino acid surfactants in aqueous solution. J. Colloid Interface Sci. 2008:**319**:526-533. DOI: 10.1016/j.jcis.2007.12.004.

[24] Perinelli DR, Casettari L, Cespi M, Fini F, Man DKW, Giorgioni G, Canala S, Lam JKW, Bonacucina G, Palmieri GF. Chemical–physical properties and cytotoxicity of N-decanoyl amino acid-based surfactants: effect of polar heads. Colloids Surf A Physicochem. Eng. Aspects. 2016;**492**:38-46. DOI: 10.1016/j.colsurfa.2015.12.009.

[25] Vijay R, Angayarkanny S, Bhaska G. Amphiphilic dodecyl ester derivatives from aromatic amino acids: significance of chemical architecture in interfacial adsorption characteristics. Colloids Surf A Physicochem. Eng. Aspects. 2008;**317**:643-649. DOI: 10.1016/j.colsurfa.2007.11.059.

[26] Bordes R, Tropsch J, Holmberg K. Role of an amide bond for self-assembly of surfactants. Langmuir. 2010; **26**:3077-3083. DOI: 10.1021/la902979m.

[27] Roy S, Das D, Dasgupta A, Mitra RN, Das PK. Amino acid based cationic surfactants in aqueous solution: physicochemical study and application of supramolecular chirality in ketone reduction. Langmuir. 2005;**21**:10398-10404. DOI: 10.1021/la051548s.

[28] Xia J, Nnanna IA, Sakamoto K. Amino acid surfactants: chemistry, synthesis and properties. In: Nnanna IA, Xia J, editors. Protein-Based Surfactants. Synthesis, Physicochemical Properties and Application. Surfactant Science Series 101. New York: Marcel Dekker; 2001; 312 p. ISBN: 0-8247-0004-X.

[29] Sreenu M, Rao BVSK, Prasad BN, Sujitha P, Chityala GK. Synthesis, surface and biological properties of sodium N-acyl isoleucines. Eur. J. Lipid Sci. Technol. 2014;**116**:193-206. DOI: 10.1002/ejlt.201300189.

[30] Bordes R, Holmberg K. Physical chemical characteristics of dicarboxylic amino acid-based surfactants. Colloids Surf A Physicochem. Eng. Aspects. 2011;**391**:32-41. DOI: 10.1016/j.colsurfa.2011.03.023.

[31] Bordes R, Tropsch J, Holmberg K. Counterion specificity of surfactants based on dicarboxylic amino acids. J. Colloid Interface Sci. 2009;**338**:529-536. DOI: 10.1016/j.cis.2009.06.032.

[32] Zhang D, Sun Y, Deng Q, Qi X, Sun H, Li Y. Study of the environmental responsiveness of amino acid-based surfactant sodium lauroylglutamate and its foam characteristics. Colloids Surf A Physicochem. Eng. Aspects. 2016;**504**:384-392. DOI: 10.1016/j.colsurfa.2016.05.097.

[33] Sanchez L, Martinez V, Infante MR, Mitjans M, Vinardell MP. Haemolysis and antihaemolysis induced by amino acid-based surfactants. Toxicol. Lett. 2007;**169**:177-184. DOI: 10.1016/j.toxlet.2007.01.003.

[34] Pinazo A, Pérez L, Lozano M, Angelet M, Infante MR, Vinardell MP, Pons R. Aggregation properties of diacyl lysine surfactant compounds: Hydrophobic chain length and counterion effect. J. Phys. Chem. B. 2008;**112**:8578-8585. DOI: 10.1021/jcisp802193p.

[35] Brito RO, Marques EF, Gomes P, Falcão S, Söderman O. Self-assembly in a catanionic mixture with an amino acid-derived surfactant: from mixed micelles to spontaneous vesicles. J. Phys. Chem. B. 2006;**110**:18158-18165. DOI: 10.1021/jp063271q.

[36] Brito RO, Oliveira IS, Araújo MJ, Marques EF. Morphology, thermal behaviour, and stability of self-assembled supramolecular tubules from lysine-based surfactants. J. Phys. Chem. B. 2013;**117**:9400-9411. DOI: 10.1021/jp400127k.

[37] Tsubone K, Ogawa T, Mimura K. Surface and aqueous properties of anionic gemini surfactants having dialkyl amide, carboxyl, and carboxylate groups. J. Surfactants Deterg. 2003;**6**:39-46. DOI: 10.1007/s4743-003-0246-5.

[38] Tsubone K, Rosen MJ. Structural effect on surface activities of anionic surfactants having *N*-acyl-*N*-methylamide and carboxylate groups. J. Colloid Interface Sci. 2001;**244**:394-398. DOI: 10.1006/jcis.2001.7995.

[39] Fan H, Han F, Liu Z, Qin L, Li Z, Liang D, Ke F, Huang J, Fu H. Active control of surface properties and aggregation behaviour in amino acid-based gemini surfactant systems. J. Colloid Interface Sci. 2008;**321**:227-234. DOI: 10.1016/j.jcis.2008.01.039.

[40] Faustino CMC, Calado ART, Garcia-Rio L. Gemini surfactant-protein interactions: effect of pH, temperature, and surfactant stereochemistry. Biomacromolecules. 2009;**10**:2508-2514. DOI: 10.1021/bm9004723.

[41] Faustino CMC, Calado ART, Garcia-Rio L. Dimeric and monomeric surfactants derived from sulphur-containing amino acids. J. Colloid Interface Sci. 2010; **351**:472-477. DOI: 10.1016/j.jcis.2010.08.007.

[42] Faustino CMC, Calado ART, Garcia-Rio L. Mixed micelle formation between amino acid-based surfactants and phospholipids. J. Colloid Interface Sci. 2011;**359**:493-498. DOI: 10.1016/j.jcis.2011.04.016.

[43] Yoshimura T, Sakato A, Esumi K. Solution properties and emulsification properties of amino acid-based gemini surfactants derived from cysteine. J. Oleo Sci. 2013;**62**:579-586. DOI: 10.5650/jos.62.579.

[44] Angayarkanny S, Vijay R, Baskar G, Mandal AB. Self-organization at the interface and in aqueous solution of a cationic gemini surfactant from the dioctyl ester of cystine. J. Colloid Interface Sci. 2012;**367**:319-326. DOI: 10.1016/j.jcis.2011.10.043.

[45] Colomer A, Pinazo A, Manresa MA, Vinardell MP, Mitjans M, Infante MR, Pérez L. Cationic surfactants derived from lysine: effects of their structure and charge type on antimicrobial and hemolytic activities. J. Med. Chem. 2011;**54**:989-1002. DOI: 10.1021/jm101315k.

[46] Pinazo A, Manresa MA, Marques AM, Bustelo M, Espuny MJ, Pérez L. Amino acid–based surfactants: new antimicrobial agents. Adv. Colloid Interface Sci. 2016;**228**:17-39. DOI: 10.1016/j.cis.2015.11.007.

[47] Faustino C, Serafim C, Ferreira I, Pinheiro L, Calado A. Solubilization power of an amino acid-based gemini surfactant towards the hydrophobic drug amphotericin B. Colloids Surf A Physicochem. Eng. Aspects. 2015;**480**:426-432. DOI: 10.1016/j.colsurfa.2014.11.039.

[48] Serafim C, Ferreira I, Rijo P, Pinheiro L, Faustino C, Calado A, Garcia-Rio L. Lipoamino acid-based micelles as promising delivery vehicles for monomeric amphotericin B. Int. J. Pharm. 2016;**497**:23-35. DOI: 10.1016/j.ijpharm.2015.11 DOI:10.1021/ed4008364 .034.

[49] Castillo JÁ, Pinazo A, Carilla J, Infante MR, Alsina MA, Haro I, Clapés P. Interaction of antimicrobial arginine-based cationic surfactants with lipossomes and lipid monolayers. Langmuir. 2004;**20**:3379-3387. DOI: 10.1021/la036452h.

[50] Colomer A, Perez L, Pons R, Infante MR, Perez-Clos D, Manresa A, Espuny MJ, Pinazo A. Mixed monolayer of DPPC and Lysine-based cationic surfactants: an investigation into the antimicrobial activity. Langmuir. 2013;**29**:7912-7921. DOI: 10.1021/la401092j.

[51] Mondal D, Zhanel GG, Schweizer F. Synthesis and antibacterial properties of carbohydrate-templated lysine surfactants. Carbohydr. Res. 2011;**346**:588-594. DOI:10.1016/j.carres.2011.01.025.

[52] Pérez L, Pinazo A, Garcia T, Lozano M, Manresa A, Angelet M, Vinardell MP, Mitjans M, Pons R, Infante MR. Cationic surfactants from lysine: synthesis, micellization and biological evaluation. Eur. J. Med. Chem. 2009;**44**:1884-1892. DOI: 10.1016/j.ejmech.2008.11.003.

[53] Shome A, Dutta S, Maiti S, Das PK. In situ synthesized Ag nanoparticle in self-assemblies of amino acid based amphiphilic hydrogelators: development of antibacterial soft nanocomposites. Soft Matter 2011;**7**:3011-3022. DOI: 10.1039/c0sm01087a.

[54] Joondan N, Jhaumeer-Laulloo S, Camul P. A study of the antibacterial activity of l-phenylalanine and l-tyrosineesters in relation to their CMCs and their interactions with 1,2-dipalmitoyl-sn-glycero-3-phosphocholine,DPPC as model membrane. Microbiol. Res. 2014;**169**:675-685. DOI: 10.1016/j.micres.2014.02.010.

[55] Tavano L, Infante MR, Abo Riya M, Pinazo A, Vinardell MP, Mitjans M, Manresa MA, Pérez L. Role of aggregate size in the hemolytic and antimicrobial activity of colloidal solutions based on single and gemini surfactants from arginine. Soft Matter. 2013;**9**:306-319. DOI: 10.1039/c2sm26670a.

[56] Coronel-León J, Pinazo A, Pérez L, Espuny MJ, Marqués AM, Manresa A. Lichenysin-geminal amino acid-based surfactants: synergistic action of an unconventional antimicrobial mixture. Colloids Surf. B Biointerfaces. 2017;**149**:38-47. DOI: 10.1016/j.colsurfb.2016.10.008.

[57] Luczynski J, Frackowiak R, Szczepaniak J, Krasowska A. Chemodegradable gemini alanine-based cationic surfactants: synthesis and antifungal activity. Chem. Lett. 2012;**41**: 1176-1177. DOI: 10.1246/cl.2012.1176.

[58] Nogueira DR, Mitjans M, Morán MC, Pérez L, Vinardell MP. Membrane-destabilizing activity of pH-responsive cationic lysine-based surfactants: role of charge position and alkyl chain length. Amino Acids. 2012;**43**:1203-1215. DOI: 10.1007/s00726-011-1176-8.

[59] Nogueira DR, Scheeren LE, Vinardell MP, Mitjans M, Infante MR, Rolim CMB. Nanoparticles incorporating pH-responsive surfactants as a viable approach to improve the intracellular drug delivery. Mat. Sci. Eng. C. 2015;**57**:100-106. DOI: 10.1016/j.msec.2015.07.036.

[60] Tavano L, Pinazo A, Abo-Riya M, Infante MR, Manresa MA, Muzzalupo R, Pérez L. Cationic vesicles based on biocompatible diacyl glycerol-argininesurfactants: physico-chemical properties, antimicrobial activity,encapsulation efficiency and drug release. Colloids Surf. B Biointerfaces. 2014;**120**:160-167. DOI: 10.1016/j.colsurfb.2014.04.009.

[61] Bračič M, Pérez L, Infante R, Kogej K, Hribernik S, Sauperl O, Pinazo A, Pons R, Infante MR. A novel synergistic formulation between a cationic surfactant from lysine and hyaluronic acid as an antimicrobial coating for advanced cellulose materials. Cellulose. 2014;**21**:2647-2663. DOI: 10.1007/s10570-014-0338-8.

[62] Ménard N, Tsapis N, Poirier C, Arnauld T, Moine L, Lefoulon F, Péan JM, Fattal E. Drug solubilization and in vitro toxicity evaluation of lipoamino acid surfactants. Int. J. Pharm. 2012;**423**:312-320. DOI: 10.1016/j.ijpharm.2011.11.030.

[63] Branco MA, Pinheiro L, Faustino C. Amino acid-based cationic gemini surfactant-protein interactions. Colloids Surf A Physicochem. Eng. Aspects. 2015;**480**:105-112. DOI: 10.1016/j.colsurfa.2014.12.022.

[64] Beliéres M, Dejugnat C, Chouini-Lalanne N. Histidine-based lipopeptides enhance cleavage of nucleic acids: interactions with DNA and hydrolytic properties. Bioconjugate Chem. 2015;**26**:2520-2529. DOI: 10.1021/acs.bioconjchem.5b00542.

[65] Yang P, Singh J, Wettig S, Foldvari M, Verral RE, Badea I. Enhanced gene expression in epithelial cells transfected with amino acid-substituted gemininanoparticles. Eur. J. Pharm. Biopharm. 2010;**75**:311-320. DOI: 10.1016/j.ejpb.2010.04.007.

[66] Sen J, Chaudhuri A. Gene transfer efficacies of novel cationic amphiphiles with alanine, β-alanine, and serine headgroups: a structure-activity investigation. Bioconjugate Chem. 2005;**16**:903-912. DOI: 10.1021/bc0500443.

[67] Rosa M, Penacho N, Simões S, Lima MCP, Lindman B, Miguel MG. DNA pre-condensation with an amino acidbased cationic amphiphile. A viable approach for liposome-based gene delivery. Mol. Membrane Biol. 2008;**25**:23-34. DOI: 10.1080/09687680701499451.

[68] Calejo MT, Kjøniksen AL, Pinazo A, Pérez L, Cardoso AMS, Pedroso de Lima MC, Jurado AS, Sande SA, Nyström B. Thermoresponsive hydrogels with low toxicity from mixtures of ethyl(hydroxyethyl) cellulose and arginine-based surfactants. Int. J. Pharm. 2012;**436**:454-462. DOI: 10.1016/j.ijpharm.2012.07.018.

Chapter 8

The Potential Application of Selected Fungi Strains in Removal of Commercial Detergents and Biotechnology

Violeta D. Jakovljević and Miroslav M. Vrvić

Additional information is available at the end of the chapter

http://dx.doi.org/10.5772/intechopen.68184

Abstract

The consumption of synthetic detergents is increasing year by year due to increasing urbanization, which reflects on higher concentration of this pollutant in the environment. In order to purify wastewaters from different pollutants, the application of new technologies such as bioremediation is necessary. From the environmental point of view, it is important to identify microorganisms that are tolerant to the presence of this pollutant. This chapter presents the experimental evaluation of ability of several fungal species, isolated from municipal sewage and industrial wastewater, in removing a high detergent concentration (anionic surfactants) from the environment as well as their potential application in biotechnology. The selected fungi were cultivated in Czapek-Dox liquid medium supplemented with commercial detergent "Merix" (Henkel, Kruševac, Serbia). Changes of physicochemical and biochemical parameters such as pH, redox potential, dry weight biomass, and enzymes activities such as alkaline protease and phosphatase were evaluated during 16 days of cultivation. The obtained results could be useful in the implementation of tested fungi in bioremediation processes and in biotechnology.

Keywords: alkaline protease, alkaline phosphatase, biodegradation, commercial detergent, fungi

1. Introduction

The detergents that we use for our daily laundry have been recognized as one of the major pollutants responsible for water pollution [1]. Detergents are synthetic organic compounds, which contain three main ingredients (%): phosphate builders (50% by weight, approximately), surface-active substances (surfactants) (between 10 and 20%), and bleaches (7%), as well as very small percentages of additives (wetting agents, optical brighteners, softeners, and enzymes).

Surfactants are the components mainly responsible for the cleaning action of detergents, whereas the additives are designated to enhance the cleaning action of surfactants [2]. Surfactants commonly used as main ingredients in commercial detergents are linear alkyl-benzene sulfonates (LASs) (anionic) and ethoxylated alcohols (AEs) (nonionic) [3]. Commercial LAS also contain coproducts called dialkyltetralinsulphonates (DATS) and iso-LAS. Furthermore, over 70 major isomers of DATS have been detected in commercial LAS [4]. The molecular structure of a synthetic detergent influences its biodegradation potential, which is the principal criterion for their ecological behavior. The biodegradation rate and acute toxicity of LAS on aquatic life are very much related to both the chain length and phenyl position of the alkyl chain [5, 6].

The main environmental impact of detergents is related to their post use effect when the wash water is discharged into sewage treatment plants (STPs) or discharged directly to the aquatic environment in areas where there is no sewage treatment [7]. These compounds can act on biological wastewater treatment processes and cause problems in sewage aeration and treatment facilities due to their high foaming, lower oxygenation potentials, and the ability to kill waterborne organisms [8]. Biodegradation processes and adsorption on active sludge remove these chemicals from wastewater to a greater or lesser extent, depending on the chemical structure of the surfactant molecule and operating conditions of the STP. Under aerobic conditions, LAS is degraded through ω-oxidation of terminal carbon in the alkyl chain followed by β-oxidation. In this process, which is known as the primary biodegradation, sulfophenyl carboxylic acids (SPACs) are forming [9]. In the next step, known as ultimate biodegradation, SPACs are transforming ultimately into CO_2, H_2O, inorganic salts, and biomass. After treatment, residual surfactants, refractory coproducts, and biodegradation products dissolved in STPs effluents or adsorbed on sludge are discharged into the environment. These chemicals through several transport mechanisms enter the hydrogeological cycle. However, even biodegradable detergents can have a toxic effect upon the living beings if they are present in quantities above the permitted level [10]. In line with their high-environmental relevance, surfactants have to meet certain requirements issued in the European detergent regulation 684/2004 [11]. According to Legislative acts, the maximal amount of detergent allowable in wastewater which effluent in public sewage is 4 mg/L, and 0.5 mg/L in natural recipient.

Traditional methods for the reduction of pollutants and environmental cleanup contain combination of physical, chemical, and biological methods [12, 13]. Bioremediation, as less expensive and eco-friendly alternative to conventional technology for decontaminating environment from wide range of pollutants by microorganisms, has been extensively studied during past two decades [14]. Mycoremediation is a type of bioremediation, which uses fungal mycelium to decontaminate or filter the toxic waste from contaminated area. Filamentous fungi have the ability to grow and transform or degrade hazardous compounds in polluted environment, as response to severe environmental conditions [15]. Because of that, they represent power useful potential in bioremediation processes. The numerous examples of fungi used in biodegradation of certain toxic pollutants (petroleum hydrocarbons, chlorofenols, pesticides, nitroaromatic explosives, etc.) have been observed [16–20]. The potential of filamentous fungi in removing of commercial detergent has been continuously studied over the past three decades by Stojanović et al. [21–23]. These studies identified a total of 15 strains of *Actinomycetes*, which have the ability to grow and metabolize synthetic detergent "Merix" (Henkel, Serbia) and its particular components: ethoxylated oleyl-cetyl alcohol and sodium tripolyphosphate at wide concentrations range 0.01–1.0%. Since the application of new technology in detergent industry leads to the development of new detergent performance, the identification and investigation of new microbial strains are necessary and justified.

The ability of filamentous fungi to produce and excrete a variety of extracellular hydrolytic enzymes is significant not only for bioremediation processes but also for fermentation industries and biotechnology. This study was focussed on two types of hydrolases: alkaline protease and phosphatase, due to their importance in different industrial areas. Proteases (EC 3.4.21-24 and 99) are one of the key enzymatic constituents in detergent formulation in which they act as protein stain removers. They represent about 60% of total worldwide sale of enzymes. The most significant commercial detergent protease additives (Savinase®, Esperase®, Alcalase®, etc.) are produced by *Bacillus* spp. [24]. In addition, *Pseudomonas* species are also recognized as sources of alkaline proteases with advantageous properties for industrial applications [25, 26]. In recent years, a great number of fungi from genus *Aspergillus, Penicillium, Rhizopus, Mucor, Humicola, Thermoascus, Thermomyces* have been identified as producers of different types of protease of biotechnological relevance. The increased demand for proteases with specific properties has led researchers to explore new sources of proteases.

***Alkaline phosphatase (ALP: EC 3.1.3.1) catalyzes the hydrolytic cleavage of phosphate monoesters under alkaline conditions and plays important roles in microbial ecology through its involvement in phosphate metabolism [27] and molecular biology [28] applications. There has been considerable effort in recent years toward the application of ALPs for bioremediation of heavy metals and radionuclides from nuclear wastes [29]. ALP can be isolated from variety of microorganisms including *Escherichia coli* [30], *Pseudomonas* [31], *Aerobactor* [32], and *Bacillus* species [33]. Usually the ALP is produced at commercial level from *E. coli* or calf intestine. However, there is no literature data about production of ALP among fungi strains so far. This study investigated potential of several filamentous fungi to produce the enzyme for the first time.

The aim of the study was isolation of fungi (micromycetes) from municipal sewage wastewater originating from households and industrial wastewater from Henkel factory (Kruševac, Serbia); selection of fungi strains that are tolerant to a high detergent concentration; cultivation of selected fungi in Czapek-Dox liquid medium supplemented with commercial detergent mark "Merix" (Henkel, Serbia), and *in vitro* investigation of their growth and metabolic activity. The investigation of these parameters is crucial for the practical application of fungi in bioremediation processes. The obtained results could be beneficiary in clarifying the potential role of fungi in bioremediation of environment contaminated with a high concentration of tested pollutant as well as in biotechnology.

2. Materials and methods

2.1. Isolation and identification of fungi

The fungi used in this work were isolated from wastewater samples, which contain commercial detergents. The wastewater samples were collected from rivers basins of Lepenica (Kragujevac, Serbia) and West Morava (Čačak, Serbia), at places where municipal wastewater discharges into the rivers. In addition, samples of wastewater were collected from river basin of Rasina (Kruševac, Serbia), downstream where the industrial wastewater of factory Henkel discharges into the river. Samples of wastewaters were taken in sterile glass bottles, transferred to the microbiology laboratory and disposed to refrigerate at 4°C. Within 24 hours, different dilutions of samples were transferred on Petri plates with malt agar and streptomycin to prevent the bacterial growth. The Petri plates were maintained at room temperature for 5 days. Positive cultures of fungi were subcultured on malt agar (MA) and potato dextrose agar (PDA) for the isolation of a pure, single colony for identification.

The identification of pure fungal cultures was carried out according to Systematic key at the Faculty of Science, University of Kragujevac, Serbia, by Prof. Branislav Ranković. For spores production, pure cultures were aseptically maintained at (28 ± 2)°C from 3 to 5 days on PDA, composed of (g/L): peeled potatoes 200, dextrose 20, and agar 15. After having sufficient population of spores, the plate were stored at (4 ± 0.5)°C with periodical subculturing in sterile conditions.

2.2. Inoculum preparation

An inoculum suspension was prepared from fresh, mature cultures of selected fungal species. The colonies were covered with 5 mL of distilled sterile water. The inoculum was achieved by carefully rubbing the colony with a sterile loop; the tube with isolate was shaken vigorously for 15 seconds with a Vortex mixer and then transferred to a sterile tube. The inoculum size was adjusted to 1.0×10^6 spores/mL by microscopic enumeration with a cell-counting hematocytometer (Neubauer chamber).

2.3. Experimental procedure and culture conditions

The fungi were grown in 250 mL Erlenmeyer flasks with 200 mL of modified Czapek Dox liquid nutrient medium the following composition (g/L): $NaNO_3$ is 3.0; K_2HPO_4 is 1.0; $MgSO_4 \times 7H_2O$ is 0.5, and sucrose is 30.0 and distilled water up to 1000 mL (control-C). The pH value of this liquid media was about 4.80 (adjusted with 0.1 M HCl). The media with addition of detergent "Merix" (Henkel, Serbia) at concentrations of 0.3% (D3) and 0.5% (D5) were prepared according to same procedure. The pH values of these media were measured and recorded as 9.35 and 9.80, respectively. All flasks were sterilized at 121°C in an autoclave for 15 minutes. After cooling the liquid media to room temperature, 1 mL spore suspension of each fungus was inoculated in liquid media in aseptic condition. Inoculated flasks were incubated on an electric shaker (Kinetor-*m*, Ljubljana) at 150 rpm and room temperature for 16 days. Summary, one positive control without detergent with spores (C), two test flask with detergent and with spores (D3 and D5), and two negative controls with detergent but without spores (ncD3 and ncD5) were used for each fungal species. Three flasks of each fungus were used for collecting samples at 3rd, 6th, 9th, 12th, and 16th day. To determine that each fungal biomass dry weight, mycelia were removed by filtration of fermentation broths, according to procedure described below. The filtrates of fermentation broths were collected by centrifugation at 12,000 g for 2 minutes. The supernatants were used as the source for determination of pH, redox potential, concentration of ASs and enzymes activities.

2.4. The measurement of dry weight (DW) biomass

At the time intervals above-mentioned, the mycelium of each fungus was filtered through filter paper (Whatman No. 1) of a known weight, washed with distilled water and dried at constant weight at 80°C. The filter paper with the mycelium was placed in the desiccator and then reweighed. Mycelium DW was calculated using the Eq. (1), and results are expressed in grams per liter (g/L) of submerged culture.

$$DW(g/L) = (Wcf - Wif) \times 5 \tag{1}$$

where DW is total biomass dry weight, Wcf is weight of culture with filter paper, and Wif is initial weight of filter paper.

2.5. The measurement of pH and redox potential values

The pH and redox potential values of the culture filtrates were measured by digital PHS-3BW microprocessor pH/mV/temperature meter model 65-1 (Bante Instruments Ltd., China) with reference electrode Ag/AgCl/3 mol/kg KCl that was initially standardized with appropriate buffer solution of pH 4.0, 7.0, and 10.0. The redox potential values are expressed in mV, and are calculated using the following Eq. (2):

$$E(mV) = E_{ems} + E_{ref} \tag{2}$$

where E_{ref} is the potential of the reference electrode (+210 mV at 25°C), E_{ems} is measured potential.

2.6. Determination of concentration of anionic surfactants (ASs) and calculation of biodegradation rate

The concentration of ASs in the detergent and fermentation broth was determined by spectrophotometrically using methylene blue (MB). The method for determining the concentration of methylene blue-active substance (MBAS) in the detergents was adapted from *Standard Methods for the Examination of water and wastewater* [34]. The method is based on the reaction presented in **Scheme 1**.

The solutions of detergent were transferred into the separatory funnels. One drop of 1% phenolphthalein solution as indicator was added to the detergents solutions, afterward 1 M NaOH was adding until obtained change in color from colorless to pink. Then, 1 M H_2SO_4 was added carefully until the solution in the funnel had become acidic, which is reflected in

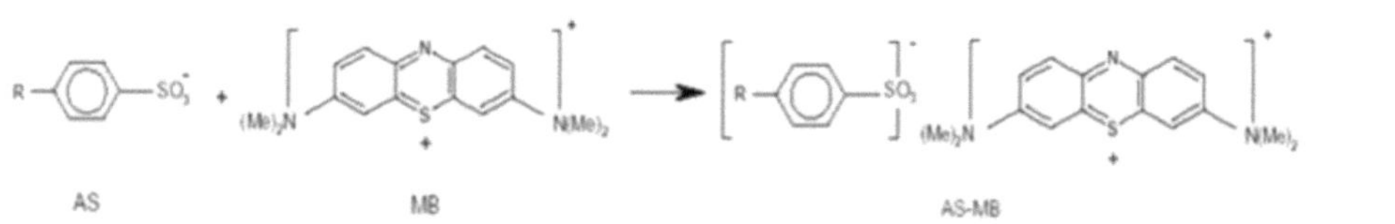

Scheme 1. Mechanism of formations an ionic pair methylene blue-active substance (AS-MB) between the anionic surfactants (AS) and the methylene blue (MB).

appearance of colorless. The procedure of extraction was continued by adding 5 mL chloroform and 13 mL methylene blue reagent in each funnel. The funnels were shaken about 30 seconds. In order for phase's separation the funnels were stored at ambient temperature for at least 30 minutes. Thereafter, the chloroform layer was decanted into a clear 100 mL Erlenmeyer bottle. The same procedure was repeated three times employing 5 mL of chloroform for each time. The chloroform layers were collected in a 100-mL Erlenmeyer bottle and then reversible transferred to the separatory funnel. To each funnel, 25 mL of wash solution (6.7×10^{-3} 1 M phosphate buffer, pH 7.1) was added; then they were shaken once more for 30 seconds and stored at ambient temperature for 30 minutes. Finally, the chloroform layer was drawn off through glass wool into a 50-mL volumetric flask. Absorbance of chloroform layer was measured using Perkin-Elmer Lambda 25 UV-Vis spectrophotometer at 652 nm against blank chloroform. The concentration of the residual surfactant present in test detergent in terms of MBAS were then plotted against the time (days) for the 16-day experimental period. The result obtained with the SDS (Fluka, Switzerland), an alkylsulfate anionic surfactant, as the referent anionic surfactant compound served as the standard. The percentage of surfactants degradation was calculated using Eq. (3):

$$\% \text{ Degradation} = 100 - [(A_{625} \text{ exp} - A_{625} \text{ blank}) / A_{625} \text{ std}] \times 100, \quad (3)$$

where A_{625} *exp* is absorbance of test sample, A_{625} *blank* is absorbance of blank sample, and A_{625} *std* is absorbance of standard sample at 625 nm.

2.7. Assay of alkaline protease activity (EC 3.4.21-24)

The protease activity was assayed by the Anson method [35]. The fermentation broth (1 mL) was mixed with 5.0 mL of substrate (0.65% casein in 25 mM tris-HCl buffer, pH 8.0), was incubated at 37°C for 30 minutes. After incubation, 1 mL of 5% trichloroacetic acid (TCA) was added to attenuate the reaction. The mixture was allowed to incubate for 30 minutes at room temperature and filtered to remove the precipitate. 5 mL of 6% Na_2CO_3 and 1 mL of diluted Folin-Ciocalteu's phenol reagent were then added to the filtrate. The solution was kept at room temperature for 30 minutes and absorbance was read at 660 nm. A standard curve was generated using tyrosine standard. One unit enzyme activity was defined as the amount of enzyme capable of producing 1 μg of tyrosine from casein in a minute under assay condition.

2.8. Assay of alkaline phosphatase activity (EC 3.1.3.1)

The activity of alkaline phosphatase (ALP) was assayed by using *ß*-glycerophosphate as substrate. The substrate solution was prepared by mixing 0.5 mL of 0.05 mol/L glycol buffer with 1.0 mM Magnesium chloride, pH 9 and 0.5 mL of substrate to both test and blank tubes. Then, 0.1 mL of enzyme solution (fermentation broth) and deionized water were added to test and blank tubes, respectively. The tubes were mixed and incubated at 37°C for 30 minutes. After incubation, 10% TCA was added to each tube in order to stop the enzyme reaction. The solution of NH_4-molybdate was added as color reagent. The amount of liberated inorganic phosphate (Pi) was quantitatively determined spectrophotometrically (Perkin-Elmer Lambda 25) at 660 nm [36]. One unit of enzyme activity (IU) was defined as the amount of enzyme that released 1 μg of inorganic phosphate per minute under the assay conditions.

2.9. Statistical analysis

Statistical analysis was performed using SPSS statistical software package (SPSS for Windows, ver. 13.0, Chicago, IL, USA). For testing the normality of distribution, means and standard deviation, student *t*-test was used. To compare the differences between growth media, Mann-Whitney and Kruskal-Wallis tests were used. Pearson's correlation coefficient was used for the measurement of the strength of the association between variables. All significance tests were two-tailed (0.05 and 0.01) and $p < 0.05$ was considered significant.

3. Results and discussion

3.1. Identification and selection of fungi species used in this study

From collected samples of wastewaters the following fungi were identified: *Aspergillus niger, Penicillium chrysogenum* (both isolated from sewage wastewater of Lepenica River, Kragujevac, Serbia), *Penicillium cyclopium* (from sewage wastewater of West Morava River, Čačak, Serbia), and *Trichoderma harzianum* and *Mucor racemosus* (both from wastewater plant of detergent industry (Henkel, Kruševac, Serbia)). The identification of fungi strains were performed using systematic keys at Faculty of Science, University of Kragujevac.

Systematic and morphological (macro- and microscopic) characterizations of isolated species are reported in **Table 1**.

Aspergillus niger (**Table 1**)—the colony color and texture varies with age. Initially, the colony is white but changes color to dark brown or black with age and conidial production. Hyphae are septate and hyaline. Conidiophores are hyaline, upright, simple, smooth-walled, length between 400 and 3000 μm, terminating in spherical to globose vesicles, 30–75 μm in diameter. *A. niger* has both metulae and phialides (biseriate), which cover the entire surface of the vesicle. The stipe measured 440–680 × 6–12 μm, smooth-walls, slightly brown in color. Conidia are 1-celled, size from 4–5 μm, very rough structure, globose, and variously in mass (brown to black color) in dry basipetal chains [37].

Penicillium chrysogenum (**Table 1**)—colony broadly spreading, blue-green to bright green, with broad white margin during the growing period, smooth velvety, usually becoming grayish or purplish brown in age with overgrowth of white or rosy hyphae; reverse yellow, with color diffusing somewhat; drop usually branches with all parts smooth; stipes 2.5–4 μm diameter; phialides ampuliform with a reduced neck, 7–10 × 2–2.5 μm; conidia elliptical to subglobose, 3–4 in μm long axis, smooth [38].

Penicillium cyclopium (**Table 1**)—colonies rather dull blue-green, with brighter zone inside the white margin, almost velvety but showing distinct fascilutation in the younger areas; reverse usually pale peach but occasionally fairly bright yellow or purplish brown; the penicillin with normally two to three stages of branching, often with branches and metulae born at the same level, stipes rough, long, 100–400 × 2.5–4 μm; phialides flask-shaped, 7–10 × 2–3 μm; conidia globose to subglobose, sometimes elliptical when first formed, smooth to very finely roughened, 3–4 μm diameter [39].

Fungi species	**Macroscopic**	**Microscopic**
Phylum: Ascomycota Class: Ascomicetes Subclass: Euromycetide Order: Eurotiales Family: Trichocomaceae Genus: *Aspergillus* *Aspergillus niger* Van Tieghem (1867)	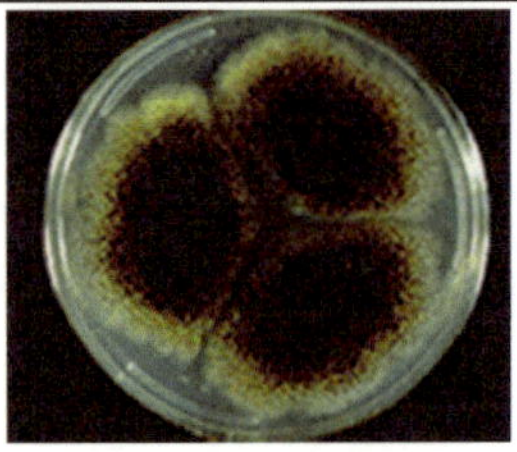	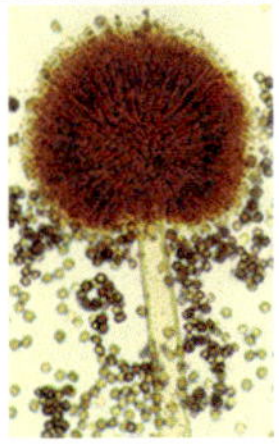
Phylum: Ascomycota Class: Ascomicetes Subclass: Euromycetidae Order: Eurotiales Family:Trichocomaceae Genus: *Penicillium* *Penicillium chrysogenum* Thom (1910)	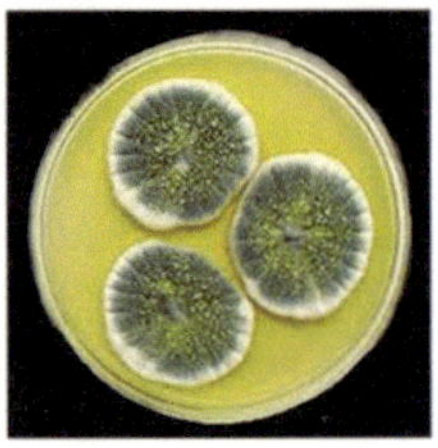	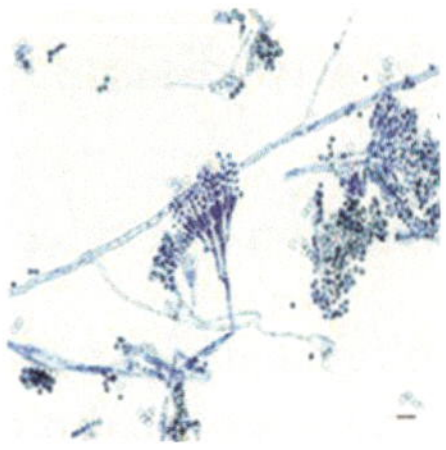
Phylum: Ascomycota Class: Ascomicetes Subclass: Euromycetidae Order: Eurotiales Family: Trichocomaceae Genus: *Penicillium* *Penicillium cyclopium* Westling (1911)		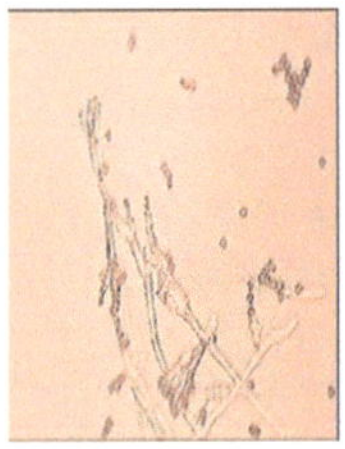
Phylum: Ascomycota Class: Sordariomycetes Order: Hypocreales Family: Hypocreaceae Genus: *Trichoderma* *Trichoderma harzianum* Rifai (1969)	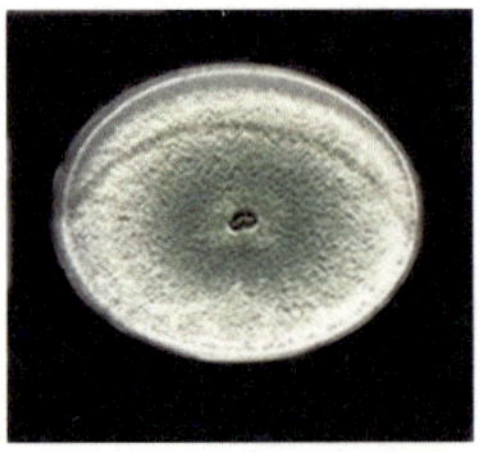	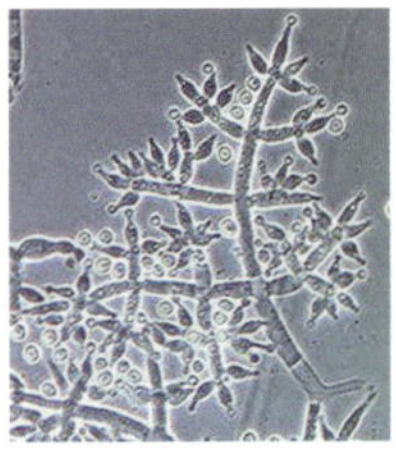
Phylum: Zygomycota Order: Mucorales Family: Mucoraceae Genus: *Mucor* *Mucor racemosus* Fresenius (1976)	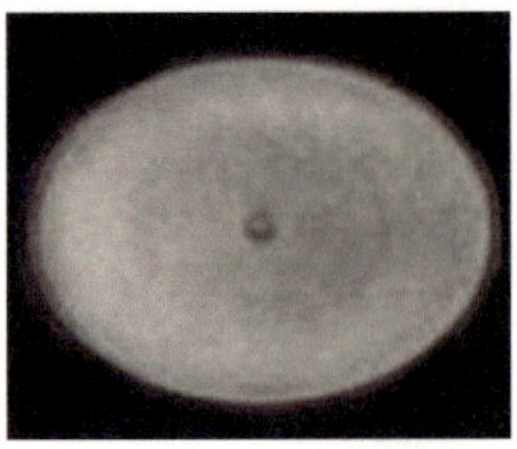	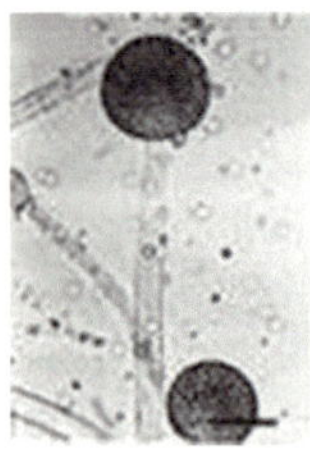

Table 1. Systematic and morphological appearance of fungi [41–45].

Trichoderma harzianum (**Table 1**)—the surface of colony is initially white or yellow, then becoming yellow-green with age. Colony texture is wooly. Hyphae are septate and hyaline. Conidiophores are hyaline, much branched, not verticillate, and may sporadically demonstrate a pyramidal arrangement. Phialides are divergent, typically flask shaped, enlarged in the middle, sharply constricted below the tip to form a narrow neck and slightly constricted at the base. Conidia are hyaline, 1-celled, smooth or roughened, range in shape from globose to ellipsoidal, born in a small terminal clusters at the tips of phialides, diameter of 3 μm [40].

Mucor racemosus (**Table 1**) is a dimorphic, facultative anaerobic zygomycete, capable of vegetative growth in either a filamentous phase or as spherical yeasts. Colonies grows rapidly at 25–30°C and quickly cover the surface of the agar. Its fluffy appearance with a height of several cm resembles cotton candy. From the front, the color is white initially and becomes dark grayish-brown or light olive-grey in time when grown on typical laboratory media. It is easily recognizable microscopically by its tall (up to 2 cm) needle-like sporangiophores and large sporangium. Sporangiophore born from aerial hyphae; stipes simpodially branched; sporangia spherical, approximately 50–300 μm in diameter; columella have ellipsoidal to pyriformal shape. Sporangiospores are hyaline, ellipsoidal, mostly diameter of 4–8 μm, with smooth wall [46].

3.2. Effect of commercial detergent on fungal biomass and growth curves

Biomass is an indicator of fungal metabolic activity and their bioremediation potential. Very important factor for biodegradation processes is physicochemical interaction between surfactants and fungal membranes and cell wall [47]. Further, the surfactants can cause inhibitory or stimulatory effect on enzymes involved in key metabolic pathways and change their metabolic activity in these two ways. Overview of the literature provides the evidences that growth of fungi depends on the type of surfactant in such a way that nonionic surfactants, Triton X-100 and Tween 80 supported, whereas anionic-type surfactant, SDS, inhibited their growth [48]. An investigation of the impact of surfactants on the growth and development of fungi is not simple process due to numerous factors such as applied concentration of surfactants, type of fungus and its genetic properties, experimental conditions, and so on influence these processes.

The current study investigated the effect of commercial-powdered detergent "Merix" (Henkel, Kruševac, Serbia) on the growth and development of five fungi species, which were quantitatively dominated in wastewaters. Previously, the maximal concentration of detergent on which fungi can grow was determined and defined as 3 mg/mL or 0.3% for all tested fungi with exception of *M. racemosus*. This fungus was able to grown at higher concentration of detergent, 5 mg/mL or 0.5%. In this case, the detergent at both concentrations was used for investigation of the growth and metabolic activity of the fungus compared to control (Czapek-Dox liquid medium). The obtained results were presented in **Figure 1**. As **Figure 1** shows, all fungi had monophasic exponential growth in C medium. The growth curves of fungi in this medium have following phases: exponential growth, stationary phase, and autolysis. A little deviation from this growth profile was observed in the C medium of *A. niger* [**Figure 1 (1)**]. This fungus had very pronounced the exponential growth phase until the 9th day, followed by the stationary phase until 16th day, without autolysis [49]. The other fungi, *P. chrysogenum*

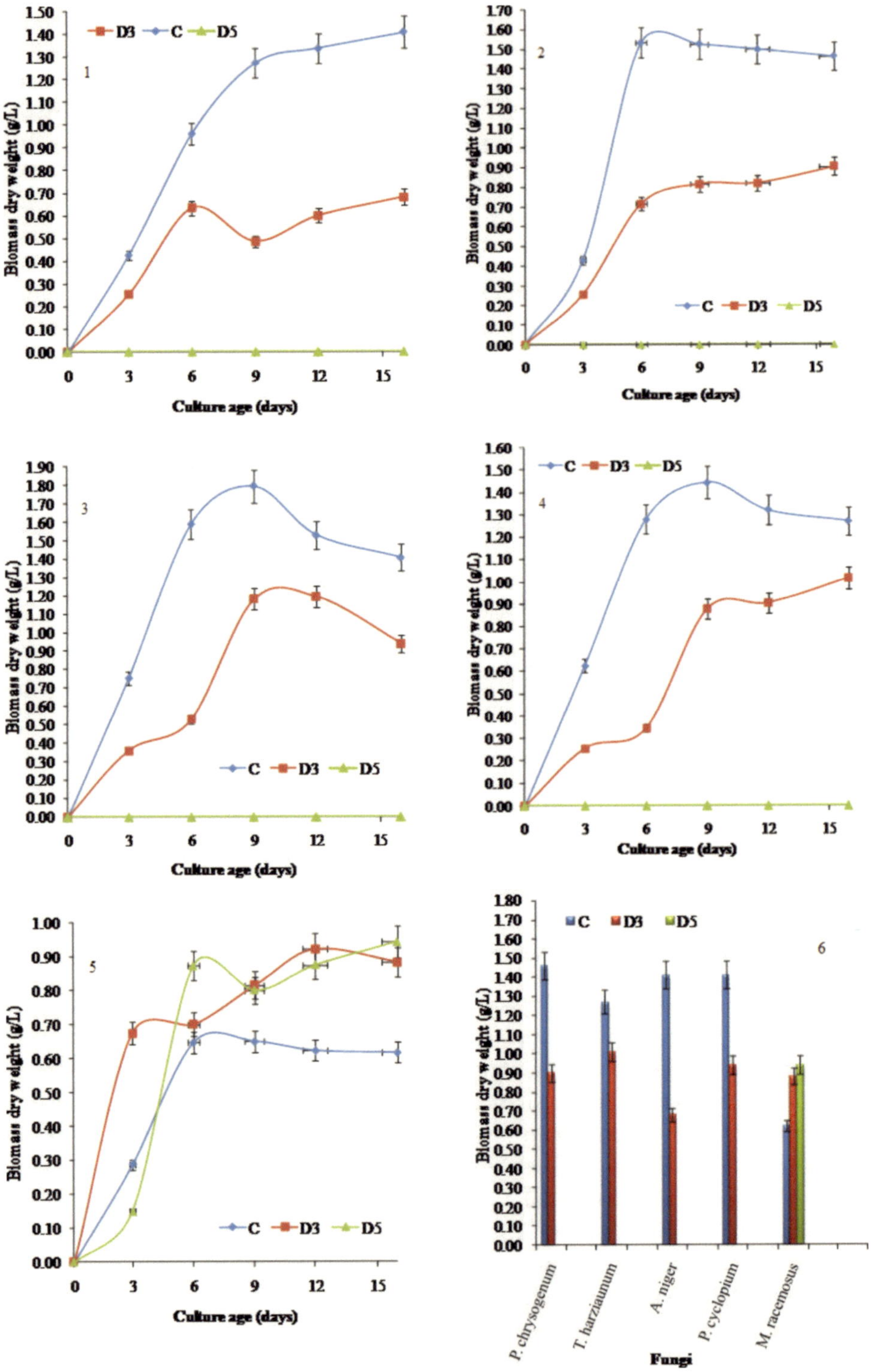

Figure 1. Fungal biomass dry weight and growth curves: *A. niger* (1), *P. chrysogenum* (2), *P. cyclopium* (3), *T. harzianum* (4), *M. racemosus* (5), and comparison of total biomass of fungi after 16-day cultivation (6).

[**Figure 1(2)**], *P. cyclopium* [**Figure 1(3)**], *T. harzianum* [**Figure 1(4)**], and *M. racemosus* [**Figure 1(5)**] had the exponential growth phase from inoculation until the 6th day and stationary phase from 6th to 9th day, when the maximal growth and biomass dry weight were observed. After stationary phase, autolysis was noted, which was reflected on the total biomass reduction. In contrast to C medium, the fungi cultivated in D3 medium had biphasic exponential growth [49–52]. The growth curve of *P. cyclopium* [**Figure 1(3)**] had cascade shape with five distinct phases: the early exponential growth (until 3rd day), the first stationary phase (from 3rd to 6th day), the second exponential phase (from 6th to 9th day), the second stationary phase (from 9th to 12th day) and autolysis (from 12th to 16th day). In the profile of *A. niger* [**Figure 1(1)**], the autolysis was observed between 6th and 9th day; afterward the fungus has slow growth until the end of experiment. The growth curves of *P. chrysogenum* [**Figure 1(2)**] and *M. racemosus* [**Figure 1(5)**] revealed the autolysis between 12th and 16th day. On the other hand, autolysis was not observed in the profile of *T. harzianum* [**Figure 1(4)**]. As we mentioned above, *M. racemosus* had the ability to grow in D5 medium and their growth curve in this medium was slightly modified in respect to D3 medium. The growth curve of the fungus in D5 medium [**Figure 1(5)**] revealed the early growth phase during the first 3 days, followed by the first exponential growth (from 3rd to 6th day), autolysis (from 6th to 9th day), and second exponential growth phase (until the 16th day) [53]. In order for better comprehension of detergent impact on the fungi growth, biomass dry weight was measured after 16 days of cultivation and was compared with the control. The results were presented in **Figure 1(6)**. The amount of biomass produced by fungi in C medium ranged in the following direction: *P. chrysogenum* > *A. niger* > *P. cyclopium* > *T. harzianum* > *M. racemosus*. This observation showed that chemical composition of Czapek-Dox liquid medium was optimal for the growth of tested fungi, except for *M. racemosus*. This finding is consistent with results of other research studies [21–23]. As the figure shows, the tested detergent at a concentration of 0.3% influenced the inhibition of biomass in the following direction: *A. niger* is 51.42%, *P. chrysogenum* is 50.0%, *P. cyclopium* is 33.38%, and *T. harzianum* is 20.0%. The inhibition of fungi growth by detergent could be explained by toxic effect of some detergent ingredients or degradation products and by autolysis. From these results, it is evident that *A. niger* and *P. chrysogenum* are the most sensitive species according to tested detergent. However, the detergent at both concentrations had stimulatory effect on the biomass production of *M. racemosus*; even the higher concentration had stronger stimulatory effect. The obtained results indicate the possible application of fungi, first of all *M. racemosus*, in bioremediation process.

3.3. Biodegradation rate of anionic surfactants incorporated in detergent and its relationship according to fungal biomass dry weight

The next step in this study was confirmation and comparison ability of the fungi to degrade anionic surfactants (ASs) of detergent in terms of their potentially application in bioremediation processes. First, it is defined percentual share of ASs in the tested detergent (about 20%) by MBAS assay. By conversion of percentage, it was obtained 600 and 1000 μg/mL of ASs in D3 and D5 media. The concentration of ASs during 16-day cultivation of fungi in liquid medium was monitored and compared with negative controls (abiotic). They were used in order to monitor the stability of detergent during the cultivation time and a process of its adsorption on the walls of glass (flasks). The obtained results were presented in **Figure 2**.

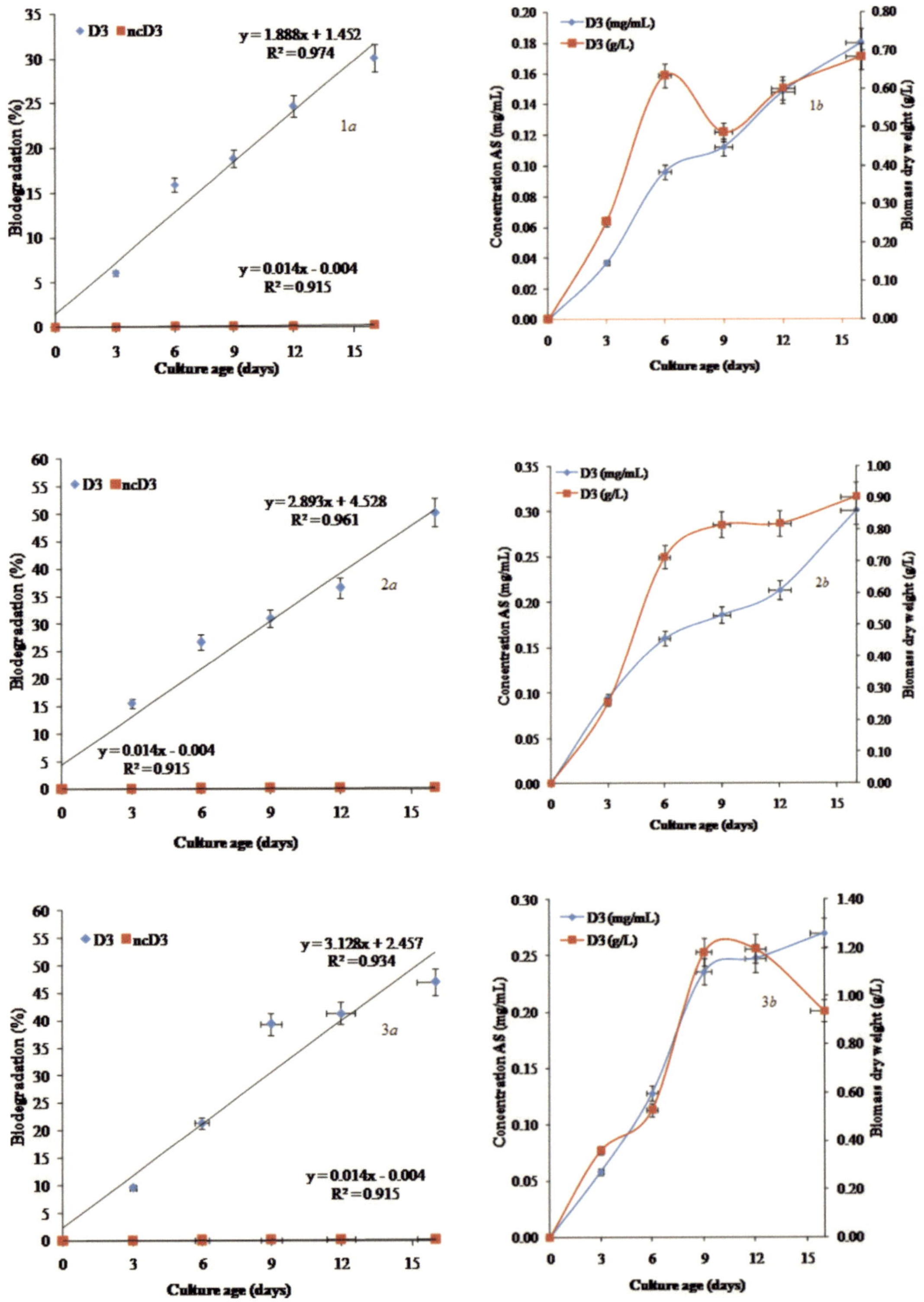
D3
ncD3
y=1.888x + 1.452
R²=0.974
1a
y=0.014x - 0.004
R²=0.915
Biodegradation (%)
Culture age (days)
D3 (mg/mL)
D3 (g/L)
1b
Concentration AS (mg/mL)
Biomass dry weight (g/L)
Culture age (days)
y=2.893x + 4.528
R²=0.961
2a
2b
y=3.128x + 2.457
R²=0.934
3a
3b

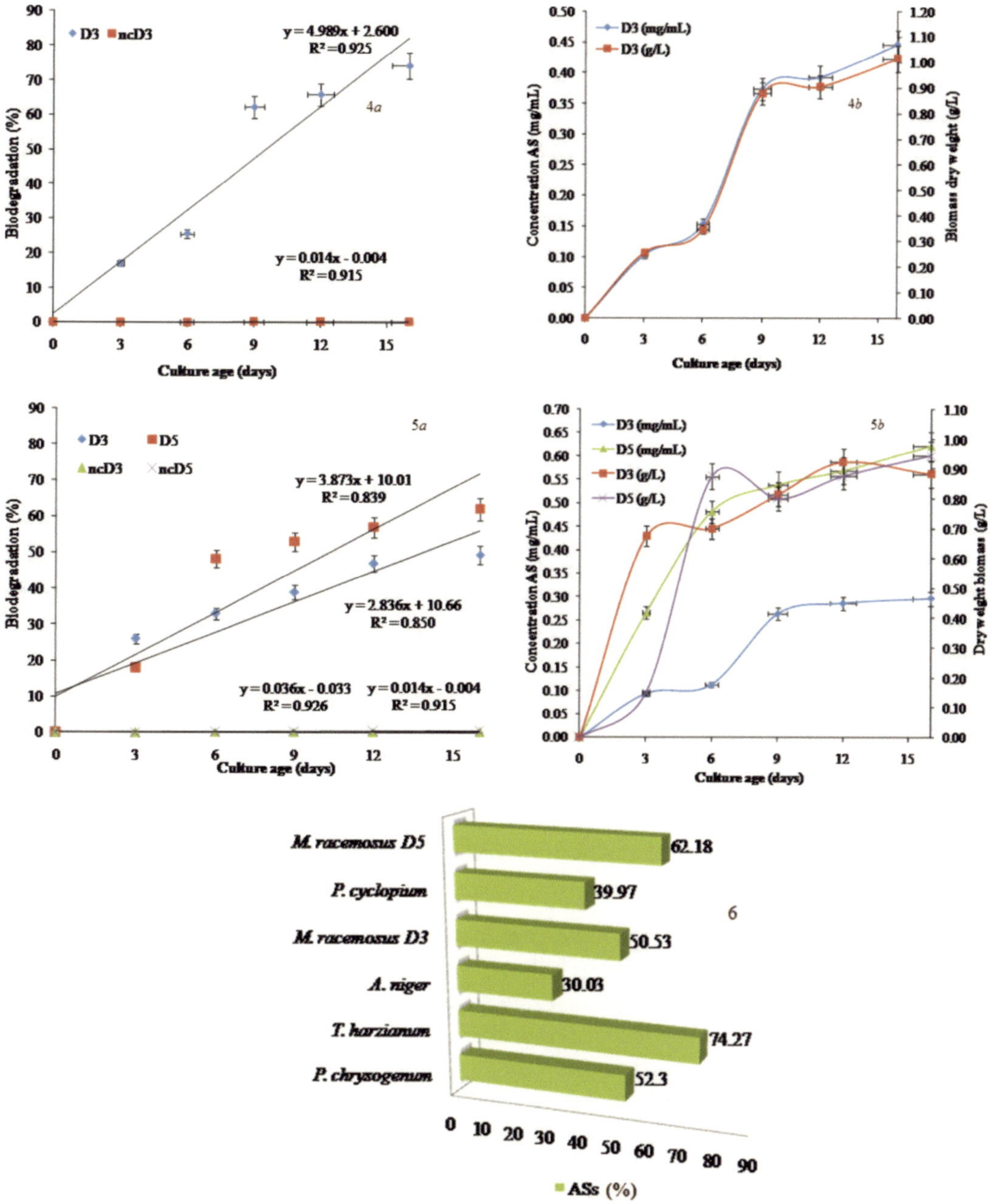

Figure 2. (a) Biodegradation rate of ASs, (b) relationship between concentration of ASs and fungal biomass dry weight: *A. niger* (1), *P. chrysogenum* (2), *P. cyclopium* (3), *T. harzianum* (4), *M. racemosus* (5), and comparison of biodegradation ability among fungi (6).

Over a period of 16 days, the tested fungi were degraded different amount of ASs, depending on fungi species and their locality (wastewater sample) [49–53]. The initial concentration of ASs in D3 medium decreased continuously with the growth and development of mycelia. The highest biodegradation rate was observed during the first exponential growth phase of *A. niger* **[Figure 2(1a)]**, *P. chrysogenum* **[Figure 2(2a)]**, *T. harzianum* **[Figure 2(4a)]**, and *M. racemosus* **[Figure 2(5a)]** or second exponential growth phase of *P. cyclopium* **[Figure 2(3a)]**. The current results are in line with the results of other authors who found that biodegradation rate of surfactants changes concomitant with cellular growth [54, 55]. **Figure 2(6)** shows a comparison of biodegradation capabilities of the fungi. Over a period of 16 days, *A. niger, P. chrysogenum, P. cyclopium, and T. harzianum* decomposed a total of 30.0, 50.2, 46.97, and 74.27% of ASs, which is equivalent to 180.20 μg/mL **[Figure 2(1b)]**, 301.0 μg/mL **[Figure 2(2b)]** and 281.80 μg/mL **[Figure 2(3b)]**, and 445.0 μg/mL **[Figure 2(4b)]** of ASs. At the same time, *M. racemosus* removed about 49.50 and 62.2% or 300.0 and 620.0 μg/mL of ASs **[Figure 2(5b)]** from D3 and D5 media, respectively.

From the equation of regression curves for each fungus (**Figure 2**) it was calculated the time needs for biodegradation at 80% of ASs (detergent). By this statistical test, the time predicted for the removing 80% of ASs, using *A. niger* **[Figure 2(1a)]**, *P. chrysogenum* **[Figure 2(2a)]**, *P. cyclopium* **[Figure 2(2a)]**, *T. harzianum* **[Figure 2(4a)]**, in applied experimental conditions, was calculated: 41.6, 26.1, 24.8, and 16.8 days, respectively. The *M. racemosus* **[Figure 2(5a)]** will be able to remove 80% of ASs for 24.3–18.3 days, depending on applied concentration. These statistical data indicate that ASs of detergent satisfying required the limit for 80% biodegradability in applied experimental conditions when *T. harzianum* and *M. racemosus* were used. Bearing this in mind their resentence to high detergent's concentrations, they could be used as test organisms in laboratory biodegradation studies. The main reason of their strong resistance to tested detergent could be their origin (industrial wastewater) and adaptation to a high concentration of detergents. However, despite the fact that they isolated from same wastewater sample, they had different response to different concentrations of detergent. Obviously, morpho-physiological characteristics of fungi have significant influence on its biodegradation capacity.

As the literature review does not provide quantitative information about the capacity of fungi to degrade ASs, the current results can only be compared with results obtained on bacterial species. The biodegradation capacity of *T. harzianum* and *M. racemosus* is very similar to the capacity of *Pseudomonas* spp., which can reduce the level of ASs up to 70% after 20 days [56]. The results of degradation of surfactants using a few bacteria strains that are available in the literature are far better compared to the current results. For example, according to Schleheck et al. [57], *Citrobacter* spp. can decompose about 90% of ASs after 35 hours of growth. Hosseini et al. [58] revealed that *Acinetobacter johnson* can use about 94% of the initial concentration of the SDS in the medium during 5-day growth. Ojo and Oso [59] investigated the capacity of two types of bacteria MH1 and MH2 to degrade LAS on alkaline pH and ambient temperature and found high biodegradation rate (93.6 and 84.6%, respectively) of LAS after 5 days. However, in above-mentioned studies with bacteria, the pure ASs were tested and their concentrations were far less compared to the current study. In our study, a commercial-type anionic detergent was used, whose composition is very complex and contains a variety of toxic substances in addition to ASs. Because of, the current results are very important from the aspect of purification of wastewaters which contain commercial detergents.

3.4. Changes of physicochemical parameters (pH and redox potential)

The normal functioning of basic cell processes and biodegradation reactions are closely related with acid-base and oxidation-reduction reactions. The acidic pH value (range 4.5–5.0) is necessary for the optimal growth of most fungi. On the other hand, the pH values between 6.5 and 8.5 are optimal for biodegradation processes in most aquatic and terrestrial systems, and pH values between 5.0 and 9.0 are considered acceptable. Bearing in mind their significant impact on mentioned processes, this study considered the changes of the pH values of media in all phases of fungal growth. The results are presented in **Table 2**.

The initial pH values recorded in C, D3, and D5 media were 4.80, 9.35, and 9.85 units, respectively. During cultivation of fungi, the pH values of C, D3, and D5 media were changing in relation to their composition, type of fungi, and the growth phases. These changes are influenced by the uptake of anions or cations from the medium by the fungal cells [60, 61] and excretion of organic acids in medium [62]. Over a period of 16 days, the pH values of the C media of *A. niger, P. chrysogenum*, and *T. harzianum* decreased toward strong acidic range. Contrarily, it was noted that the pH values of *M. racemosus* and *P. cyclopium* increased toward slightly acidic (neutral) range. The most significant changes in the pH value of C media were recorded during the exponential growth phase of fungi. From 3rd to 9th day, the pH values of *A. niger, P. chrysogenum*, and *T. harzianum* were dropped from 4.80 to 2.53, 3.42 and 4.56 units, respectively. The pH values of *P. cyclopium* and *M. racemosus* were increased from 4.80 to 6.92 and 6.13 units, respectively. These significant changes recorded in the pH values can be interpreted by intensive metabolism of fungi. Thereafter, during the stationary phase and autolysis, the changes in the pH value of the C media were less pronounced [40–53].

The pH values of the D3 media decreased throughout the cultivation time, except in phase of autolysis. The most significant changes in the pH values were observed during the first exponential growth phase of *P. chrysogenum* (from 9.12 to 6.07 units) followed by *M. racemosus* (from 9.40 to 6.24 units), *A. niger* (from 9.13 to 6.49 units), *T. harzianum* (from 9.05 to 6.07 units) or second exponential growth phase of *P. cyclopium* (from 8.75 to 6.33 units). Similar to the C medium, the changes of pH values were less pronounced during the stationary phase and autolysis [49–53]. The obtained results are consistent with results of other authors who confirmed decreasing of pH media during extensive mycelium development of *Glomus intraradices* [63], *Fusarium oxysporum* [64], etc. The obtained results suggest that fungi have different mechanisms for regulation of environmental pH, which depend on the initial pH. It could be speculated that the presence of detergent in the medium, which can be considered as strong alkali agent, probably induces the expression of different sets of genes, compared to C medium, as reflected in the regulation of external pH. Based on literature data, fungi can response to alkaline pH by two possible mechanisms: proteolytic activation of PacC transcription factors (*A. niger, C. albicans, S. cerevisiae*, etc.) or calcium-mediated pathway.

The *Eh* clearly influence the development of microorganisms. Each microorganism type is adapted to specific *Eh* conditions and is characterized by its ability to develop within a wide of narrow *Eh* range. The concentration levels of oxidants or reductants have an impact on the enzymatic activity via effects on three-dimensional conformation. Many investigations demonstrate the influence of the *Eh* value on the activity of some enzymes such as ADP-glucose pyrophos-

pH (Units)					Redox potential (mV)		
Fungi	**Day**	**C**	**D3**	**D5**	**C**	**D3**	**D5**
A. niger	3	3.92 ± 0.10	9.13 ± 0.26	-	390 ± 4	91 ± 2	-
	6	3.12 ± 0.15	6.49 ± 0.20	-	434 ± 2	238 ± 0.4	-
	9	2.53 ± 0.10	7.06 ± 0.10	-	457 ± 2	207 ± 5	-
	12	2.73 ± 0.18	5.63 ± 0.15	-	441 ± 5	309 ± 3	-
	16	2.52 ± 0.12	5.59 ± 0.28	-	435 ± 5	392 ± 5	-
P. chrysogenum	3	5.26 ± 0.10	9.12 ± 0.25	-	312 ± 0.1	91 ± 0.4	-
	6	3.77 ± 0.15	6.07 ± 0.20	-	403 ± 5	261 ± 2.5	-
	9	3.42 ± 0.23	6.36 ± 0.29	-	397 ± 3	246 ± 2	-
	12	3.44 ± 0.19	6.05 ± 0.25	-	401 ± 10	271 ± 5	-
	16	3.61 ± 0.15	7.02 ± 0.16	-	388 ± 8	210 ± 1	-
P. cyclopium	3	5.78 ± 0.14	7.89 ± 0.23	-	281 ± 0.1	101 ± 0.5	-
	6	6.68 ± 0.18	7.89 ± 0.25	-	229 ± 0.5	111 ± 0.5	-
	9	6.92 ± 0.10	7.29 ± 0.19	-	213 ± 2	247 ± 1	-
	12	5.63 ± 0.15	7.37 ± 0.15	-	283 ± 2.5	279 ± 1	-
	16	6.21 ± 0.20	6.48 ± 0.25	-	272 ± 2	243 ± 2	-
T. harzianum	3	5.40 ± 0.15	8.95 ± 0.26	-	299 ± 1	90 ± 2	-
	6	4.82 ± 0.20	8.75 ± 0.24	-	336 ± 0.5	97 ± 0.5	-
	9	4.56 ± 0.10	6.33 ± 0.15	-	341±2.5	261 ± 0.1	-
	12	4.67 ± 0.18	5.90 ± 0.15	-	335 ± 0.5	290 ± 2	-
	16	4.44 ± 0.12	6.48 ± 0.28	-	344 ± 2	301 ± 2	-
M. racemosus	3	5.24 ± 0.10	6.24 ± 0.25	9.36 ± 0.14	313 ± 1.6	250 ± 2	79 ± 2
	6	6.13 ± 0.15	6.37 ± 0.20	6.46 ± 0.10	261 ± 1.8	245 ± 2.5	240 ± 1.4
	9	6.01 ± 0.23	6.14 ± 0.29	6.89 ± 0.05	262 ± 2.5	257 ± 1	217 ± 2
	12	5.87 ± 0.19	6.36 ± 0.25	6.31 ± 0.12	270 ± 0.2	294 ± 2	257 ± 2.2
	16	5.80 ± 0.15	5.50 ± 0.16	5.62 ± 0.08	272 ± 2	297 ± 2	290 ± 2.5

Table 2. Changes in the pH and redox potential values of media.

phorylase [65], the activity and binding of α-glucan, water dikinase (SEX1) to starch granules [66], the activity of β-amylase [67]. The *Eh* values within a range from +100 to +350 mV indicate well-aerated conditions that are optimal for biodegradation processes. However, the addition of detergent in medium caused a significant shift in *Eh*, as **Table 2** shows. The initial *Eh* values of C, D3, and D5 media (before inoculation) were 340, 80, and 60 mV, respectively. From inoculation until the 16th day, the changes in *Eh* values of nutrient media of fungi were expressed with different intensity respect to the type of media, fungi species, and growth phase. During the growth of *A. niger, P. cyclopium, T. harzianum*, and *M. racamosus*, the *Eh* value decreased in C medium.

In contrast, the *Eh* value of *P. chysogenum* increased during most time of cultivation, with exception in first 3 days. In D3 medium, the *Eh* value increased during the growth of all fungi species, without exception. Statistically significant changes were observed during the exponential growth phase whereas during stationary and autolysis phase these changes were insignificant [51, 52].

3.5. Fungal alkaline protease (EC 3.4.21-24) activity

Alkaline proteases have been maximally exploited in food, leather, silk, detergent industries, and waste management. The use of alkaline protease as active ingredient in laundry detergent is the single largest application of this enzyme [68]. From this aspect, isolation and characterization of new promising microbial strains is a continuous process [69]. For the practical application of alkaline proteases in detergents industry, the following conditions are important: their compatibility with the detergent, efficiency at lower temperatures, and stability.

The current study investigated the effect of tested commercial detergent on alkaline protease activity of selected fungi species, and results are presented in **Table 3**. In C medium, the maximum enzyme activity was produced by *P. cyclopium* (0.73 IU/mL) followed by *P. chrysogenum* (0.31 IU/mL), *T. harzianum* (0.27 IU/mL), *A. niger* (0.18 IU/mL), and at least *M. racemosus* (0.15 IU/mL). The addition of detergent in the C medium influenced the changes of enzyme activity depending on the fungi species. The detergent at a concentration of 0.3% showed slight inhibitory effect on alkaline protease activity of *P. cyclopium* (for 12.3%) and strong inhibitory effect of enzyme activity of *P. chrysogenum* (for 89.87%). The detergent at concentrations of 0.3 and 0.5% inhibited enzyme activity of *M. racemosus* for 81.7 and 21.57%, respectively. On the other hand, the activity of alkaline protease of *A. niger* and *T. harzianum* was enhanced in presence of tested detergent for 372 and 128.0%, respectively [49, 51].

Overview of the literature provides contradictory results about the impact of pure surfactants or commercial detergents on activity and stability of alkaline proteases of microbial origin. Choudhary and Jain [70] have reported the detergent compatibility of the alkaline protease of *Aspergillus* sp., but enzyme was not able to retain maximum activity more than 1 hour of incubation. Rani et al. [71] found that alkaline protease from *Aspergillus flavus* AS2 retained 56–92% activity in presence of commercial detergents in following range: Rin (India) < Surf Excel (India) < Tween 80 < Ariel (India) <Tween 20. According to Sankeerthana et al. [72], alkaline proteases from *A. niger* and *A. flavus* retained about 75–70% activity in the most of tested commercial detergents with maximum activity (65–85%) obtained in Surf Excel (India). Niyonzima and More [73] observed that protease of *Aspergillus terreus* gr. was 100% stable and compatible for 2 hours at 60°C with all the detergents except for Super wheel (India), and retained of 84–89% activity after 24 hours at 60°C in following order: Super wheel < More choice < Ariel < Henko < Surf excel (all from India). The alkaline protease was also active and retained 79.1–83.2% and 55.8–75.1% of activity in the presence of tested detergents at 28 and 90°C, respectively, after 24 hours. The different proteolytic activities of fungi in the presence of detergents could be the consequence of experimental conditions (medium composition, aeration, temperature, etc.) as well as morphological characteristics and genetic basis of fungi. Therefore, examination of the above mentioned effects on each individual type of fungus is very important and justified. Finally, this study clearly indicates that alkaline protease of *A. niger*, *T. harzianum*, and *P. cyclopium* could be used as an additive in formulation of tested detergent.

Alkaline protease activity					Alkaline phosphatase activity		
Fungi	Day	C	D3	D5	C	D3	D5
A. niger	3	0.18 ± 0.15	0.09 ± 0.26	-	12.60 ± 0.15	1.65 ± 0.26	-
	6	0.001 ± 0.20	0.67 ± 0.24	-	0.07 ± 0.20	17.74 ± 0.24	-
	9	0.002 ± 0.10	0.001 ± 0.15	-	21.57 ± 0.10	24.14 ± 0.15	-
	12	0	0.001 ± 0.15	-	0	24.31 ± 0.15	-
	16	0.13 ± 0.12	0 ± 0.28	-	2.57 ± 0.12	11.19 ± 0.28	-
P. chrysogenum	3	0.004 ± 0.10	0.005 ± 0.25	-	5.10 ± 0.10	9.79 ± 0.25	-
	6	0.03 ± 0.15	0.001 ± 0.20	-	0.21 ± 0.15	8.40 ± 0.20	-
	9	0.15 ± 0.23	0.03 ± 0.29	-	25.00 ± 0.23	1.17 ± 0.29	-
	12	0.31 ± 0.19	0.01 ± 0.25	-	3.45 ± 0.19	8.23 ± 0.25	-
	16	0.29 ± 0.15	0.01 ± 0.16	-	0.10 ± 0.15	18.29 ± 0.16	-
P. cyclopium	3	0.26 ± 0.14	0.03 ± 0.23	-	31.71 ± 0.14	18.36 ± 0.23	-
	6	0.01 ± 0.18	0.64 ± 0.25	-	96.58 ± 0.18	9.76 ± 0.25	-
	9	0.73 ± 0.10	0.03 ± 0.19	-	22.26 ± 0.10	19.14 ± 0.19	-
	12	0.22 ± 0.15	0.09 ± 0.15	-	12.57 ± 0.15	4.79 ± 0.15	-
	16	0.003 ± 0.23	0.01 ± 0.20	-	0.35 ± 0.23	15.21 ± 0.20	-
T. harzianum	3	0.23 ± 0.15	0.05 ± 0.26	-	0.55 ± 0.15	9.57 ± 0.26	-
	6	0.27 ± 0.20	0.59 ± 0.24	-	0.07 ± 0.20	0.02 ± 0.24	-
	9	0.01 ± 0.10	0.03 ± 0.15	-	0.55 ± 0.10	21.37 ± 0.15	-
	12	0 ± 0	0.63 ± 0.15	-	10.21 ± 0.18	4.87 ± 0.15	-
	16	0.03 ± 0.12	0.01 ± 0.28	-	4.21 ± 0.12	26.24 ± 0.28	-
M. racemosus	3	0.001 ± 0.10	0.03 ± 0.25	0.10 ± 0.01	2.43 ± 0.10	0.07 ± 0.25	2.41 ± 0.21
	6	0.15 ± 0.15	0 ± 0	0.03 ± 0.01	73.34 ± 0.15	4.86 ± 0.20	8.98 ± 1.45
	9	0.14 ± 0.23	0.001 ± 0.29	0.001 ± 0	26.64 ± 0.23	28.64 ± 0.29	24.48 ± 1.92
	12	0.15 ± 0.19	0.002 ± 0.25	0.07 ± 0.02	9.00 ± 0.19	28.56 ± 0.25	0.43 ± 0.02
	16	0.002 ± 0.15	0.002 ± 0.16	0.013 ± 0.01	6.54 ± 0.15	2.43 ± 0.16	16.26 ± 0.16

Table 3. Activity of alkaline protease and phosphatase of fungi.

3.6. Activity of alkaline phosphatase (EC 3.1.3.1) of fungi

Alkaline phosphatase (ALP) enzyme hydrolyzes the phosphomonoesters from number of organic molecules like ribonucleotides, deoxyribonucleotides, proteins, alkaloids, phosphate esters, and anhydrides of phosphoric acid [74] ALP enzymes are involved in various biological processes (cell cycle, differentiation, etc.) and industries; therefore have a wide range of applications [75]. Since the relevant literature provide the information about production of the enzyme only using bacterial strains, the current study investigated the potential of selected fungi to produce ALP. The obtained results are presented in **Table 3**.

In C medium, the maximum enzyme activity was produced by *P. cyclopium* (96.58 IU/mL) followed by *M. racemosus* (73.84 IU/mL), *P. chrysogenum* (25.0 IU/mL), *A. niger* (21.57 IU/mL), and *T. harzianum* (10.21 IU/mL). The addition of detergent in growth medium influenced the inhibition of enzyme activity of *P. cyclopium* (81.18%), *P. chrysogenum* (for 26.86%), and *M. racemosus* (for 61.33 to 66.85%; at applied concentration of 0.3 and 0.5%, respectively). The inhibition of ALP activity by detergent is understood, considering the specific action of the enzyme on *ß*-glycerophosphate and the composition of the growth media [52, 53]. According to Aseri et al. [76], ALP hydrolyses more easily monoesters originating from the carbohydrate metabolism than ester bonds in the alkyl chain of surfactants. The study of Koffiet al. [77] found that SDS has a strong inhibitory effect (about 98%) on phosphatase activity. The current results are in agreement with observations of mentioned authors with exception of *A. niger and T. harzianum*. As **Table 3** shows, the ALP activity of *A. niger and T. harzianum* was slightly (for 12.70%) or significantly (for 156.86%) enhanced by detergent. Finally, this study showed for the first time that fungi grown in Czapek-Dox liquid medium, in applied experimental conditions, can be considered as significant source of ALP. Moreover, the addition of a commercial detergent in liquid medium with *A. niger and T. harzianum* can be used as a strategy for improving the enzyme activity. The knowledge obtained in this study on ALP can have considerable effort in application of tested fungi in biotechnology and waste management, and provides a good base for further investigation in this manner.

4. Conclusions

The main conclusion of this study is that all fungi showed the ability to degrade a high concentration of tested detergent during experimental period of 16 days. The fungi *T. harzianum* and *M. racemosus* had the best biodegradation ability, which is expected since they were isolated from industrial wastewater of Henkel Factory (Kruševac, Serbia). Second conclusion, the alkaline protease and phosphatase activities of *A. niger* and *T. harzianum* were significantly enhanced by detergent. On the other hand, the alkaline protease of *P. cyclopium* and alkaline phosphatase of *P. chrysogenum* retained a high percentage of activities in the presence of detergent. The obtained results could have practical application of tested fungi in bioremediation processes and in biotechnology.

Acknowledgements

This research was financially supported by Serbian Ministry of Education, Science and Technology Development (Grant numbers III 43004).

Author details

Violeta D. Jakovljević[1]* and Miroslav M. Vrvić[2]

*Address all correspondence to: jakovljevicvioleta@gmail.com

1 Faculty of Science, Institute for Biology and Ecology, University of Kragujevac, Kragujevac, Serbia

2 Faculty of Chemistry, Department for Biochemistry, Belgrade, Serbia

References

[1] Ghai VU. Soap Nut Detergent—The Best HE Detergent. [Internet]. 2010. http://www.ezine articles.com/ [Accessed: 29-November-2010)

[2] Okpokwasili GO, Nwabuzor CN. Primary biodegradation of anionic surfactants in laundry detergents. Chemosphere. 1988;**17**:2175-2182. doi.org/10.1016/0045-6535(88)90165-8. DOI: 10.1016/0045-6535(88)90165-8#doilink

[3] Ying GG. Fate, behaviour and effects of surfactants and their degradation products in the environment. Environmental International. 2006;**32**:417-431. DOI: 10.1016/j.envint.2005.07.004

[4] Trehy ML, Gledhill WE, Orth RG. Determination of linear alkylbenzene sulfonates and dialkyltetralinsulfonates in water and sediment by gas chromatography/mass spectrometry. Analytical Chemistry. 1990;**62**:2581-2586. DOI: 10.1021/ac00222a012

[5] Nomura Y, Ikebukuro K, Yokoyama K, Takeuchi T, Arikawa Y, Ohno S, Karube I. Application of a linear alkylbenzene sulfonate biosensor to river water monitoring. Biosensors & Bioelectronics. 1998;**13**:1047-1053. doi.org/10.1016/S0956-5663(97)00077-8. DOI: 10.1016/S0956-5663(97)00077-8#doilink

[6] Swisher RD. Biodegradation of ABS in relation to chemical structure. Journal of Water Pollution Control Federation. 1963;**35**(7):877-892. http://www.jstor.org/stable/25034871

[7] Goel G, Kaur S. A study on chemical contamination of water due to household laundry detergents. Journal of Human Ecology. 2012;**38**:65-69. http://www.indiaenvironmentportal.org.in

[8] Eichhorn P, Rodriguez SV, Baumann W, Knepper T. Incomplete degradation of SDS in Brazilian surface waters and pursuit of their polar metabolites in drinking waters. Science of the Total Environment. 2002;**284**:123-134. doi.org/10.1016/S0048-9697(01)00873-7. DOI: 10.1016/S0048-9697(01)00873-7#doilink

[9] Cook AM. Sulfonated surfactants and related compounds: Facets of their desulfonation by aerobic and anaerobic bacteria. Tenside Surfactants Detergent. 1998;**35**(1):52-56

[10] Berna JL, Ferrer J, Moreno A, Prats D, Ruiz F. The fate of LAS in the environment. Tenside Surfactants Detergents. 1989;**26**:101-107. http://www.lasinfo.eu/images/Documents/fate_of_las_in_the_environment.pdf

[11] European Parliament Regulation (EC) No 648/2004 of the European Parliament and the Council of 31 March 2004 of detergents. EC No 648/2004, In Official Journal of the European Union, 28 April 2004. http://www.wipo.int/edocs/lexdocs/laws/en/eu/eu073en.pdf

[12] Araujo R, Casal M, Cavaco-Paulo A. Application of enzymes for textile fibers processing. Biocatalysis and Biotransformation. 2008;**26**:332-349. doi.org/10.1080/10242420802390457

[13] Mo JH, Lee YH, Kirn J, Jeong JY. Treatment of day aqueous solutions using nanofiltration polyamide composite membranes for the day wastewater reuse. Dyes and Pigments. 2008;**76**:429-439. doi.org/10.1016/j.dyepig.2006.09.007. DOI: 10.1016/j.dyepig.2006.09.007#doilink

[14] Bhatnagar S, Kumari R. Bioremediation: A sustainable tool for environmental management—a review. Annual Review and Research in Biology. 2013;**3**:974-993. http://imsear.hellis.org/handle/123456789/162312

[15] Jeon CO, Madsen EL. *In situ* microbial metabolism of aromatic hydrocarbon environmental pollutants. Current Opinion in Biotechnology. 2012;**24**:1-8. doi.org/10.1016/j.copbio.2012.09.001. DOI: 10.1016/j.copbio.2012.09.001#doilink

[16] Lisowska K, Szemraj J, Rózalska S, Dlugonski J. The expression of cytochrome P-450 and cytochrome P-450 reductase genes in the simultaneous transformation of corticosteroids and phenanthrene by *Cunninghamella elegans*. FEMS Microbiology Letters. 2006;**261**:175-180. DOI: 10.1111/j.1574-6968.2006.00339.x

[17] Slaba M, Szewczyk R, Bernat P, Dlugonski J. Simultaneous toxic action of zinc and alachlor resulted in enhancement of zinc uptake by the filamentous fungus *Paecilomyces marquandii*. Science of the Total Environment. 2009;**407**:4127-4133. doi.org/10.1016/j.scitotenv.2009.03.023. DOI: 10.1016/j.scitotenv.2009.03.023#doilink

[18] Rózalska S, Szewczyk R, Dlugonski J. Biodegradation of 4-*n*nonylphenol by the non-ligninolytic filamentous fungus *Gliocephalotrichum simplex*: A proposal of a metabolic pathway. Journal of Hazardious Materials. 2010;**180**:323-331. doi.org/10.1016/j.jhazmat.2010.04.034. DOI: 10.1016/j.jhazmat.2010.04.034#doilink

[19] Rózalska S, Pawlowska J, Wrzosek M, Tkaczuk C, Dlugonski J. Utilization of 4-n-nonylphenol by *Metarhizium* sp. isolates. Acta Biochimica Polonica. 2013;**60**:677-682. http://www.actabp.pl/pdf/4_2013/677.pdf

[20] Bernat P, Długoński J. Comparative study of fatty acids composition during cortexolone hydroxylation and tributyltin chloride (TBT) degradation in the filamentous fungus *Cunninghamella elegans*. International Biodeterioration & Biodegradation. 2012;**74**:1-6. doi.org/10.1016/j.ibiod.2012.07.001. DOI: 10.1016/j.ibiod.2012.07.001#doilink

[21] Stojanović J, Jakovljević V, Matović I, Mijušković Z, Nedeljković T. The influence of detergents, sodium tripoly-phosphates and ethoxyled oleyl-cetyl alcohol on metabolism of the fungi *Penicillium verrucosum* Peyronel. Acta Veterinaria (Beograd). 2010;**60**:67-77. DOI: 10.2298/AVB1001067S

[22] Stojanović J, Jakovljević V, Matović I, Gajović O, Mijušković Z, Nedeljković T. Influence of detergent on metabolic activity of fungi *Aspergillus niger*. Natural Science. 2011;**3**:466-470. DOI: 10.4236/ns.2011.36064 DOI:10.4236/ns.2011.36064#_blank

[23] Stojanović J, Milićević J, Gajović O, Jakovljević V, Matović I, Mijušković Z, Nedeljković T. The effects of detergent, sodium tripoly-phosphate and ethoxyled oleyl-cetyl alcohol on metabolic parameters of the fungus *Trichothecium roseum* Link. Archive of Biological Sciences. 2011;**63**:1001-1006. DOI: 10.2298/ABS1104001S

[24] Grbavčić S, Bezbradica D, Izrael-Živković L, Avramović N, Karadžić I, Kneževic-Jugović Z, Milosavić N. Production of lipase and protease from an indigenous *Pseudomonas aeruginosa* strain and their evaluation as detergent additives: Compatibility study with detergent ingredients and washing performance. Bioresource Technology. 2011;**102**: 11226-11233. doi.org/10.1016/j.biortech.2011.09.076. DOI: 10.1016/j.biortech.2011.09.076#doilink

[25] Rahman RNZA, Geok LP, Basri M, Salleh AB. Physical factors affecting the production of organic solvent-tolerant protease by *Pseudomonas aeruginosa* strain K. Bioresource Technology. 2005;**96**:429-436. doi.org/10.1016/j.biortech.2004.06.012. DOI: 10.1016/j.biortech.2004.06.012#doilink

[26] Gupta A, Roy I, Khare SK, Gupta, MN. Purification and characterization of a solven stable protease from *Pseudomonas aeruginosa* PseA. Journal of Chromatography A. 2005;**1069**:155-161. doi.org/10.1016/j.chroma.2005.01.080. DOI: 10.1016/j.chroma.2005.01.080#doilink

[27] Torriani A. From cell membrane to nucleotides: The phosphate regulon in *Escherichia coli*. Bioessays. 1990;**12**:371-376. DOI: 10.1002/bies.950120804

[28] Suzuki C, Ueda H, Tsumoto K, Mahoney WC, Kumagai I, Nagamune T. Open sandwich ELISA with VH-/VL-alkaline phosphatase fusion proteins. Journal of Immunological Methods. 1999;**224**:171-184. doi.org/10.1016/S0022-1759(99)00020-4. DOI: 10.1016/S0022-1759(99)00020-4#doilink

[29] Chaudhuri G, Chatterjee S, Venu-Babu P, Ramasamy K, Thilagaraj WR. Kinetic behaviour of calf intestinal alkaline phosphatase with pNPP. Indian Journal of Biochemistry & Biophysics. 2013;**50**:64-71. http://hdl.handle.net/123456789/16064

[30] Torriani A. Alkaline phosphatase subunit and their dimerization *in vivo*. Journal Bacteriology. 1968;**96**:1200-1207. http://jb.asm.org/content/96/4/1200.full.pdf

[31] Friedberg I, Avigad G. Some properties of alkaline phosphatase of *Ps. flourescens*. European Journal Biochemistry. 1967;**1**:193-198. DOI: 10.1111/j.1432-1033.1967.tb00063.x

[32] Wolfenden R, Spence G. Depression of phosphomonoesterase and phosphodiesterase activities in *Aerobacter aerogenes*. Biochimica et Biophysica Acta (BBA)—Enzymology. 1967;**146**:296-298. doi.org/10.1016/0005-2744(67)90099-X. DOI: 10.1016/0005-2744(67)90099-X#doilink

[33] Takeda K, Tsugita A. Phosphoesterase of *B. subtilis*. II crystallization and properties of alkaline phosphatase. Journal of Biochemistry. 1967;**61**:231-241. https://www.jstage.jst.go.jp/article/biochemistry1922/61/2/61_2_231/_pdf

[34] Eaton AD, Clesceri LS, Rice EW, Greenberg AE, editors. American Public Health Association (APHA) Standard Methods for the Examination of Water and Wastewater. Washington: American Water Works Association; Water Pollution Control Federation; 2005. http://www.efn.unc.edu.ar/otros/bibliocentro/index_archivos/LS44.pdf

[35] Anson ML. The estimation of pepsin, trypsin, papain, and cathepsin with hemoglobin. Journal of General Physiology. 1938;**20**:79-89. https://www.ncbi.nlm.nih.gov/pmc/articles/PMC2213732/pdf/79.pdf

[36] Heinonen JK, Lahti RJ. A new and convenient colorimetric determination of inorganic orthophosphate and its application to the assay of inorganic pyrophosphatase. Analytical Biochemistry. 1981;**113**:313-317. doi.org/10.1016/0003-2697(81)90082-8. DOI: 10.1016/0003-2697(81)90082-8#_blank#Persistent link using digital object identifier

[37] Tieghem, P. van. Description d'une nouvelle espèce d'*Aspergillus: A. niger*. Annales des Sciences Naturelles Botanique. 1867;**8**:240-244

[38] Thom C. Cultural studies of species of *Penicillium*. U.S.D.A. Bureau of Animal Industry Bulletin. 1910;**118**:1-107

[39] Westling R. Über die grünen Spezies der Gattung *Penicillium*. Arkiv før Botanik. 1911; **11**(1):1-156

[40] Rifai MA. A revision of the genus *Trichoderma*. Mycological Papers. 1969;**116**:1-56

[41] Fungi of Great Britain and Ireland [Internet]. 2017. Available from: http://fungi.myspecies.info/sites/fungi.myspecies.info/files/394230-Aspergillus-nMEAa.jpg

[42] Mold and Bacteria Consulting Laboratories (MLB) Inc., Ontario, Canada [Internet]. 2004. Available from: https://www.moldbacteria.com/wp-content/uploads/penicillium-chrysogenum-spores.jpg

[43] MycoBank. International Mycological Association [Internet]. 2016. Available from: http://www.mycobank.org/BioloMICS.aspx?TargetKey=14682616000002126&Rec=5769

[44] Harman GE. *Trichoderma spp.*, including *T. harzianum, T. viride, T. koningii, T. hamatum,* and other *spp. Deuteromicetes, Moniliales* (asexual classification system. Cornel University College of Agriculture and Life Sciences, New York. 2000. [Internet]. Available from: https://biocontrol.entomology.cornell.edu/pathogens/trichoderma.php

[45] Schipper MAA. On *Mucor circinelloides, Mucor racemosus* and related species. Studies in Mycology 1976; **12**:1-40. Available from: http://www.bcrc.firdi.org.tw/fungi/fungal_detail.jsp?id=FU200802070052

[46] Fresenius G. Beiträge zur Mykologie. Heinrich Ludvig Brömmer Verlag, Frankfurt; 1850. 38 p.

[47] Alifyah Y, Kagalwala, Kavitha K. Effects of surfactant (Sodium Lauryl Sulphate) on *Hydrilla verticillata*. International Journal of Life Sciences Biotechnology and Pharma Research. 2012;**1**:128-138. http://www.ijlbpr.com/uploadfile/2015/0413/20150413053903337.pdf

[48] Garon D, Krivobok S, Wouessidjewe D, Seigle-Murandi F. Influence of surfactants on solubilization and fungal degradation of fluorene. Chemosphere. 2002;**47**:303-309. doi.org/10.1016/S0045-6535(01)00299-5. DOI: 10.1016/S0045-6535(01)00299-5#doilink

[49] Jakovljević VD, Vrvić MM. Capacity of *Aspergillus niger* to degrade anionic surfactants and coproduce the detergent compatible enzymes. Applied Biochemistry & Microbiology. 2016;**52**:183-189. DOI: 10.1134/S0003683816020083

[50] Jakovljević VD, Vrvić MM. Potential of *Penicillium cyclopium* Westling for removing of anionic surfactants and biotechnology. Applied Biochemistry and Microbiology. 2015;**51**:704-711. DOI: 10.1134/S000368381506006X

[51] Jakovljević VD, Stojanović JD, Vrvić MM. The potential application of fungus *Trichoderma harzianum* Rifai in biodegradation of detergent and industry. Chemical Industry and Chemical Engineering Quarterly. 2015;**21**:131-139. DOI: 10.2298/CICEQ140414017J

[52] Jakovljević VD, Milićević JM, Stojanović JD. Detergent-like stressor and nutrient in metabolism of *Penicillium chrysogenum*. Biotechnology & Biotechnological Equipment. 2014;**28**:41-53. DOI: 10.1080/13102818.2014.901674

[53] Jakovljević VD, Milićević JM, Stojanović JD, Vrvić MM. The ability of fungus *Mucor racemosus* Fresenius to degrade high concentration of detergent. Chemical Industry and Chemical Engineering Quarterly. 2014;**20**:587-595. DOI: 10.2298/CICEQ130922002J

[54] Schleheck D, Cook AM. ω-Oxygenation of the alkyl side chain of linear alkylbenzene sulfonate (LAS) surfactant in *Parvibaculum lavamentivorans* T. Archives of Microbiology. 2005;**183**:369-377. http://nbn-resolving.de/urn:nbn:de:bsz:352-opus-67247

[55] Velan M, Sheeba Varma S, Gnanambigai P, Brinda Lakshmi M. Biodegradation of toluene in the contaminated soil by *Mycoplana* sp. MVMB2. International Journal of Chemical and Environmental Engineering. 2012;**3**:318-323. http://citeseerx.ist.psu.edu/viewdoc/download?doi=10.1.1.682.7745&rep=rep1&type=pdf

[56] Jerabkova H, Blanka K, Nahlik J. Biofilm of pseudomonas C12B on glass support as catalytic agent for continuous SDS removal. International Biodeterioration & Biodegradation. 1999;**44**:233-241. DOI: 10.1016/S0964-8305(99)00084-0

[57] Schleheck D, Lechner M, Schonenberger R, Suter MJ-F, Cook AM. Desulfonation and degradation of the disulfodiphenylethercarboxylates from linear alkyl-diphenyletherdisulfonate surfactants. Applied and Environmental Microbiology. 2003;**69**:938-944. DOI: 10.1128/AEM.69.2.938-944.2003

[58] Hosseini F, Malekzadeh F, Amirmozafari N, Ghaemi N. Biodegradation of anionic surfactants by isolated bacteria from activated sludge. International Journal of Environmental Science and Technology. 2007;**4**:127-132. DOI: 10.1007/BF03325970

[59] Ojo OA, Oso BA. Isolation and characterization of synthetic detergent-degraders from wastewater. African Journal of Biotechnology. 2008;**7**:3753-3760. DOI: 10.5897/AJB08.530

[60] Griffin DH. Fungal Physiology. New York: John Wiley & Sons; 1994. pp. 375-398

[61] Moore-Landecker E. Fundamentals of Fungi. Upper Saddle River: Pretice Hall; 1996. pp. 15-18

[62] Orlowski M. Mucor dimorphism. Microbiological Reviewers. 1991;**55**:234-258. https://www.ncbi.nlm.nih.gov/pmc/articles/PMC372813/pdf/microrev00033-0052.pdf

[63] Bago B, Vierheilig H, Piche Y, Azcon-Aguilar C. Nitrate depletion and pH changes induced by the extraradical mycelium of the arbuscular mycorrhizal fungus *Glomus intraradices* grown in monoxenic culture. New Phytologist. 1996;**133**:273-280. DOI: 10.1111/j.1469-8137.1996.tb01894.x

[64] Srivastava S, Pathak N, Srivastava P. Identification of limiting factors for the optimum growth of *Fusarium oxysporum* in liquid medium.Toxicology International. 2011;**18**:111-116. DOI: 10.4103/0971-6580.84262. DOI: 10.4103%2F0971-6580.84262#pmc_ext

[65] Ballicora MA, Frueauf JB, Fu Y, Schürmann P, Preiss J. Activation of the potato tuber ADP-glucose pyrophosphorylase by thioredoxin. Journal of Biological Chemistry. 2000;**275**:1315-1320. DOI: 10.1074/jbc.275.2.1315

[66] Mikkelsen R, Mutenda KE, Mant A, Schürmann P, Blennow A. Alpha-glucan, water dikinase (GWD): A plastidic enzyme with redox-regulated and coordinated catalytic activity and binding affinity. Proceedings of the National Academy of Sciences of the USA. 2005;**102**(5):1785-1790. DOI: 10.1073/pnas.0406674102

[67] Balmer Y, Koller A, del Val G, Manieri W, Schürmann P, Buchanan BB. Proteomics gives insight into the regulatory function of chloroplast thioredoxins. Proceedings of the National Academy of Sciences of the USA. 2003;**100**:370-375. DOI: 10.1073/pnas.232703799

[68] Nehra KS, Dhillon S, Chaudhary K, Singh R. Production of alkaline protease by *Aspergillus* species under submerged and solid state fermentation. Indian Journal of Microbiology. 2002;**42**:43-47. http://cat.inist.fr/?aModele=afficheN&cpsidt=13980089

[69] Kumar A, Sachdev A, Balasubramanyam SD, Saxena AK, Nain L. Optimization of conditions for production of neutral and alkaline protease from species of *Bacillus* and

Pseudomonas. Indian Journal of Microbiology. 2002;**42**:233-236. http://cat.inist.fr/?aModele=afficheN&cpsidt=14568782

[70] Choudhary V, Jain PC. Screening of alkaline protease production by fungal isolates from different habitats of Sagar and Jabalpur district (M.P). Journal of Academia and Industrial Research. 2012;**1**:215-220. http://www.jairjp.com/

[71] Rani MR, Prasad NN, Sambasivarao KRS. Optimization of cultural conditions for the production of alkaline protease from a mutant *Aspergillus flavus* As2. Asian Journal of Experimental Biological Sciences. 2012;**3**:565-575. http://shodhganga.inflibnet.ac.in/bitstream/10603/8300/12/12_chapter%205.pdf

[72] Sankeerthana C, Pinjar S, Jambagi RT, Bhavimani S, Anupama S, Sarovar B, Inamdar SR. Production and partial characterization of protease from *Aspergillus flavus* using rice mill waste as a substrate and its comparision with *Aspergillus niger* protease. International Journal of Current Engineering and Technology. 2013;Special Issue **1**:143-147. http://inpressco.com/wp-content/uploads/2013/09/Paper29143-147.pdf

[73] Niyonzima FN, More SS. Purification and characterization of detergent compatible protease from *Aspergillus terreus* gr. 3 Biotech. 2015;**5**:61-70. DOI: 10.1007/s13205-014-0200-6

[74] Mahesh M, Guleria Neha, Rajesh TS, Somashekhar R, Puttaiah ET. Isolation and characterization of extracellular thermostable alkaline phosphatase enzyme from *Bacillus* spp. International Journal of Applied Biology and Pharmaceutical Technology. 2010;**1**:21-33. http://imsear.hellis.org/handle/123456789/157626

[75] Rani K, Datt S, Rana R. Brief review on alkaline phosphatases: An overview. International Journal of Microbiology and Bioinformatics. 2012;**2**:1-4. http://bretjournals.com/sites/default/files/issues-pdf/ijmb201504605_1428121897.pdf

[76] Aseri GK, Neelam J, Tarafdar JC. Hydrolysis of organic phosphate forms by phosphatases and phytase producing fungi of arid and semiarid soils of India. American-Eurasian Journal of Agricultural & Environmental Sciences. 2009;**5**:564-570. https://www.idosi.org/aejaes/jaes5(4)/17.pdf

[77] Koffi D, Faulet B, Gonnety J, Bédikou M, Kouame LP, Zero Bi IA, Niamke SL. Biochemical characterization of phosphatase, *ß*-galactosidase and α-mannosidase activities of seeds of an oleaginous cucurbit: *Lagenaria siceraria* (Molina) Standl blocky-fruited cultivar. Sciences & Nature. 2010;**7**:221-235. http://www.ajol.info/index.php/scinat/article/viewFile/59966/48235

Section 2

Characterization of Surfactants

Chapter 9

Electrochemistry of Surfactants

Eduardo Nicolás Schulz, Erica Patricia Schulz and Pablo Carlos Schulz

Additional information is available at the end of the chapter

http://dx.doi.org/10.5772/67975

Abstract

The application of different electrochemical techniques to surfactant systems, namely polarography and cyclic voltammetry, differential capacitance, chronocoulometry and electrochemical impedance spectroscopy, is reviewed.

Keywords: cyclic voltammetry, polarography, differential capacitance, chronocoulometry, electrochemical impedance spectroscopy, monolayers, micelles, microemulsions

1. Introduction

This chapter addresses the application of several electrochemical methods to the study of surfactant assemblies in both monolayers adsorbed on solid surfaces and free aggregates such as micelles, vesicles and microemulsions. The reviewed techniques are polarography and cyclic voltammetry for free aggregates and differential capacitance, chronocoulometry and electrochemical impedance spectroscopy for adsorbed monolayers.

In some cases, the reliability of the results can be verified with other techniques such as light scattering. However, when alternative methods cannot be applied such as in concentrated or coloured systems, these techniques give complementary and valuable information.

In this work, only some specific details concerning the use of these methods on surfactant systems are explained, while for a general description of the techniques, the reader is referred to electrochemistry books.

2. Polarographic and voltammetric methods applied to surfactant solutions

Polarographic and voltammetric methods allow for the determination of the diffusion coefficient of an electroactive probe. If the electroactive species is attached to a micelle, the measured diffusion coefficient is that of the micelle (D_M), which is related to the size and the shape of the micelles. Therefore, useful information can be derived from these experiments such as the aggregation number or changes in the shape with the concentration. The procedures are simple and can be applied when light scattering cannot be used, such as in coloured, absorbent or very concentrated samples or in systems showing very low refringence index contrast.

The diffusion coefficient (D), together with other properties, such as the intrinsic viscosity, may give information about the particle dimensions. The aggregation number can be obtained from the particles' dimensions and the partial molar volume of the surfactant. If the density of the surfactant is known, the aggregate weight may be computed.

The diffusion coefficient gives a hydrodynamic radius R_H. If the aggregate is a sphere, this radius is that of the particle. Otherwise, R_H is that of a sphere whose hydrodynamic behaviour is equal to that of the actual particle.

The diffusion coefficient is a function of the solute concentration and the temperature. There are two kinds of diffusion coefficients. The mutual or of pair (D_m) diffusion coefficient is obtained measuring the rate of reduction of an imposed concentration gradient of the solute.

On the other hand, the lone particle or self-diffusion coefficient, (D_T) [1] is obtained by following one or several tagged particles through the matrix formed by the other untagged particles and components in a solution with uniform concentration.

Both D_T and D_m quantify different physical processes and have different dependence on concentration. The D_M obtained with the techniques studied here is a self-diffusion coefficient and thus the Einstein's equation applies [2], which is not appropriate for diffusion over very small distances [3]:

$$D_T = k_B T / f_T \tag{1}$$

The coefficient f_T depends on the concentration. There is a theoretical dispute about the role of the '*dynamic friction*', i.e. the increase of f_T arising from direct interactions such as collisions. Several authors [4] are of opinion that the dynamic friction affects f_T Mazo [5] demonstrated that the dynamic friction may have a greater effect on D_T in micellar solutions with scarce swamping (or support) electrolyte.

The accuracy of Eq. (1) to predict the diffusion of macroparticles has been experimentally verified [6].

Stokes stated that the diffusional frictional coefficient for a sphere with radius a_0 moving through a continuous medium with viscosity η is given by $f_T = 6\pi\eta a_0$. Introducing this in Eq. (1), the Einstein-Stokes equation for spherical particles is obtained:

$$D = k_B T / 6pha_o \tag{2}$$

http://dx.doi.org/10.5772/67975

2.1. The determination of the micelle dimensions from self-diffusion coefficients

The hydrodynamic radius (a_0) of the micelles, computed with the Stokes-Einstein equation (2), is influenced by two factors: the effect of the intermicellar interactions and the possible change in size and shape of micelles when the surfactant and/or the supporting electrolyte concentration changes.

The intermicellar interactions may be computed considering a hard sphere and Coulomb repulsions, and a van der Waals attraction through a model proposed by Pusey [6]. The interaction is represented by a unique parameter defined by the effective radius of the hard sphere a_{eff}, and the hydrodynamic radius a_0 obtained from the Stokes-Einstein equation, giving:

$$D_M = D_{M.0}\ [1 + k_D(c - CMC)] \tag{3}$$

where D_M and $D_{M,o}$ are the micelle diffusion coefficient at concentration c and without the effect of interactions, respectively, and

$$k_D = \left[0.5 + 2\ (1 + x)^2(1 + 4x) - 15/8\ \ (1 + x)^{-1}\right]\ \nu \tag{4}$$

where $x = a_{eff}\ /a_0 - 1$ and v is the partial specific volume of micelles. In this model, $a_{eff} \approx a_0 + \kappa^{-1}$, κ^{-1} being the Debye length. In SDS rod-like micelles at 25°C in 0.1 M NaCl $a_{M,o} = (9.6 \pm 0.2) \times 10^{-7}\ cm^2 \cdot s^{-1}$ and $k_D = 15 \pm 1\ cm^3 \cdot g^{-1}$, when c is measured in $g \cdot cm^{-3}$.

In order to compute the ionic strength and then the Debye length, the concentration of free counterions and surfactant ions in equilibrium with micelles must be known. In absence of actual data from ion-selective electrodes, the free surfactant ions concentration [S] is usually taken as the CMC ([S] = CMC) and the free counterions concentration [X] as [X] = CMC + (c − CMC) α, where α is the ionization degree of micelles, which is usually almost invariant in a homologous series.

The ionization degree has been extensively reported in literature and may be readily obtained from conductivity measurements. It is usually assumed that the free surfactant concentration and α are invariant at any concentration above the CMC but it has been demonstrated with ion-selective electrode measurements that this is not always true. In general, α and [S] do not remain constant with concentration. For instance, in disodium n-decane phosphonate solutions the free surfactant and the counterion concentrations increase, while the micelle ionization degree strongly decreases when increasing the concentration above the CMC [7]. The same behaviour of α was observed in amiodarone micelles [8]. In cationic surfactants [S] strongly decreases and counterions concentration [X] monotonically increases at concentrations above the CMC [9–11]. The same behaviour was observed in sodium dehydrocholate micelar solutions [12]. However, in some anionic surfactants, such as SDS [9] sodium perfluorooctanoate [13], and *n*-alkane phosphonic acids [14, 15] the [S] and [X] values above the CMC are almost constant and equal to the CMC. The proper procedure is to measure [S] and [X] by using ion-selective electrodes at each concentration, c. The contribution of micelles to the ionic strength is negligible and may be ignored [16].

If the micellised surfactant molar partial volume V ($cm^3 \cdot mol^{-1}$) is known, the volume fraction of micelles can be computed as $v = V \cdot c_M/1000$, where c_M is the micellised surfactant concentration on a monomer basis, $c_M = c - [S]$. If V is not known, it may be computed from tables of group contributions to the partial molar volume and procedures from literature [17].

If no experimental surfactant molar partial volumes are available, a good estimation may be obtained from the equation [18]:

$$V_{S,m} = V_{CH3} + (n_C - 1)\, V_{CH2} + V_{ph} + n_W\, V_W... \quad (5)$$

where $V_{CH2} = 0.02669 + 0.0000143t$ nm^3 and $V_{CH3} = 0.05108 + 0.0001311t$ nm^3, t being the temperature in °C [19], V_w is the volume of the hydration water molecule in the Stern layer (=0.01038 nm^3 [20]); n_w is the number of water molecules per hydrated micellised surfactant molecule (which is an approximately constant value in an homologous series); V_{ph} is the volume of the polar head group computable on the basis of its structure.

It is necessary to compare the value of a_0 obtained from Eq. (2), with $D_{M,0}$from Eq. (3) ($= a_{M,o}$) with the length of the completely extended surfactant molecule that may be estimated with the equation [20]: l_s (nm) $= 0.13n_C + 0.1704 + 2r_{ph}$, where n_C is the number of carbon atoms in the hydrocarbon chain and r_{ph} is the radius of the hydrated polar headgroup that can be estimated from its structure or from the size of a related ion. For instance, for the carboxylate ($-COO^-$) group, r_{ph} is estimated to be 0.168 nm using the limiting equivalent formiate ion conductance (λ_o): $r_{ph} = ZeF/6\pi\eta\lambda_o$, where F is the Faraday constant and Z the ion charge. For formiate, this yields $r_{ph} = 0.168$ nm [21].

As a refinement, the effective length of the surfactant molecule can be calculated as ρl_s, employing the chain flexibity factor ρ (ρ ≤ 1); ρ is approximately 0.75 for sodium dodecyl sulphate (SDS) [22].

If $a_{M,0} \leq l_s$, it may be assumed that the micelles are spherical. Thus, a_0 is its actual radius and the volume is $V_{sph} = 4\pi a_{M,0}^3/3$.

The micelles are generally not spherical for high surfactant and/or supporting electrolyte concentrations. If $l_s < a_o$, micelles cannot be spherical but they can be rod-like or disk-like. Moreover, the aggregates may not be micelles but vesicles, microemulsion droplets or liposomes. This may be elucidated with turbidity measurements, even with a common photospectrometer. Except for coloured surfactants, micelles are optically transparent, while vesicles and microemulsions show some absorbance. If the system is composed of micelles, they may be rod-like or disk-like.

In this case the hydrodynamic radius is not the true radius of the micelle, but that of a sphere having the same hydrodynamic behaviour. They generally are prolate or (rarely) oblate ellipsoids [21]. Non-spherical micelles are commonly rod-like and may be assumed as prolate ellipsoids and analysed with Eq. (5). In this case it may be taken $b = l_s$, and then the length of the rod may be computed as $L = 2a$.

When micelles are rod-like or disk-like, the hydrodynamic radius is not the true radius of the micelle, but that of a sphere having the same hydrodynamic behaviour. They generally are

http://dx.doi.org/10.5772/67975

prolate or (rarely) oblate ellipsoids [21]. Non-spherical micelles are commonly rod-like and may be regarded as prolate ellipsoids and analysed with Eq. (6), considering the transversal radius $r = l_s$ and the length of the rod $L = 2a$.

Eq. (6) can be numerically solved to obtain the length of a micelle (L) of hydrodynamic radius a_0 [23]:

$$a_0 = \frac{L}{2\sigma - 0.19 - \frac{8.24}{\sigma} + \frac{12}{\sigma^2}} \tag{6}$$

where $\sigma = \ln(L/r)$ and r is the transversal section radius ($r = l_s$).

Disk-like micelles are uncommon [24]. They may be treated as oblate revolution ellipsoids with principal axes a ($a = l_s$) and b ($b > a$), that can be obtained with Eq. (7) [23]. The volume of the micelle is $V_d = \pi b^2 l_s$.

$$a_0 = \frac{a\left(b^2/a^2 - 1\right)^{1/2}}{\tan^{-1}\left[\left(b^2/a^2 - 1\right)^{1/2}\right]} \tag{7}$$

Rod-like micelles are rigid if L is less than 100 nm, whilst longer micelles are flexible. The flexibility has been compared with that of a caterpillar [25].

The volume of a rod-like micelle may be calculated as a cylinder with radius $l_s$and length L - l_s, capped in its extremes by hemispheres with radius l_s [26]: i.e. $V_{rod} = \pi l_s^2 (L - 2l_s) + 4\pi l_s^3/3$.

The partial molar volume of the micellised surfactant ($PMV_{s,m}$) can be obtained from literature or computed from solution density measurements or with the contributions of the different groups taken from literature [17]. The hydration water of micelles must be added to the surfactant molecule volume in the calculations.

The aggregation number n can be estimated as $V_{micelle}/v_s$, where v_s is the molecular volume of the surfactant, $v_s = PMV_{s,m}/N_A$; N_A is the Avogadro's number. The aggregation numbers obtained in this way are in good concordance with those obtained with other methods as light scattering, even with rod-like micelles [21].

It must be taken into account that the above equations are based on simplified models, although they are good approximations. So, it is possible to obtain, from hydrodynamic measurements, the dimensions of an equivalent particle that behaves hydrodynamically as the actual particle [27]. However, in general the approximation is good.

An illustrative application of the above procedures is the study by polarography of the effect on the size, shape and diffusion of disodium n-decane phosphonate micelles when adding two different electrolytes [28]. Under equal conditions, the addition of NaCl produces micelles with an aggregation number one order of magnitude larger than those produced when adding NaOH. This has been attributed to an increase in the effective charge per micellised head group caused by the reaction of OH^- ions with the hydrolysed head groups, mainly present as$-PO_3H^-$ in the micelle Stern layer. This is an uncommon effect since co-ions do not normally affect the size and shape of the micelles.

It can be seen by cross checking the results from voltammetric and polarographic methods with other techniques that the information obtained is very reliable.

2.2. Some applications

Spherical micelles have self-diffusion coefficients of the order of 1.5×10^{-6}–0.6×10^{-6} $cm^2 \cdot s^{-1}$, whereas for rod-like micelles it is about 10^{-8} $cm^2 \cdot s^{-1}$ [21, 28]. The changes in the structure of the aggregates when modifying the system conditions, such as temperature, concentration or added salts, can be followed through the determination of D.

Long rod-like micelles entangle and the diffusion coefficient drops sharply. The length at which sodium hexadecanoate micelles entangle was determined by plotting (in logarithmic scales) the aggregation number n as a function of the counterion concentration [X] (**Figure 1**).

A change in slope indicates the entanglement and the aggregation numbers obtained at higher concentrations were unrealistically high [21].

Cyclic voltammetry has been used to test some assumptions commonly accepted in the study of mixed micelles [29]. The dependence of n on the composition of the surfactant mixture and the total concentration of the catanionic mixed micelles of sodium oleate and hexadecyltrimethylammonium bromide has been analysed under thermodynamic and steric considerations, including the affinity of water molecules with the double bond of the chain of oleate ions. Results suggest that the mixed micelles' composition also will change with c. Therefore, the techniques based on the assumption that the composition of the mixed micelles does not change with concentration must be used with precaution.

As shown in **Figure 2**, the diffusion coefficient of mixed micelles as a function of the total composition and the concentration of the system has a complex behaviour. In **Figure 3**, the largest dimension of the micelles is plotted as a function of the concentration and the composition of the mixtures, showing the evolution from spheres to stiff rods, then to flexible rod-like micelles and finally to entangled micelles. As already mentioned, when the micelles are entagled the values of L are unrealistic due to the restricted movement. Besides, these huge micelles probably include more than one probe molecule and some of these probe molecules may not probably access to the electrode.

Polarography has been also used to study concentrated microemulsions that cannot be studied by light scattering [30]. Zana and Mackay [31] demonstrated that these methods may not only be used to obtain the size of aggregates but also to study the inter-aggregates interactions and the partition of electroactive substances between the aggregates and the solvent.

Polarography and cyclic voltammetry can also be used to determine the critical micelle concentration (CMC). However, this method has no advantage over other simpler methods. Moreover, the inclusion of a hydrophobic probe in the system may induce the formation of micelles at concentrations below the CMC of the pure surfactant.

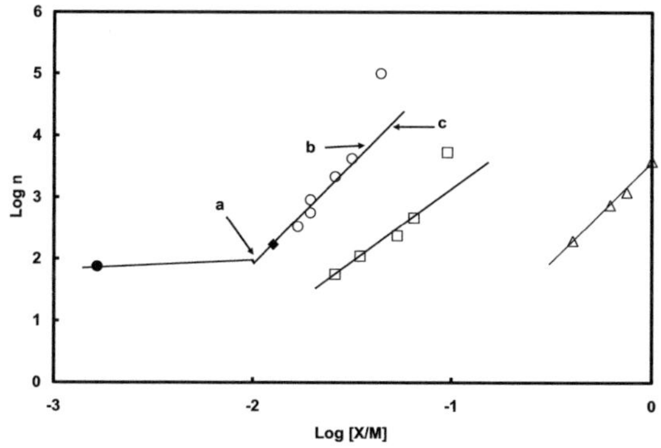

Figure 1. Log n vs. log [X] (counterion concentration), for □ : sodium dodecanoate; ∘: sodium hexadecanoate, Δ: potasium dodecanoate, a: maximum concentration for spherical micelles, b: upper limit for stiff rod-like micelles, c: upper limit for non-entangled rod-like micelles, Redrawn from [21].

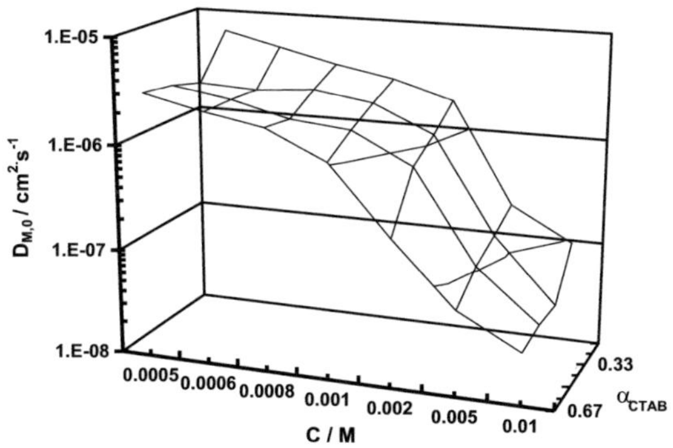

Figure 2. The values of $D_{M,0}$ as a function of concentration and the mole fraction of hexadecyltrimethylammonium bromide (α_{CTAB}.) [29].

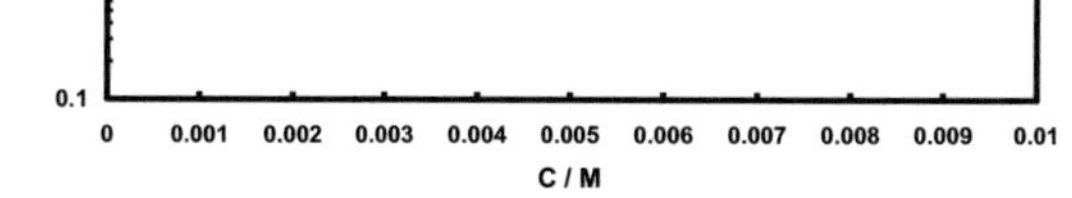

Figure 3. The largest dimension of micelles L as a function of concentration and $\alpha_{CTAB} = 0$ (◊,◆), 0.33 (□,■), 0.5 (Δ,▲) and 0.667 (O, ●). Open symbols represent rod-like micelles, while closed symbols correspond to spherical ones [29].

2.3. Conditions to study micelles

It is necessary to tag the micelles with an electroactive probe in order to study their diffusion coefficient. Hoyer and Novodoff [32] were pioneers using polarography and employing solid cadmium dodecanoate as a probe since the wave of Cd^{+2} is not in the region of the studied surfactant.

The viscosity η used in the Stokes-Einstein equation to determine D of micellar systems is that of the intermicellar solution, which is approximately equal to that at the critical micelle concentration (CMC). If the CMC is low and there are no added salts, this viscosity is close to that of pure water.

When the electrode reaction is controlled by the mass transport, the diffusion current allows the determination of the diffusion coefficient of the electroactive species and it is that of the carrier when the electroactive species is attached to a micelle, a droplet of microemulsion or a vesicle. Polarography, cyclic voltammetry, lineal scanning voltammetry, chronocoulombimetry, amperometry and spinning disk voltammetry are the most common techniques employed not only to obtain information on diffusion, but on the kinetics and energetics of adsorption and the electrode reaction.

Both methods, polarography and voltammetry, measure the intensity of the diffusion current generated by the discharge of an electroactive particle (i_D). Then, as the electroactive particle contribution to the charge transport through the cell must be negligible, a swamping or support electrolyte is commonly used which eliminates the electroactive probe transport current contribution. This may be a problem when ionic surfactant micelles are studied because their size and shape is usually altered by the nature and the concentration of added salts. However, the intermicellar solution has enough concentration of monomeric ions to act as swamping electrolyte (except for very low CMC) because the low concentration of micelles and their large size compared with that of monomeric surfactant ions and counterions. The addition of an electrolyte does not alter their size and shape of non-ionic micelles provided if the salt concentration is not very high, much above that needed to ensure the conditions for the i_D measurement.

2.4. Experimental details

2.4.1. Electrodes

Saturated calomel electrode (SCE) (potential $E_{SCE} = 0.241$ V at 25°C) and the Ag/AgCl with saturated KCl ($E_{Ag/AgCl} = 0.197$ V at 25°C) are usually employed as reference electrodes. The working electrode is commonly of quicksilver, platinum or carbon (vitreous or pyrolytic), but other electrodes may be used. Polarography uses a mercury electrode (dropping or with a static drop). Voltammetry uses hanging quicksilver drop or solid electrodes. The electrode surface lies between 0.01 and 0.10 cm^2. Microelectrodes or ultramicroelectrodes with areas of some square microns have been used in high high-resistivity W/O emulsions [33].

2.4.2. Time scale

The time scale involved in continuous current polarography is 1−10 s (droplet falling time); in cyclic voltammetry and lineal scanning voltammetry is between 10^{-4} and 1 s (scanning time);

http://dx.doi.org/10.5772/67975

in chronoculombimetry, amperometry and potentiometry from 10^{-3} to 10 s (transition time) and in spinning disk voltammetry, 10^{-3}–0.1 s (spinning speed).

2.4.3. The techniques

Figure 4 shows a continuous current polarogram with spinning disk electrode.

Figure 5 shows a cyclic voltagram for a reversible process, in which the half-wave potential is $E_{1/2} = (E_a + E_c)/2$ and the diffusion current i_D corresponds to E_c.

The following equations may be used for reversible and irreversible processes at 25°C:

Continuous current polarography, Ilkovič equation:

$$i_D = K\,n\;D^{1/2}\,C\,m^{2/3}\,t^{1/6} \tag{8}$$

where D is the diffusion coefficient in cm^2/s; *n* is the number of transferred electrons; *m* is the flux speed of mercury in $mg \cdot s^{-1}$; C is the concentration of the electroactive probe (not that of the micelles) in $mol \cdot cm^{-3}$ and *t* the dropping time in s. The current is given in Ampère. The factor *K*

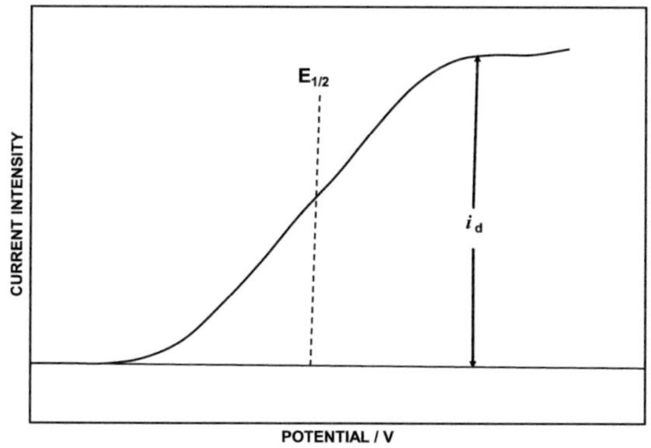

Figure 4. Continuous current polarogram with spinning disk electrode.

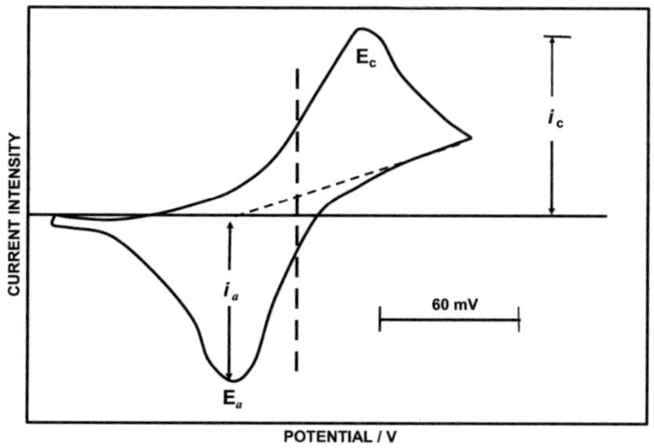

Figure 5. Cyclic voltagram for a reversible process.

is 708.1 for instantaneous current measurements and 607 for average current. Notice that for a given cell and experimental conditions, $Km^{2/3}t^{1/6}$ is constant and may be determined measuring the i_D for an electroactive probe such as Cd^{+2}, with $n = 2$ and $D_{Cd+2} = 69 \times 10^{-7}\ cm^2 \cdot s^{-1}$ in absence of surfactant.

Another procedure is to plot i_D as a function of the concentration of the electroactive probe in water and in surfactant solution and determine the corresponding slopes S_w and S_m of both straight lines. The ratio of the slopes is $S_m/S_w = (D_M/D_{probe})^{1/2}$, where D_M is the micelle self-diffusion coefficient

Spinning disk voltammetry, Levich equation:

$$i_D = 0.620\, n\, F\, A\, D^{2/3}\, \omega^{1/2}\, v^{-1/6}\, C \qquad (9)$$

where A is the electrode area (cm^2); F is the Faraday constant (Coulomb · mol ^{1}); ω the spinning speed (radian · s^{-1}) and v is the kinematic viscosity ($cm^2 \cdot s^{-1}$).

The Randles-Sevcik equation may be used for reversible processes in cyclic or linear scanning voltammetry:

$$i_D = 2.69\ 10^5\, n^{2/3}\, A\, D^{1/2}\, \nu^{1/2}\, C \qquad (10)$$

where v is the scanning speed in $V \cdot s^{-1}$.

For microelectrodes:

$$i_D = 4\, n\, F\, D\, C\, r \qquad (11)$$

where r is the electrode radius [34, 35].

2.4.4. *Probes*

An ideal electroactive probe attaches to the aggregate and does not dissolve in the interparticle solution, i.e. the probe must be soluble in the micelle and water insoluble. In order to avoid any modification of the size and shape of micelles, there must be less than one probe molecule per micelle. The diffusion current must be caused only by micelle translational diffusion, with its attached probe directed to the electrode surface.

The hydrophobicity of the electroactive probe plays an important role in their inclusion in the aggregate [36]. It is necessary that the micelle does not carry more than one probe molecule in order to obtain proper i_D values. The probe/micelle ratio can be less than unity.

The probe must be always electroactive, i.e. it must be able to exchange electrons with the electrode whatever the location in the micelles. This condition is almost always fulfilled when solubilized in micelles and microemulsions droplets, but it may not be the case in macroemulsions or vesicles [37, 38].

When a probe is dissolved in a very hydrophobic region of an aggregate, where it is not available for the electrons transfer, it does not remain electroactive. This is the case for 1-dodecyl-cianopyridinium in SDS micelles (but not in hexadecyltrimethylammonium bromide

(HTAB) micelles) [39]. The half-wave potential is also affected: −0.6 V in SDS and −1.30 V in HTAB. However, there are cases as the methylferrocene that remains electroactive although it dissolves in the hydrophobic core of the HTAB micelles where it is relatively inaccessible to water and oxygen [40].

The D value depends on both, the surfactant and the probe concentration. D is computed using the analytic concentration of the probe, which is only valid if all the probes in the micelles are discharged when the micelles reach the electrode [41]. The Ilkovič equation can be used to check this: if i_d is not linear with the probe concentration, the effective concentration of the probe [P_{eff}] is reduced. If a Poisson distribution of probe molecules among micelles is assumed then:

$$[P_{eff}] = [P]\left(1 - \exp\left(-[P]/[M]\right)\right)/([P]/[M]) \tag{12}$$

where [P] and [M] are the probe and the micelles concentrations (in micelle moles per litre), respectively. The diffusion current is proportional to [P_{eff}] instead of [P]. If not all the probes are discharged at the electrode, i_d, and consequently D, will diminish when the probe concentration is augmented and the surfactant concentration remains constant. This inconvenient is usually avoided using probes that do not partition between micelles and water and with $[P]/[M] \leq 1$. This latter condition also ensures that the size and shape of micelles are not affected by the inclusion of the probe molecule in the micelle [31, 42]. This has been also verified in microemulsions [30]. However, in very concentrated systems with rod-like micelles, the presence of several probe molecules in the same micelle may modify its size and shape [43].

The electroactive probe may be directly added to the surfactant solution in an appropriate amount [44]. This procedure was used with Cd^{+2} in sodium dodecylsulphate [65], disodium n-dodecane phosphonate [28] and sodium hexadecanoate [21]. The cadmium ions adsorb at the anionic micelle Stern layer forming a water insoluble compound solubilized by micelles.

The first probe used was Cd^{+2}, with $n = 2$ [32], which is useful for anionic micelles.

Water insoluble anthraquinone dyes (1,4-diamineanthraquinone and 1,4,5,8 tetraamineanthraquinone) have been used to tag non-ionic micelles [45]. The reactions of both dyes are reversible with $n = 2$ and the diffusion coefficients determined by polarography are of the order of $3 \times 10^{-7}\ cm^2 \cdot s^{-1}$. These values are consistent with those expected for non-ionic micelles. The half-wave potentials were not affected by changes in the concentrations of the probe or the surfactant.

Ferrocene and tetrahydrofulvene were used in HTAB, hexadecyltrimethylammonium chloride (HTAC) and SDS. In all cases the probe was associated with micelles [46]. Ferrocene solubility in 0.1 M NaCl aqueous solution is 5×10^{-5} M [47] and $D_{ferrocene}$ is $6.7 \times 10^{-6}\ cm^2 \cdot s^{-1}$. Cu^{+2} [48], several Fe^{+2} complexes [49] and N-alkyl-p-cyanopyridinium [31] have been also employed.

For cationic micelles, 1-dodecyl-4-cyanopyridinium iodide ($C_{12}PI$) and 1-hexadecyl-4-pyridinium iodide ($C_{16}PI$) [31] can be used as probes. $C_{12}PI$ is partitioned between water and micelles, whilst $C_{16}PI$ is almost water insoluble ($4.55 \pm 0.1\ 10^{-4}$ M at 23°C) and dissolves completely in micelles. These probes are useless with an anionic surfactant (such as SDS) because they are not accessible to the electrode electrons. Another homologous, the $C_{18}PBr$ has a water solubility of $4.05 \pm 0.1 \cdot 10^{-5}$ M and $E_{1/2} = -0.63$ mV ($n = 1$) and -1.05 mV ($n = 1$) [42].

If the probe does not partition between water and micelles, i_D is independent of the surfactant concentration. Otherwise, both i_D and $E_{1/2}$ change with surfactant concentration.

Other probes are:

1-nitropyrene ($PyNO_2$): water solubility at 25°C: (5.9 ± 2) × 10^{-7} mol · dm^{-3}, solubility in 2.78 × 10^{-3} mol · dm^{-3} TTAB: (4.4 ± 1) × 10^{-6} mol · dm^{-3}. Half-wave potential in cationic surfactants $E_{1/2}$ = −0.61 mV (n = 4) and −1.20 mV (n = 2), in anionic surfactants (SDS): $E_{1/2}$ = −0.26 mV (n = 1), −0.68 mV (n = 1), −1.18 mV (n = 2).

1-pyrenecarboxylaldehide (PyCHO):): water solubility at 25°C: ((3.4 ± 0.5) × 10^{-6} mol · dm^{-3}, solubility in TTAB 2.78 × 10^{-3} mol · dm^{-3}: (6.75 ± 1) × 10^{-4} mol · dm^{-3}. $E_{1/2}$ = −1.12 mV, n = 1 [42].

9-nitroanthracene (ANO_2): water insoluble. $E_{1/2}$ = −0.70 mV, n = 1

9-anthracenecarbonitrile (ACN): water insoluble. $E_{1/2}$ = −1.30 mV, n = 1 and −1.57 (n = 1) [42].

2,2′-dinitrobiphenyl: it undergoes a reduction in up to four steps in anionic, cationic or non-ionic surfactants. It is water insoluble but it seems that it is partitioned between the intermicellar solution and micelles, in a proportion of 77% in SDS, 95% in HTAB and 96% in Tween 80. In water-ethanol solution its diffusion coefficient is 1.1 × 10^{-5} $cm^2 \cdot s^{-1}$ [50].

All these probes may be dissolved in organic solvents (ethanol, toluene or benzene). Then the solvent is evaporated and the surfactant solution is added and sonicated to allow the probe to be solubilised in micelles. In some cases the probe is added to the surfactant solution and then sonicated. Generally 2–4 hours of sonication is enough, but in some cases up to 10 hours were needed to ensure the correct solubilisation.

2.4.5. *Surfactants whose counterion is electroactive*

The measured diffusion coefficient is formed by the contributions of the counterions attached to the micelles, those released by the micelle ionization and those belonging to the dissociation of the unmicellised monomers. Adriamanampisoa and Mackay [51] performed voltammetric measurements on cadmium dodecylsulphate, and analysed the measured diffusion coefficient ($D_{measured}$) as:

$$D_{measured} = \left[1 - \beta - \beta(CMC/c)^2 + 2\beta\,(CMC/c)\right] D_x + \beta\, D_M\,(1 - CMC/c)^2 \qquad (13)$$

where c is the total concentration; β = 1 − α = m/n, the number of counterions (m) bounded to the micelle having an aggregation number n; D_x and D_M are the diffusion coefficients of the unmicellised counterions and of the micelles, respectively. It is assumed in this equation that β and the concentration of unmicellised surfactant ions and counterions are constant above the CMC, what has been proved not to be generally true. Eq. (13) can be rearranged so that a plot of $(D_X - D_{med})^{1/2}$ vs. c^{-1} must give a straight line:

$$(D_x - D_{measured})^{1/2} = -\beta^{1/2}\, CMC\,(D_x - D_M)^{1/2}\, c^{-1} + \beta^{1/2}\,(D_x - D_M)^{1/2} \qquad (14)$$

Or it can be also rearranged to obtain a straight line whose intercept is D_X and the slope is $\beta(D_M - D_X)$ when plotting $D_{measured}$ vs. $(1 - CMC/c)^2$:

$$D_{measured} = \beta\,(D_M - D_x)(1 - CMC/c)^2 + D_x \tag{15}$$

The authors found that the model failed at low supporting electrolyte concentration what was attributed to a significant contribution of the micelles to i_D. It is also possible that part of the cadmium attached to the micelles is not accessible to the electrode. Besides, the exchange between Na^+ and Cd^{+2} between the intermicellar solution and the micelle Stern layer may also affect the measured diffusion coefficient.

2.4.6. *Electroactive surfactant ion*

Saji et al. [52] deduced the following expression for the measured diffusion coefficient:

$$D_{measured} = [(CMC/c) - (CMC/c)^2]\,D_p + D_M(1 - CMC/c)^2 \tag{16}$$

where D_P is the diffusion coefficient of the monomeric electroactive surfactant ion.

2.4.7. *Oxygen elimination*

Oxygen interferes with the measurements. Generally, oxygen is eliminated by slowly bubbling nitrogen or argon at most by 12 minutes, and the bubbling is maintained while the measurement is performed [42]. Alternatively, when the foam formation is excessive, the sample may be placed in a two-neck flask and bubbled 1–2 minutes, then the flask is closed and the sample is left in an oxygen-free atmosphere. Once the foam is reduced, the procedure is repeated. Normally three cycles suffice to obtain a sample ready to measure.

2.4.8. *Support electrolyte*

Many studies on micelar or O/W microemulsions systems are performed with the addition of supporting electrolyte such as NaCl, KBr or KCl. This allows the suppression of the contribution of the aggregates tagged with the electroactive probe to the migration current while maintaining their diffusion current contribution.

In systems with ionic surfactants, provided the CMC is not very low, it is usually not necessary to add supporting electrolyte since the non-micellised surfactant ions and counterions act as supporting electrolyte. As the micelles are bigger than these ions and their concentration is low, their charge transport number is generally negligible. However, in some systems, the contribution of the micelles to the total conductivity of the system may be significant [9].

In order to study non-ionic micelles it is necessary to add supporting electrolyte which are usually adsorb on the micelles [53]. However, except in very high concentrations well above those needed to ensure conductivity, the non-ionic micelles are not affected by the swamping electrolyte.

When supporting electrolyte is added, it must be taken into account that it may affect the size and shape of micelles (especially ionic ones), the adsorption of the surfactant to the electrode surface and the electrostatic interactions among micelles [54].

2.5. The adsorption of surfactants on the electrode surface

At concentrations well below the CMC lone molecules adsorb. However, at a concentration called the critical hemimicellisation concentration (CHMC) the adsorption is by aggregates called hemimicelles (for SDS, CHMC ≈ CMC/20) [55]. Above the CHMC the coverage of the electrode surface increases rapidly until it is saturated. Mono- and multi-layers may be formed, depending on the surfactant concentration and the applied potential to the electrode [54]. Since the studies on micelles are obviously above the CMC, the electrode surface is commonly covered by one or more surfactant layers. Anionic surfactants will desorb only at very negative potentials and the cationic ones at very positive potentials. Depending on the applied potential, the layer may re-orientate changing its density or reverse the orientation of their polar groups and chains. The adsorbed layer may affect the discharge of the electroactive species on the electrode [56]. The surfactant adsorbed layer may displace an absorbable electroactive species or may promote its incorporation to the adsorbed layer. If the surfactant is ionic, the layer may produce an additional electrostatic barrier against the electroactive ionic species. Ionic surfactants may also affect the structure of the electrode ionic double layer.

Adsorbed cationic surfactants usually extend the anodic range in aqueous solutions. The hydrophobic conducting film on the electrode inhibits water to reach the electrode surface [57, 58]

The presence of the adsorption layers does not seem to affect the reactions of electron transfer of the electroactive probes carried by aggregates [59].

The adsorbed layers may affect the potential but if they are thin their effect on the diffusion of the electroactive species to the electrode is negligible. However, it is convenient to check this in each case [54]. If the employed technique is not sensitive to the details about how the electrons are transferred in the electrode reaction, this does not affect the determined D value [54]. As examples, the Ilkovič and the Levich equations may be used for both reversible and irreversible reactions, whereas that of Randles-Sevcik must only be employed in reversible or quasi-reversible processes.

2.5.1. The half-wave potential

$E_{1/2}$ of the electroactive species is presumed to be different when the probe is attached to micelles and when it is dissolved in water. The availability of the electron coming from the electrode is modified by the surface potential of the micelles and the micro-environment of the probe in the micelle. If $E_{1/2,w}$ and $E_{1/2,M}$ are the half-wave potential values for the probe in water and in micelles, respectively, the experimentally measured half-wave potential, in case of a probe partition between water and micelles will be:

$$E_{1/2} = \frac{E_{1/2,W}}{1 + K\,C_M} + \frac{E_{1/2,W}\,K\,C_M}{1 + K\,C_M} \qquad (17)$$

where C_M is the concentration of the micelles and *K*, the probe distribution constant between micelles and water. This equation assumes that the probe exchange between water and micelles is so fast, when compared with the electronic transference speed, that the reduction of the free and attached probes is seen as a unique wave in polarography. This situation is actually observed.

http://dx.doi.org/10.5772/67975

Since i_D depends not only on the partition but on the size and shape of micelles, whereas $E_{1/2}$ is essentially only dependent on the partition, this later may be used to obtain the partition constant of the probe between micelles and intermicellar solution, on the supposition that *K* depends only slightly on the surfactant amount in the intermicellar solution.

2.5.2. *Effect of probe partition between micelles and the intermicellar solution*

If the electroactive probe is only solubilized by micelles, or its water solubility is extremely low, the measured diffusion coefficient is that of the micelle (D_M).

If the probe is distributed between micelles and intermicellar solution, the measured diffusion coefficient is higher, because $D_M < D_{P,w}$ ($D_{P,w}$ being the diffusion coefficient of the probe in water). This must be taken into account to obtain correct $D_M$values [31].

Eq. (17) applies when it is assumed that the interchange of probe molecules between micelles and intermicellar solution is fast in comparison with the electronic transference [60]:

$$D_{measured} = x_W\, D_{P,W} + x_M\, D_M \tag{18}$$

where D $_{measured}$ is the measured diffusion coefficient, x_w and x_M are the probe mole fraction in water and micelles, respectively. Supposing that the partition constant of the probe between water and micelles (K = $[P_w]/[P_M$, [P]: probe concentration) is independent of the probe concentration, then:

$$D_{measured} = (D_M + K\, D_{P,W})/(1 + K) \tag{19}$$

This assumption holds if the probe or micelles concentration is low, and the probe is preferentially solubilized in micelles.

If the exchange velocity between micelles and intermicellar solution is lower that the electron transfer process, then the equilibrium between probes in water and in micelles inside the electrode diffusion layer cannot be obtained and the equation to be used is [60]:

$$D_{measured} = \left(x_W\, D_{P,W}{}^Z + x_M\, D_M{}^Z\right)^{1/Z} \tag{20}$$

where Z = ½ for polarography and cyclic voltammetry, and Z = 2/3 for spinning disk voltammetry. For microelectrodes, Z = 1 [34, 35].

This situation is the most frequent [61]:

$$D^{1/2}_{measured} = \frac{i_D}{708.1\, n\, m^{2/3}\, t^{1/6}\, c} = \frac{D_M{}^{1/2}\, K\, C_M \,+\, D_{P,W}{}^{1/2}}{1 + K\, C_M} \tag{21}$$

where C_M is the micellised surfactant concentration (on a monomer basis). It is supposed that the diffusion of the probe dissolved in the intermicellar solution and that of micelles are independent. $D_{P,w}$ may be experimentally determined in absence of surfactant.

Provided that $D_M$is independent of the surfactant concentration (which is not generally true), a plot of $D_{measured}$ vs. the surfactant concentration allows the determination of *K* and D_M.

To determine the relation between the reduction time (t_{red}) and that of the diffusion (t_{dif}), the diffusion current i_D is plotted against the probe concentration [P]. If this plot is linear, $t_{red} >> t_{dif}$ and Eq. (18) may be used. Otherwise, $t_{red} << t_{dif}$ and Eq. (20) must be used [42]. In some cases, linearity is only obtained for some [P]/[M] ratios.

In the derivation of the preceding equations it has not been considered the possibility that probe molecules may be attached to different micelle loci. It has been also assumed that the partition constant is independent of the probe concentration, similarly to the partition of a solute between two immiscible liquids. However, in some cases it has been observed a dependence of the diffusion coefficient with [P] for methylviologen and ferrocene [62]. This phenomenon has been studied as the equilibrium of multiple union sites. Eq. (21) is obtained when it is considered that the probe is strongly bounded to the micelle and that there are μ probe molecules per micelle:

$$D = D_{P.W}\,[1 + [M]\,K_M\,[P]^{\mu-1}] + D_M\,[M]\,K_M\,[P]^{\mu-1}/[1 + [M]\,K_M\,[P]^{\mu-1}] \quad (22)$$

where $K_M = \mu K'$, K′ being the equilibrium constant for the union of the μ probe molecules to the micelle.

3. Electrochemical study of the adsorption of surfactants on solid interfaces

Surfactant monolayers on a solid substrate have the potential to modify both the chemical and electrochemical behaviour of the interface between the solid and electrolyte [63]. These films have many applications in areas such as flotation, oil recovery, detergency [64], and templating of metallic nanoparticles [65, 66]. In this respect, nanoparticle (NP) synthesis in microemulsions has been a hot topic since the early 1980s, when the first colloidal solutions of platinum, palladium and rhodium metal nanoparticles were prepared [67]. Since then, a huge variety of nanoparticles has been synthesized in water in oil, and water in supercritical fluid solutions.

Coming back to the formation of surfactant monolayers at solid surfaces, this is also a subject of intensive research, with topics ranging from the influence of the surfactant's molecular structure [68, 69], to the hydrophilicity-hydrophobicity of the substrate surface [70], and the ionic strength and the nature of the counter ion [71]. The charging of the solid surface also has a significant impact on the surface assembly of non-ionic and ionic surfactants [72]. The electrical state of a surface plays a key role in determining the morphology of surfactants at solid interfaces and, unlike other shape determining factors such as the surfactant packing parameter [73], the electrical parameter can readily be adjusted *in situ*, providing a tuneable means to control films of soft condensed matter [74]. It is for this reason that electrochemistry provides the ideal set of tools to study the effect of charge on the behaviour of adsorbed surfactant molecules. By choosing a conductive substrate (such as gold), one has the unique opportunity to investigate the influence of charge density and hence the electrostatic field on the surface aggregation of surfactant molecules [75–80].

3.1. Differential capacitance measurements

The differential capacitance is defined as the derivative of the double layer's charge (q) with respect to the electric field (E) at a constant chemical potential (μ) (see Eq. (23))

$$C_d = \left(\frac{dq}{dE}\right)_\mu \tag{23}$$

In the electrical double layer, the capacitance depends upon the area of the electrode ($\boldsymbol{A}$), the thickness of the double layer (d) the relative permittivity of the solution ($\boldsymbol{\varepsilon_r}$), and the permittivity in a vacuum, ($\boldsymbol{\varepsilon_0}$):

$$C_d = \frac{\varepsilon_r \varepsilon_0 A}{d} \tag{24}$$

In practice, the differential capacitance is measured using an electrochemical cell with a lock-in amplifier (LIA) that produces a sinusoidal voltage that is superimposed on the static electric potential, and analyses the oscillating current response. Since a LIA can measure signals with different phase shifts separately, using Eq. (25), one can obtain the C_d value from the real and imaginary current components [81]:

$$C_d = \frac{I_{\mathrm{Im}}}{2\pi f V_{ac}} \left\{ 1 + \left(\frac{I_{\mathrm{Re}}}{I_{\mathrm{Im}}}\right)^2 \right\} \tag{25}$$

where f is the frequency of the alternating current, V_{ac} is the amplitude of the sinusoidal voltage and the real current ($\boldsymbol{I_{\mathrm{Re}}}$) and the 'imaginary' current ($\boldsymbol{I_{\mathrm{Im}}}$) are the currents measured in-phase and out-of-phase with the voltage, respectively.

Measurements of C_d therefore provide information on the permittivity of the layer adsorbed at the interface as well as the thickness of such a layer

3.2. Chronocoulometry measurements

Chronocoulometry (CC) allows the measurement of the change in charge density, σ_M, as a function of the potential. If the molecule of interest either does not conduct or is insoluble, then the measurements must be performed in an electrolyte solution. A measurement wherein current is measured with respect to time is called a current transient. At a certain potential, ($\boldsymbol{E_{\mathrm{des}}}$), all of the molecules will have been desorbed from the surface. The surface charge at this potential is the surface charge of the bare electrode (σ_M). After stepping the potential by $\boldsymbol{\Delta}E$ to a potential of interest ($\boldsymbol{E_i}$), where the molecules are adsorbed and waiting for equilibrium to be established, the relative surface charge of the electrical double layer can be measured by subsequently desorbing the molecules and integrating the current that flows during the desorption step. This process is repeated for a sequence of potentials $\boldsymbol{E_i}$. A plot of charge density as a function of applied potential can be produced as a result.

In a plot of σ_M *vs* E the difference in area between the base electrolyte curve and the molecule of interest curve is equal to the surface pressure at that potential. This is because a surface pressure is equal to the difference between the surface energies of a system with and without the surface-bound molecule [82].

If the potential of zero charge (***pzc***) of the electrode is known (often measured by differential capacitance in a weak electrolyte solution such as 5 mM KPF_6), the measured surface charge ($\Delta\sigma_M$) at that potential can be used to calculate the absolute surface charge at the desorption potential by means of:

$$\Delta\sigma_M(pzc) = \sigma_M(pzc) - \sigma_M(E_{des}) = -\sigma_M(E_{des}) \tag{26}$$

Thus, the absolute surface charge at the potential of interest ($\sigma_M(E_i)$) can be calculated as:

$$\Delta\sigma_M(E_i) = \sigma_M(E_i) - \sigma_M(E_{des}) \tag{27}$$

When $E_i < E_{des}$, only the base electrolyte should contribute to $\Delta\sigma_M$, and the data should therefore resemble a curve of the base electrolyte alone.

By means of numerical integration one can obtain the area between the CC curve and that of the base electrolyte; that area is the surface pressure (π), usually in $mN \cdot m^{-1}$, when the surfactant adsorbs on the electrode's surface. The surface pressure is closely related to the excess free energy of the system (G^E) thus giving information on how much stable (or unstable) is the surface by having the surfactant adsorbed.

3.3. Electrochemical impedance spectroscopy

The classical electrochemical techniques use measurements of currents, electrochemical potentials, and charges as a function of time, which can in turn be related to the electrochemical potential. In contrast to this, electrochemical impedance spectroscopy (EIS) presents the signal as a function of the frequency at a constant potential. This can pose a problem to electrochemists, since we are used to thinking in terms of time, not frequencies. Another issue with EIS is that it requires a certain amount of knowledge in mathematics, in particular of Laplace and Fourier transforms, along with complex numbers. The following section is meant as a brief introduction to EIS along with its applications in systems with adsorption, such as the adsorption of surfactants on electrochemical interfaces.

The general definition of impedance is given by Eq. (28) as follows:

$$\hat{Z}(s) = \frac{L[E(t)]}{L[i(t)]} = \frac{\overline{E}(s)}{\overline{i}(s)} \tag{28}$$

where $\hat{Z}$ (s) is the operational impedance, and has units of resistance (Ω), Ldenotes the Laplace transform, s is the frequency, E the electrochemical potential and $i$the current density. The parameter s can be complex of the form $s = \sigma + j\omega$,or real $s = \sigma$, as in the classical Laplace transform. The impedance of each electrical circuit element is detailed in **Table 1** [83]. For each electrical component one can write the corresponding impedance and then, by applying

Element	Operational impedance	Ac impedance
R	R	R
C	$1/(sC)$	$1/(j\omega C)$
L	sL	$j\omega L$

Table 1. Impedance of linear electrical elements in an electrical circuit [83].

Kirchhoff's laws, calculate the total impedance of the electrical circuit. In the case of ***ac*** impedance, i.e. when the potential perturbation is sinusoidal, one uses the Fourier transform (***FT***), as shown in Eq. (29):

$$\widehat{Z}\ (j\omega) = \frac{F[E(t)]}{F[i(t)]} = \frac{\overline{E}(j\omega)}{\overline{i}\,(j\omega)} \tag{29}$$

Where the parameter s in this case imaginary and of the form $\boldsymbol{s} = \boldsymbol{j\omega}$. For further information about ***FT*** or ***LT***, the reader is referred to any of the complex variable calculus books that cover in detail these transforms. Another quantity usually employed is the admittance, which is the inverse of the impedance:

$$\widehat{Y}\ (s) = \frac{1}{\widehat{Z}\ (s)} \tag{30}$$

Regarding applications with surfactants in electrochemistry, EIS has been widely used for the study of the inhibition of corrosion by surfactant coatings [84, 85], the investigation of ionic surfactant selective electrodes [86], the effect of anionic and cationic surfactants in the performance of batteries [87] and the study of electrochemical reactions in surfactant films [89], such as the study of O_2 reduction by haemoglobin in a film of didodecyldimethylammonium bromide [88], among other applications. By analysing the experimental EIS data and creating an equivalent electrical circuit for the reactions being studied, one can get a unique insight on the mechanism for those reactions.

The electrochemical techniques briefly described in this chapter can provide an insight on the mechanics of the adsorption of surfactants on solid electrochemical interfaces, as well as those of surfactant aggregates in solution. Electrochemistry possesses the advantage that by simply tuning the electrode potential one can create different conditions for the study of these systems, thus providing a powerful tool for the probing of surfactant systems.

Acknowledgements

ENS would like to acknowledge the Marie Sklodowska Curie FP7 programme 629771 – PHASE (Phosphonic Acids Surface Electrochemistry) for their financial support for the last 2 years. ENS and EPS would also like to acknowledge CONICET – Argentina for their financial support in their careers as researchers in that institution.

Author details

Eduardo Nicolás Schulz[1]*, Erica Patricia Schulz[2] and Pablo Carlos Schulz[2]

*Address all correspondence to: nschulz@uns.edu.ar

1 University of Birmingham, School of Chemistry, Edgbaston, Birmingham, United Kingdom

2 Chemistry Department, Universidad Nacional del Sur and Instituto de Química del Sur (INQUISUR; CONICET), Bahía Blanca, Argentina

References

[1] Phillies GDJ. Interpretation of micelle diffusion coefficients. J. Colloid Interface Sci. 1982; **86**: 226–233.

[2] Phillies GDJ. Effects of intermolecular interactions on diffusion. II. Three component solutions. J. Chem. Phys. 1974; **60**: 983–989.

[3] Phillies GDJ. Effects of intermolecular interactions on diffusion. I. Two component solutions. J. Chem. Phys. 1974; **60**: 976–982.

[4] Marqsee JA, Deutch JM. Concentration dependence of the self-diffusion coefficient. J. Chem. Phys. 1980; **73**: 5396–5397.

[5] Mazo RM. The theory of the concentration dependence of the self-diffusion coefficient. J. Chem Phys. 1965; **43**: 2873–2877.

[6] Phillies GDJ, Benedek GB, Mazer NM. Diffusion in protein solutions at high concentrations: a study by cuasielastic light scattering spectroscopy. J. Chem. Phys. 1976; **65**: 1883–1892.

[7] Pusey PM. Macromolecular diffusion, in: Cumming HZ, Pike ER, editors, Photon Correlation and Light Beating Spectroscopy, New York, Plenum Press; 1975. p. 387.

[8] Schulz PC. Potentiometric Studies on disodic *n*-decane phosphonate micellar aqueous solutions. Colloids & Surfaces. 1988/89;**34**:69–74.

[9] Benedini L, Messina PV, Manzo RH, Allemandi DA, Palma SD, Schulz EP, Frechero MA, Schulz PC. Colloidal properties of amiodarone in water at low concentration. J. Colloid Interface Sci. 2010; **342**: 407–414.

[10] Schulz PC, Hernández Vargas ME, Puig JE, Do Micelles Contribute to the Total Conductivity of Ionic Micellar Systems? Latin American Appl. Res. 1995; **25**: 153–159.

[11] Morini MA, Minardi RM, Schulz PC, Puig JE, Hernandez-Vargas ME. Basicity of dodecyltrimethylammonium hydroxide. Colloids Surf. A: Physicochem. Eng. Aspects. 1995; **103**: 37–40.

[12] Morini MA, Schulz PC, Puig JE. Counterion specificity of the micelle surface and its implications on micellar catalysis. Colloid Polym. Sci. 1996;**274**:662–668.

[13] Schulz PC, Messina PV, Morini MA, Vuano B. Potentiometric studies on sodium dehydrocholate micelles. Colloid Polym. Sci. 2002; **280**: 1104–1109.

[14] López-Fontán JL, Sarmiento F, Schulz PC. The aggregation of sodium perfluorooctanoate in water. Colloid Polym. Sci. 2005; **283** (8)**:** 862–871.

[15] Minardi RM, Schulz PC, Vuano B. The Aggregation of n-dodecanephosphonic acid in water. Colloid Polym. Sci. 1996; **274/11**: 1089–1093.

[16] Di Anibal CV, Morini MA, Verdinelli V, Rodríguez JL, Minardi RM, Schulz PC, Vuano B. Critical micelle concentration of tridecane, tetradecane and hexadecane phosphonic acids and their mono and disodic salts. Colloids Surf. A: Physcicochem Eng Aspects. 2009; **380**: 276–281.

[17] Morini MA., Schulz PC. Zeta potential of dodecyltrimethylammonium hydroxide micelles in water. Colloid & Polym. Sci. 1997; **275**: 802–805.

[18] Lepori L, Gianni P. Partial molar volumes of ionic and nonionic organic solutes in water: simple additivity scheme based on the intrinsic volume approach. J Solutn. Chem. 2000; **29**(5): 405–447.

[19] Schulz PC. Volúmenes molares parciales de n- alcano fosfonatos disódicos. Anales Asoc. Quím. Argentina. 1981; **69**: 185–194.

[20] Gruen DW. QA model for the chains in amphiphilic aggregates. 1. Comparison with a molecular dynamics simulation of a bilayer. J. Phys. Chem. 1986; **89**: 146–153.

[21] Schulz PC. Steric Fitting of the spherical micelle size. Colloid & Polym. Sci. 1991; **269**: 612–619.

[22] Schulz PC, Zapata-Ormachea M. Sodium hexadecanoate micellar aggregation number. Colloid & Polymer Sci. 1994**;** **272**: 1259–1263.

[23] Tanford C. Theory of micelle formation in aqueous solutions. J. Phys. Chem. 1974; **78**: 2469–2479.

[24] Young CY, Missel PJ, Mazer NA, Benedek, GB. Deduction of the micelle shape from angular dissimetry measurements of light scattering from aqueous sodium dodecyl sulfate solutions at high sodium chloride concentration. J. Phys. Chem. 1978; **82**(12): 1375–1378.

[25] Schulz EP, Piñeiro A, Rodríguez JL, Minardi RM, Frechero M, Schulz PC. Intermediate structures for higher level arrangements: catching disk-like micelles in decane phosphonic acid aqueous solutions. J Phys. Chem B. 2013; **117**: 6231–6240, and references therein.

[26] Imae T, Ikeda S. Sphere-rod transition of micelles of tetradecyltrimethylammonium halides in aqueous sodium halides solutionand flexibility and entanglement of long dodlike micelles. J. Phys. Chem.1986; **90**: 5216–5223.

[27] Schulz PC. Steric fitting if the rodlike micelle size. J. Colloid Interface Sci. 1991; **152**: 333–337.

[28] Yang JT. The viscosity of macro-molecules in relation to molecular conformation, in: Anfinsen CB, Anson ML, Bailey K, Edsall JT, editors. Advances in Protein Chemistry, vol. XVI, New York, Academic Press; 1951. pp. 323–400.

[29] Schulz PC. Effect of NaOH and NaCl on the disodium n-decane phosphonate micellar aggregation number. Colloid Polymer Sci. 1995; **273/3**: 288–292.

[30] Miraglia DB, Schulz EN, Rodriguez JLM, Schulz PC, Salinas D. Effect of the concentration and composition on the size and shape of micelles of sodium oleate -cetyltrimethylammonium bromide mixtures. J Colloid Interface Sci. 2010; **351**: 197–202.

[31] Mackay R, Dixit N, Agarwal N, Seiders R. Diffusion measurements in microemulsions. J. Dispersion Sci. Technol. 1983**; 4:** 397–407.

[32] Zana R, Mackay RA. Polarographic measurement of micellar diffusion coefficients. Langmuir. 1986; **2**: 109–113.

[33] Hoyer HW, Novodoff J. The polarographic determination of micellar diffusion coefficients. J. Colloid Intrerface Sci. 1968; **26**(4): 490–492.

[34] Howell JO, Wrightman RM. Ultrafast voltammetry of anthracene and 9-10-diphenylanthracene. J. Phys. Chem. 1984; **88**: 3915–3918.

[35] Wang,Z, Owlia A, Rusling JF. Reduction of methylviologen in a resistive wat4er-in-oil microemulsion. J. Electroanal. Chem.1989; **270**: 407–413.

[36] Chen JW, Georges J. Applicartion of carbon-fiber microelectrodes in the study of a non-conductive inverse microemulsions: interpretation of the diffusion coefficients. J. Electroanal. Chem. 1986; **210**: 205–211.

[37] Kaifer AE, Bard AJ. Micellar effects on the reductive electrochemistry of methylviologen. J. Phys. Chem. 1985; **89**: 4876–4880.

[38] Texter J, Beverly T, Templar ST, Matsubara T. Partition of para-phenylenediamines in oil-in-water emulsions. J. Colloid Interface Sci. 1987; **120**: 389–403.

[39] Berthod A, in: Mackay RA, Texter J, editors. Electrochemistry in Colloids and Dispersions, New York, VCH Publishers; 1992. p. 137.

[40] Mackay RA, Dixit NS, Hermansky C, Kertes AS. Conductivity and diffusion measurements on micellar and o/w microemulsion systems: a comparative study. Colloids Surfaces. 1986**;** **21**: 27–39.

[41] Hashino K, Saji T, Ohsawa Y, Aoyagui S. The solubilization sit4e of methylferrocene in cationic micelle. Bull. Chem. Soc. Japn. 1984; **57:** 1685–1686.

[42] Shinozuka N, Hayano S. Electrochemical investigations in surfactant media, in: Mittal KL, editor. Solution Chemistry of Surfactants. Vol. 2, New York, Plenum Press; 1949. pp. 599–626.

[43] Verrall RE, Milioto S, Giraudeau A, Zana R. Polarographic measurements of micellar diffusion coefficients: new results and implicationsof surfactant adsorption at the mercury–solution interface. Langmuir. 1989; **5**: 1242–1249.

[44] Offen H, Dawson D, Nicoli D. Probe effects on micelllar size. J. Colloid Interface Sci. 1981; **80**: 118–122.

[45] Novodoff J, Rosano HL, Hoyer HW. Diffusion coefficents of cadmium tagged sodium alkyl sulfate micelles. J. Colloid Interface Sci. 1972; **38**(2): 424–430.

[46] Hayano S, Shinozuka N. Polarographic studies of the solubilization of disperse dyes. Bull. Chem. Soc. Japan. 1969; **42**: 1469–1472.

[47] Georges J, Desmettre S. Electrochemical oxidation of hydrophobic compounds in aqueous micelar solution and oil-in-warter emulsion. Electrochim. Acta. 1986; **31**: 1519–1524.

[48] Georges J, Desmettre S. Electrochemistry of ferrocene in anionic, cationic and nonionic micelar solutions. Ffect of the micelar solubilization on the half-wave potentials. Electrochim. Acta. 1984; **29**: 521–525.

[49] MacKay R, Dixit N, Agarwal, R, Inorganic reactions in microemulsions, in: Holt SL, editor. Inorganic Reactions in Organized Media. Vol. 177, New York, ACS Symposium Series; 1982. pp. 179–194.

[50] Ko YC, Ree J, Chung KH. Electrochemical behavior of iron (II) chelates sodium dodecyl sulfate in H_2SO_4 aqueous solution. Bull. Korean Chem. Soc. 1997; **18**(1): 113–116.

[51] Avranas A, Sazou D. Electroreduction of 2, 2′-dinitrophenil in anionic, cationic and nonionic micellar systems. J. Colloid Interface Sci. 1994; **164**: 309–317.

[52] Adriamanampisoa R, Mackay RA. Voltammetry of cadmium cdodecylsulfate micelles. Langmuir. 1994; **10**: 4339–4343.

[53] Saji T, Hoshino K, Aoyagui S. Reversible formation and disruptionof micelles by controlof the rodox state of the head group. J. Am. Chem. Soc. 1985; **107**: 6865–6868.

[54] Morini MA, Messina PV, Schulz PC, The interaction of electrolytes with non-ionic surfactant micelles. Colloid Polym Sci. 2005; **283**: 1206–1218.

[55] Mackay RA. Electrochemistry in association colloids. Colloids Surfaces A: Physicochem. Eng. Aspects. 1994; **82**: 1–28.

[56] Harwell JH, Haskins TC, Schecter RS, Wade W. Pseudophase separation model for surfactant adsorption: isomerically pure surfactants. Langmuir. 1985; **1**: 251–262.

[57] Schmid RW, Reilley CN. Concerning the effect of surface-active substances on polarographic currents. J. Am. Chem. Soc. 1958; **80**: 2087–2094.

[58] Franklin TC, Sidarous L. The effects of surfactantsomn the electro-oxydation of benzhydrol in emulsion and micelle systems. J. Electrochem. Soc. Electrochem. Sci. Technol.1977; **124**: 65–69.

[59] Franklin TYC, Nodimele R, Adeniyi WK, Hunt D. The anode oxidation of galena in a cationic surfactant aqueous sodium hydroxide emulsion. J. Electrochem. Soc.1988; **153**: 1944–1947.

[60] Frank SG, Zograff G. Determination of micellar weight fort di-sodiumk sulfosuccinates in ahydrous and hydrous hydrocarbons solution. J. Pharm. Sci. 1969; **58:** 993–997.

[61] Dayalan E, Qutubuddin S, Texter J, in: Mackay RA, Texter J, editors. Electrochemistry in Colloids and Dispersions, New York, VCH Publishers; 1992. p. 119.

[62] Texter J. Two-component diffusionand reactiomin rotating disk voltammetry. Can fast kinetics force equilibration in the diffusion layer? J. Electroanal. Chem. 1991; **304**: 257–262.

[63] Rusling JF, Shi SN, Kumosinski TF, Diffusion of micelle-bound molecules to electrodes in solutions of ionic surfactants. Anal. Chem.1988; **60**: 1260–1267.

[64] Queffélec C, Petit M, Janvier P, Knight DA, Bujoli B. Surface modification using phosphonic acids and esters. Chem. Rev. 2012; **112:** 3777–3807.

[65] Swalen JD, *et al.* Molecular monolayers and films. A panel report for the Materials Sciences Division of the Department of Energy. Langmuir. 1987; **3:** 932–950.

[66] Firouzi A, *et al.* Cooperative organization of inorganic-surfactant and biomimetic assemblies. Science. 1995; **267:** 1138–1143.

[67] Huo Q, Hansma PK, Morse D E, Stucky GD, Aksay IA. Gemini surfactants at solid–liquid interfaces: control of interfacial aggregate geometry. Langmuir. 1997; **13(24):** 6382–6387.

[68] Boutonnet M, Kizling J, Stenius P, Maire G. The preparation of monodisperse colloidal metal particles from microemulsions. Colloids and Surfaces. 1982; **5:** 209–225.

[69] Manne S, Gaub HE. Force microscopy: Measurement of local interfacial forces and surface stresses. Curr. Opin. Colloid Interface Sci. 1997; **2:** 145–152.

[70] Qiu H, Hudson ZM, Winnik MA, Manners I. Micelle assembly. Multidimensional hierarchical self-assembly of amphiphilic cylindrical block comicelles. Science. 2015; **347:** 1329–32.

[71] Ducker WA, Grant LM. Effect of substrate hydrophobicity on surfactant surface–aggregate geometry. J. Phys. Chem. 1996; **100:** 11507–11511.

[72] Wanless EJ, Ducker WA. Weak influence of divalent ions on anionic surfactant surface-aggregation. Langmuir 1997; **13:** 1463–1474.

[73] Chandar P, Somasundaran P, Turro NJ. Fluorescence probe studies on the structure of the adsorbed layer of dodecyl sulfate at the alumina—water interface. J. Colloid Interface Sci. 1987; **117:** 31–46.

[74] Israelachvili JN, Mitchell DJ, Ninham BW. Theory of self-assembly of hydrocarbon amphiphiles into micelles and bilayers. J. Chem. Soc. Faraday Trans. 2. 1976; **72:** 1525.

[75] Chen M, Burgess I, Lipkowski J. Potential controlled surface aggregation of surfactants at electrode surfaces—a molecular view. Surf. Sci. 2009; **603:** 1878–1891.

[76] Burgess I, *et al.* Direct Visualization of the potential-controlled transformation of hemimicellar aggregates of dodecyl sulfate into a condensed monolayer at the Au(111) electrode surface. Langmuir. 1999; **15:** 2607–2616.

[77] Xu S, Chen M, Cholewa E, Szymanski G, Lipkowski J. Electric-field-driven surface aggregation of a model zwitterionic surfactant. Langmuir. 2007; **23:** 6937–6946.

[78] Retter U. One-dimensional nucleation−growth−collision in the formation of surface hemimicelles of amphiphiles. Langmuir. 2000; **16:** 7752–7756.

[79] Retter U, Avranas A. On anion-induced formation of hemicylindrical and hemispherical surface micelles of amphiphiles at the metal/electrolyte interface. Langmuir. 2001; **17:** 5039–5044.

[80] Retter U, Tchachnikova M, Avranas A. On packing constraints in the formation of surface micelles of amphiphiles at the metal/electrolyte interface-the necessity of introducing a generalized packing parameter. J. Colloid Interface Sci. 2002; **251:** 94–100.

[81] Retter U, Tchachnikova M. On the formation of surface micelles at the metal | electrolyte interface. J. Electroanal. Chem. 2003; **550–551:** 201–208.

[82] Zamlynny V. Electrochemical and spectroscopic studies of pyridine surfactants at the gold-electrolyte interface[PhD thesis]. University of Guelph; 2002.

[83] Richer J. Measurement of physical adsorption of neutral organic species at solid electrodes. J. Electrochem. Soc. 1986; **133:** 121.

[84] Lasia A. Electrochemical Impedance Spectroscopy and its Applications, New York, Springer; 2014. ISBN 978-1-4614-8932-0.

[85] Kertit MS, Gouttaya HM, Nciri B, Bensouda Y, Perez L, Infante MR, Elkacemi K. Corrosion inhibition of iron in 1 M HCl by some gemini surfactants in the series of alkanediyl-α,ω-bis-(dimethyl tetradecyl ammonium bromide). Prog. in Organic Coatings. 2001; **43 (4)**: 267–273.

[86] Ma H, Chen S, Yin B, Zhao S, Liu X. Impedance spectroscopic study of corrosion inhibition of copper by surfactants in the acidic solutions. Corrosion Science. 2003; **45(5)**: 867–882.

[87] Gabrielli C, Hemery P, Letellier P, Masure M, Perrot H, Rahmi M-I,Turmine M. Investigation of ionic surfactant-selective electrodes by EIS. Electrochimica Acta. 2002; **47(13–14)**, 2117–2126.

[88] Ghavami RK, Rafiei Z, Mojtaba TS. Effects of cationic CTAB and anionic SDBS surfactants on the performance of Zn–MnO2 alkaline batteries. J. of Power Sources. 2007; **164(2)**:934–946.

[89] Mimica D, Ringuedé A, Agurto C, Bedioui F, Zagal J. Biomimetic electroreduction of O2 by hemoglobin in a surfactant film: Preliminary electrochemical impedance spectroscopy insight. Electroanalysis. 2004; **16(19)**: 1632–1636.

Chapter 10

SAXS and SANS Techniques for Surfactant Characterization: Application in Corrosion Science

Deepak Dwivedi and Kateřina Lepková

Additional information is available at the end of the chapter

http://dx.doi.org/10.5772/intechopen.69290

Abstract

This chapter presents advances in the characterization of surfactants and detergents using small angle X-ray scattering (SAXS) and small angle neutron scattering (SANS) techniques. Surfactant molecules have been extensively used for corrosion prevention as part of commercial corrosion-inhibitor formulations. It is generally established that the interactions between surfactant molecule and metallic substrate play a key role in the formation of a corrosion-protective film. It is therefore essential to develop understanding about the nature of surfactant and detergent molecules in bulk solutions prior to formation of a surface film, as well as the mechanisms of their interactions with metallic substrates. These properties and interactions determine the properties of the surface film, including its persistency, and in turn define its protectiveness against corrosion. X-ray and neutron reflectivity methods are important investigating tools that could be used to characterize surfactant interactions with metallic substrates. These techniques have recently been utilized to investigate adsorption energies and contact angles between molecules or particles and variable substrates. This chapter addresses basic principles of these techniques and discusses their application for surfactant and detergent studies in corrosion science. Several case studies are presented and provide outlook for future prospects in this field of science.

Keywords: corrosion inhibitor, corrosion, surfactant, small angle neutron scattering, small angle X-ray scattering, aggregation, surface film formation, critical micelle concentration

1. Introduction

Corrosion of metals or alloys is a prime concern for various industries, such as oil and gas, steel and chemical to name few. To mitigate corrosion, application of corrosion inhibitors is a common and effective practice.

Adsorption of inhibitor molecules on substrate depends on various factors such as inhibitor concentration, pH and temperature of the corrosion media, etc. It is known that properties of metallic substrates, such as for example carbon steel, influence adsorption of organic surfactants. These properties have been recently reviewed [1]. It is very important for corrosion scientists to understand mechanistic insights and fundamentals associated with adsorption of surfactants on metallic substrates.

Film-forming surfactant molecules have many advantages, such as low cost, easy production, and high inhibition efficiency [2]. Surfactant molecules consist of hydrophilic head group and hydrophobic tail group, but the mechanism by which they adsorb to the substrate and form protective films has not been fully understood to date. Surfactant molecules tend to aggregate and interact with other molecules available in system, for example other components in commercial corrosion inhibitor formulations.

Film formation of organic surfactants has been largely characterized using for example scanning electron microscopy, X-ray photoelectron spectroscopy and atomic force microscopy. Some methods only provide information about fully formed film, but do not allow for analysis of possible interactions of the surfactant molecules with other components present in the corrosive media. Small angle X-ray scattering (SAXS) and small angle neutron scattering (SANS) could potentially characterize surfactant molecules in solution even in the precursor state (if it is available with other molecules in solution) and provide information about the surface film properties. This chapter provides background technical information on the two emerging characterization methods, SAXS and SANS, as well as discusses prospects of their use to characterize organic surfactants in corrosion science.

2. Principles of small angle neutron scattering (SANS) and small angle X-ray scattering (SAXS)

2.1. Small angle neutron scattering (SANS)

Phenomenon such as reflection or diffraction can be observed with neutron. Neutron behaves as a wave (same as electron) and is accepted in quantum mechanics. The interaction of neutron with nucleus is one way of its interaction with matter. Alternately, unpaired electrons also interact with neutron (specifically momentum of neutron). The scattering of neutron depends on the interaction potential between nucleus and neutron. Wavelength of neutron falls in the order of 1A° and it is worth to note that interaction potential, which is represented by V(r), falls in the order of 10^{-15} m. Therefore, nucleus incites scattering.

http://dx.doi.org/10.5772/intechopen.69290

Let us discuss few important terms that are encountered in neutron scattering. Scattering length can be defined as a complex number, but it should be noted that imaginary component of a complex number can only be considered if the atom is a heavy nuclei, such as boron, and shows high absorption coefficient. One should also be aware that nuclear scattering length is different from magnetic scattering length. Magnetic scattering length needs to be considered in magnetization studies because neutron momentum interacts with unpaired electrons. This fundamental phenomenon was responsible for the developments of SANS technique, which is widely used to characterize magnetic materials.

Interaction of neutron beam with matter is possible through two ways: (a) interaction between neutron beam and nucleus and (b) momentum of neutron with unpaired electron (magnetic interaction). Therefore, magnetic information about a surface film and a substrate can be obtained through SANS. It is particularly important for corrosion applications, where SANS could be used for determination of various iron oxide phases (i.e. corrosion products) owing to the magnetic nature of iron. Corrosion of iron usually promotes formation of these phases, often characterized with X-ray diffraction spectroscopy (XRD). It is important to point out, with regards to analysis of XRD and SANS data, that SANS analysis is more complex compared to a conventional XRD.

Elastic scattering describes a phenomenon of no change of neutron during scattering. The opposite phenomenon is known as inelastic scattering where the energy of neutron changes during scattering. Total scattering cross section is another technical parameter, particularly important to corrosion scientists performing experiments with SANS. Total scattering cross section is a ratio of total number of neutrons scattered per second to the number of neutron incident per second. The scattering data quality depends on this parameter as it suggests how strong the neutron scattering signal will be.

Let us discuss briefly small angle neutron scattering (SANS) technique. SANS is an instrument that uses neutron for sample characterization, and a monochromator is used to obtain monochromatized neutron beam. The velocity of neutron beam is controlled through rotating velocity selector made from an absorbing blade. Neutron beam passes through velocity selector and is controlled in terms of velocity along with monochromatization (order of 10%). Neutron beam divergence is limited by collimator, diameter of which is changed according to an application. The collimation length varies between 1 and 10 m. Flat samples must be used and mounted perpendicular to the beam direction so that the entire sample thickness is available to neutron scattering. In corrosion applications, thin and flat samples are suitable for SANS analysis. Using thin metallic foil (1–1.5 mm) is advisable in order to avoid multiple scattering and absorption of neutron beam. A position-sensitive detector (PSD) is usually used to determine incident neutron position, and the length can be changed as per the experiment requirement (5–20 m). PSD is typically positioned into an evacuated tank to avoid additional scattering occurring due to air (nitrogen). Schematic of SANS experimental setup is shown in **Figure 1** [3].

In SANS, coherent elastic neutron beam interacts with a sample and scattering vector, which is also known as a wave vector or a momentum vector, and is an important factor in a SANS experiment. The scattering vector is denoted by q and can be defined as:

$$q = K_s - K_i \tag{1}$$

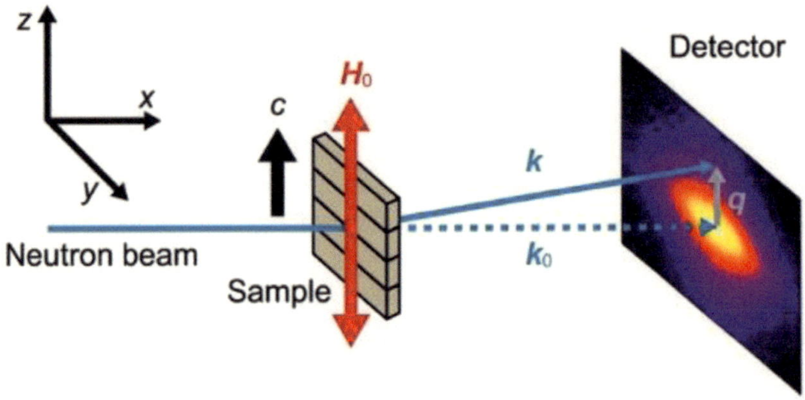

Figure 1. Schematic of SANS experimental setup [3]. Reprinted with permission. Copyright 2016 Nature Publishing Group.

where K_s is the scatter wave vector and K_i is the incident wave vector.

Mathematically, q can be represented as:

$$q = \frac{4\pi}{\lambda} \sin \theta \tag{2}$$

where 2θ is the angle between K_s and K_i. The unit of q is A^{o-1} or nm^{-1} and λ is the wavelength of neutron.

In direct space, q can be represented as:

$$q = \frac{2\pi}{d} \tag{3}$$

where d is the interplanar spacing in the crystal.

It is recommended to the reader to refer to a specialized literature for detailed information [4].

Time of flight (ToF) SANS is an advanced technique that requires special detector for detecting large dynamic range in q (scattering vector). It is worth to mention that for different λ of neutron, different q is depicted even at the same scattering angle, and this phenomenon describes the range of q. It is possible to cover large q range through the ToF instrument. It is feasible to change a steady state instrument into ToF SANS by introducing a ToF detector. SANS data analysis has been reported by Brulet et al. [5] and provides fundamental insight into the process.

Neutron-based technique such as SANS is considered a non-destructive technique that does not introduce any changes to the sample. It is therefore recognized as an important characterization technique in corrosion science where in situ monitoring of the corrosion and corrosion-inhibition processes is required in mechanistic studies. SANS can also be useful for various corrosion-related scientific fields, such as analysis of surfactants, nanoparticles, microparticles, polymers, corrosion inhibitors and thin films.

2.2. Small angle X-ray scattering (SAXS)

Small deviation of radiation from its incident direction occurs due to interaction between radiation and a homogeneous matter. This small deviation from the incident direction is known as small angle scattering. This scattering can also be explained as a wave or an object. Background information for this phenomenon can be obtained from other sources [6, 7].

It is fundamentally established that scattering power depends on electron density. This means that electron density is directly proportional to the scattering phenomenon. SAXS analysis is suitable for a range of samples and is most often used to analyse two-phase systems. An application of SAXS in a multiphase system requires assumption of a two-phase system only. In such case, one phase consists of much higher electron density than the other phases (considered as one combined phase) [8]. This methodology of a two-phase assumption results in a rare application of SAXS for multiphase systems [9, 10].

SAXS has been utilized for the analysis of precipitation phenomenon in metals and alloys, crystal growth, particles' behaviour in solution, and so on. For example, Wang et al. [11] studied biocorrosion of artificial hip implant. Orthopaedic prostheses were lubricated by pseudosynovial fluid that contains protein, which results in corrosion and wear of the hip (implant). It was observed that a nanocrystalline layer formed on Co-Cr-Mo alloy surface. Particle size distribution of this layer was measured through SAXS. The application of SAXS has not been limited to stainless steel or carbon steel, but was used also for corrosion of Mg alloys [12–15].

It should be highlighted here that coupling SAXS and SANS data could provide valuable information about complex systems, such as corrosion. Combination of SAXS with transmission electron microscopy (TEM) was successfully applied in the analysis of gold nanoparticles [16] and also for metals/alloys with the objective to identify rod-shaped precipitates [17]. In summary, one can even use SAXS to obtain useful information about morphological features that usually belong in the region of micro- to subnanometer. SAXS may require combinatory data from TEM analysis for a multiphase system and thus appears less favourable. SAXS however allows for obtaining data in situ, which is not possible from TEM analysis alone.

Experimental setup for SAXS measurements is technically very similar to that used for SANS (**Figure 1**). It is advisable for the reader to recall few technical parameter definitions discussed earlier in the SANS section. SAXS intensity [18, 19] is a function of scattering vector q, which is presented as a function of θ and wavelength of photon that scattered from the sample, and is mathematically described by Eq. (2). It is, however, important to note that the λ used in neutron (SANS) and X-ray (SAXS) are different. Some other factors, such as electron density, particle size and collimation of beam resolution [20], are important for corrosion science and for application of surfactant molecules in corrosive media. Let us, for example, consider particle size in the analysis of SAXS data. It was mentioned earlier in this chapter that scattering is a function of electron density. Amplitude of scattering intensity

A(q) usually depends on radial electron density of particle ρ(r), besides being influenced by volume of the particle. When particles are in solution, buffer solution scattering must be separated (subtracted) from scattering that originates from the system of study, that is the analysed samples. This method is known as contrast method and utilizes the difference of average electron density ρ(r) of particles from the electron density of solvent in which particles are dispersed. It is therefore necessary to acquire SAXS profile from blank buffer solution apart from measurements conducted with particles in solvent. One should keep in mind that dialyzing a sample for macromolecules is highly desirable. Corrosion scientists should note that the presence of salts results in increased background response. For example, a signal recorded from a brine solution (sodium chloride-containing medium) with a corrosion inhibitor will combine responses from both the brine solution and the inhibitor molecules.

In corrosion inhibition, aggregation of molecules is a common phenomenon that affects scattering results. Using diluted solutions of corrosion inhibitor is expected to reduce the aggregation, but could result in insufficient scattering signal as large clusters of particles or agglomerates will give a stronger scattering signal.

Data obtained from SAXS measurements can provide information on size, shape and surface of agglomerates or particles. A schematic representing the type of information that is available from SANS data, through various data-processing methods, is presented in **Figure 2** [21]. Discussion on the data analysis, specific for corrosion scientists is provided later in this chapter, in application section. Readers who are interested in obtaining additional details may refer to the review published by Pauw et al. [22] which addresses in detail SAXS profile collection and processing.

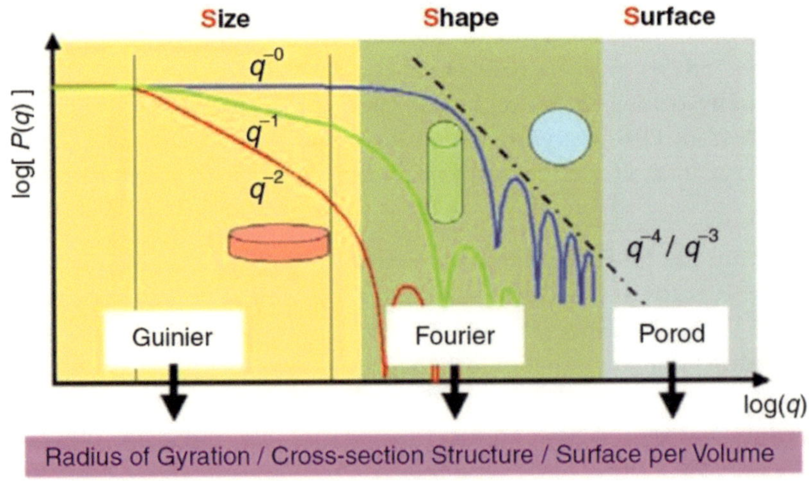

Figure 2. Schematic of the information which is possible to extract through various data-processing techniques in SAXS. Copyright 2015, Lauren Boldon et al. [21].

http://dx.doi.org/10.5772/intechopen.69290

3. Application of SANS and SAXS techniques for analysis of surfactants in corrosion science

3.1. Application of SANS technique

SANS can provide significant mechanistic insights into the corrosion phenomenon specifically related to applications of corrosion inhibitors. In particular, SANS can be applied to determine the interaction between various components used in commercial corrosion-inhibitor formulations. It is known, for example, that organic surfactants used in the corrosion-inhibitor formulations form micelles in the solution that are believed to adsorb to metallic substrate and protect against corrosion. It is however not known whether any interactions between these organic surfactants and other compounds in the corrosion inhibitor formulations take place in the corrosive media. Such interactions could potentially influence their adsorption to the steel surface and their efficiency to inhibit corrosion. Additionally, SANS could provide information about the precursor state of inhibitor molecules including nucleation of these molecules at steel, which depends on the degree of super saturation. The super saturation level and consequently nucleation are expected to be affected by different chemical compounds present in corrosion inhibitor formulations. It is therefore desirable for corrosion scientists to evaluate behaviour of inhibitor molecules both before and after their adsorption to a metallic substrate in order to develop mechanistic understanding of the corrosion-inhibition process.

Processing and analysis of SANS spectra is graphically described in **Figure 2**. Inverse Fourier Transform (IFT) is one of the methods used to analyse scattering data, including describing the pair distribution function p(r). It is important to mention here that if there are any oscillations in the curve between p(r) vs. r (inter atomic radius), then it is mandatory to consider intermicellar interactions as part of the data analysis. This is best demonstrated with the help of data reported by Hassan et al. [23]. In this work, the X_{PTHC}(molar ratio between salt and surfactant) was varied from 0 to 0.6 (**Figure 3**). In the case of X_{PTHC}= 0.5 and 0.6, no negative oscillations were observed at the spectra, meaning the interparticle interactions do not have to be included in this analysis. On contrary, in the case of X_{PTHC}= 0, 0.3 and 0.4, negative oscillations appear at the spectra and confirm the presence of intermicellar interactions. Another feature in **Figure 3** is the intensity of oscillations, especially the negative ones. Larger negative oscillation can be easily observed for X_{PTHC}= 0.3 compared to X_{PTHC}= 0.4 and indicates larger sized aggregates. It is worth noting that anisotropic micelles can be examined directly using IFT spectra with high scattering vector (q).

Following the IFT analysis, generalized indirect Fourier transform (GIFT) can be employed to provide information about growth of micelles, particularly the growth mechanism. GIFT usually separates two scattering phenomenon known as intraparticle scattering (form factor) and interparticle scattering (structure factor). The GIFT analysis could be combined with a mathematical approximation model, such as Ornstein-Zornike equation that is used for evaluating structure factor and for describing resultant waves (phase and amplitude) from atoms in unit cell. The most common model is rescaled mean spherical approximation (RMSA). RMSA

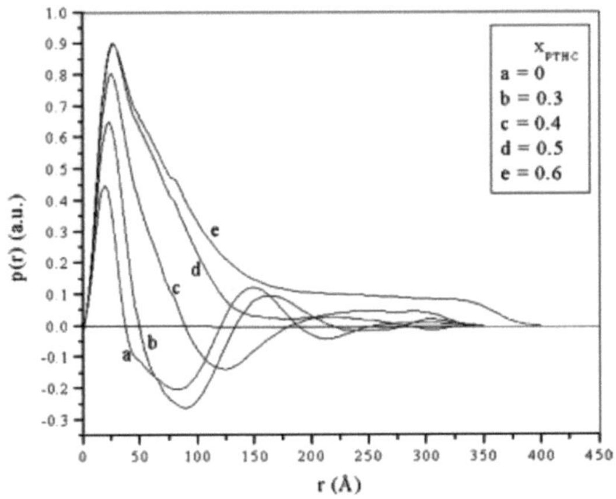

Figure 3. Pair distance distribution functions, p(r), through IFT analysis of SANS spectra for 50 mM SDS at different values of X_{PTHC} [23]. Reprinted with permission. Copyright 2003 Elsevier Science (USA).

provides significant advantage in reducing computational complexities. Alternatively, one can use Rogers and Yong (RY) closure. Readers should keep in mind that results derived from different closure applications will vary as they are dependent on the method of analysis used. Details of these approximation models can be found in a SANS tool box [24].

Guinier analysis is a common terminology in scattering analysis. Since Porod approximation takes high scattering vector to analyse low q limit of scattering spectra, Guinier analysis becomes useful. There are different formulas available for particles of different shapes. For example, radius of gyration of sphere would be $R_g^2 = 3/5\ R^2$ and R_g can be extracted slope of In I (Q) vs. Q^2 curve. But one should note that $R_g << 1$. We recommend a book by A. J. Jackson for further details on the data analysis [25].

Apart from the examination of interaction among surfactant molecules, SANS can also be used to extract information about electrostatic interactions and type of growth, for example uniaxial growth of surfactant assemblies. Shape of the micelles could also be identified with SANS and correlated with for example TEM analysis., while combined application of SANS and zeta potential measurements could help to relate particle charge with aggregation number. Application of Yukawa potential with RMSA closure provides similar information to the latter. Determining particle properties would be helpful in mechanistic investigations as defining the effect of surface charge on particle growth could reveal whether any specific charge promotes the growth of a surfactant assembly.

Let us discuss the application of SANS using case studies involving surfactant molecules in corrosion applications. Hassan et al. [23] studied a surfactant molecule, namely sodium dodecyl sulphate (SDS) and its interaction with cationic hydrophobic salt known as P-toluidine hydrochloride (PTHC). This experiment depicted the interaction among these molecules. Repulsive intermicellar interactions were observed and confirmed through the appearance of a well-pronounced peak in SANS spectra at low salt concentration. The experiment was conducted with neutron wavelength 6 A^{o-1}, path length 2 mm with scattering vector range

(q) 0.005–0.6 $A^{\circ -1}$. As mentioned in the principle section on SANS technique, background correction is necessary before final data analysis. In this study, the authors recorded spectra of empty test cell (quartz). Background could also be determined through Porod correction [25]. To avoid aggregation of inhibitor molecules during the experiment, the molar ratio of PTHC to SDS was kept below 0.6 or the solution became turbid. In general, molar ratio of inhibitor to surfactant will depend on the type of corrosion inhibitor, and turbidity will vary for solutions with a cationic or an anionic surfactant. These solution properties need to be considered by corrosion scientists when designing their experiments.

Mobin et al. [26] investigated SDS and cetyltrimethyl ammonium bromide (CTAB) for their synergistic effect with L-methionine to inhibit corrosion of mild steel. Through potentiodynamic polarization measurements, the authors concluded that the efficiency of L-methionine as a corrosion inhibitor was higher in the presence of SDS compared to CTAB. L-methionine was more effective with CTAB than with SDS. As both compounds, CTAB and SDS, can be analysed with SANS, it can be envisaged that the application of SANS would provide additional information on the mechanisms of the interactions of L-methionine with CTAB and SDS.

Surfactants can also be combined with protein to protect against corrosion [27]. Sodium dodecyl sulphate (SDS) was used as an anionic surfactant with zein, which is a corn protein and is insoluble in water. It was observed that SDS was less effective in inhibiting corrosion when used as a sole molecule, compared to its combined use, that is SDS-protein and zein-SDS. It was suggested that the enhanced zein-SDS efficiency resulted from the formation of zein-SDS complex. The interactions between zein and SDS are yet to be described. SANS could be a suitable technique to evaluate this phenomenon where the approach by Khan et al. [28] can be adopted. The authors applied SANS technique to study the interaction between protein, nanoparticles and surfactants. Nanoparticle used in this study was silica with sodium dodecyl sulphate surfactant along with the bovine serum albumin protein. Each system was studied separately. In the nanoparticle-protein system, protein did not adsorb on nanoparticle and formed aggregates. No physical interaction was observed in the nanoparticle-surfactant system, but surfactant interacted with protein and formed necklace-like structure probably due to the formation of a protein-surfactant complex as elucidated by Roy and Sukul [27]. It is worth to note that the test solution was D_2O instead of H_2O as D_2O provides better contrast than H_2O.

Figure 4 shows a complex SANS spectra for a three-component system [29]. The three-component system contained 1 wt% BSA protein with 50 mM SDS and 1 wt% silica nanoparticles in 0.2 M NaCl in D_2O.SANS was applied with neutron wavelength 6 A°, $\Delta\lambda/\lambda$= 9% (fwhm) with q ranges between 0.007 and 0.30 $A^{\circ -1}$. Quartz cell was used with 1–2 mm path length, and a non-linear least square fitting was applied. **Figure 4** clearly depicts that the combined system behaved differently than each individual system (see insets of **Figure 4**). In the mixed system, low q region contained a lot of scattering intensity whereas this was not the case for individual components, such as nanoparticle, protein and surfactant. Interestingly, high Q region was almost the same for the individual and combined systems. Aggregation of nanoparticles was observed in the nanoparticle-protein system. This is caused by the lack of

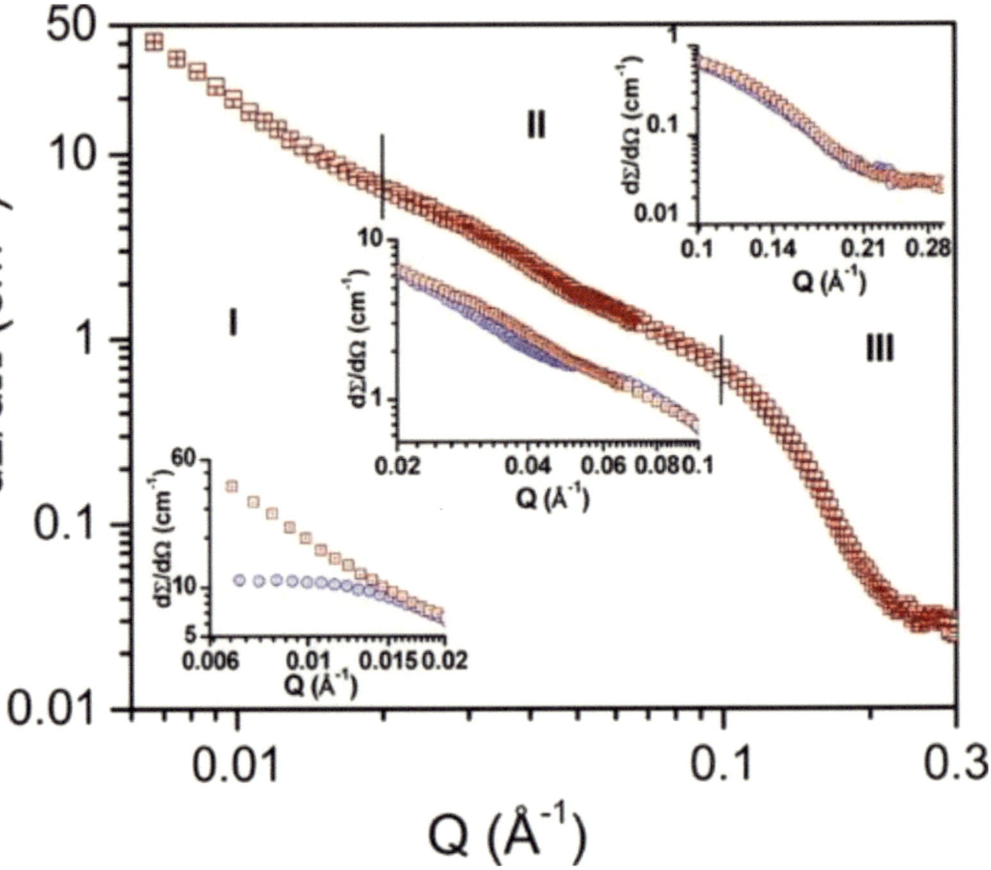

Figure 4. SANS data from a three-component system. Reprinted with permission from [29]. Copyright 2013 American Chemical Society.

adsorption of protein in nanoparticle due to depletion of forces. In other words, there was not enough driving force to govern the process of adsorption. Similarly, no physical interaction between nanoparticle and SDS was observed whereas interaction between SDS and protein was noted.

Bergstroem and Garamus [30] studied cationic surfactants, hexadecyltrimethylammonium bromide (CTAB), dodecyltrimethylammonium bromide (DTAB) and didodecyldimethylammonium bromide (DDAB) with variable chain lengths. The authors observed significant differences in the growth behaviour in various surfactant mixtures. It was noticed that DTAB formed oblate spheroidal micelle with [NaBr]=0.1 M. On contrary, addition of CTAB resulted in a transformation into prolate spheroidal micelles. Similar growth rate of CTAB/DTAB mixture was observed in all directions, whereas DDAB/DTAB mixture grew only in the length direction of micelle. The DDAB/DTAB micelles later transformed to form a bilayer structure. Another interesting outcome was the effect of salt addition on the transition point of micelle to bilayer. Furthermore, an addition of [NaBr] salt in DDAB/DTAB mixture resulted in aggregation of DTAB. This is relevant to corrosion science as corrosive media often contain various salts, such as NaCl in sea-salt simulations. The presence of salts could affect the behaviour of a surfactant in the solution and vary the micelle aggregation.

Singh et al. [31] used SANS to study interaction of CTAB surfactant with Pluronic F 88, a compound used as a corrosion inhibitor for Zn in alkaline battery in suspension form. It was found that Pluronic formed super molecular assembly with CTAB. Pluronic's micelle hydrophobic core was occupied with CTAB (hydrophobic chain). Interestingly, an increment in the CTAB concentration affected the electrostatic interaction, which was found to increase due to the change in the rotational correlational time of anionic probe. The pluronic-CTAB interaction led to formation of super molecular assembly with hydrophobic chain of CTAB placed at the hydrophobic core of pluronic micelle. Pluronic was established as a useful corrosion inhibitor for Zn and therefore further analysis of pluronic acid with DTAB, TTAB and other surfactants could be explored as it holds potential for corrosion inhibition applications.

Similarly, micelles of cationic octadecyltrimethylammonium chloride (OTAC) and anionic ammonium dodecyl sulphate (ADS) surfactants were investigated through SANS for their structural transitions [32]. It was observed that 100 mM ADS exhibited spherical shape with core shell, whereas oblate ellipsoid shapes were observed at concentrations above 100 mM. SANS was also employed to elucidate interaction of non-ionic surfactant with silica nanoparticle and decaethylene glycol monododecyl ether. The experiment was conducted using 1 wt.% nanoparticles and 0–1 wt.% surfactant. The non-ionic surfactant was found to adsorb on nanoparticles in the absence of electrolyte (NaCl), whereas in the presence of electrolyte it was not adsorbed, leading to the aggregation of nanoparticles [33]. This study could be useful for corrosion scientists as nanoparticle interactions with non-ionic surfactants are relevant to applications of corrosion-protective coatings containing nanoparticles. Nanoparticles are typically synthesized with the help of various capping agents and there is an opportunity to select a surfactant as a capping agent suitable for corrosion-protective coatings.

SDS surfactant was also used in combination with Al^{3+} ($Al(NO_3)_3.9H_2O$) and D_2O as a solvent. SANS analysis confirmed the formation of clusters as shown in **Figure 5** [34]. Including other modern characterization techniques into the study could provide mechanistic insights into the formation (e.g. shape) and composition of the formed clusters. Transmission electron microscopy with energy dispersive spectroscopy and electron energy loss spectroscopy and secondary ion mass spectrometry could be the possible options.

In this section, we have discussed the possibility of SANS application for surfactants that have been used as corrosion inhibitors and outlined future experiments concerning the use of SANS. Despite most studies being conducted in solutions (electrolytes), it should be noted that SANS can also be useful for investigating atmospheric corrosion studies, such as wet-dry oxidation [35].

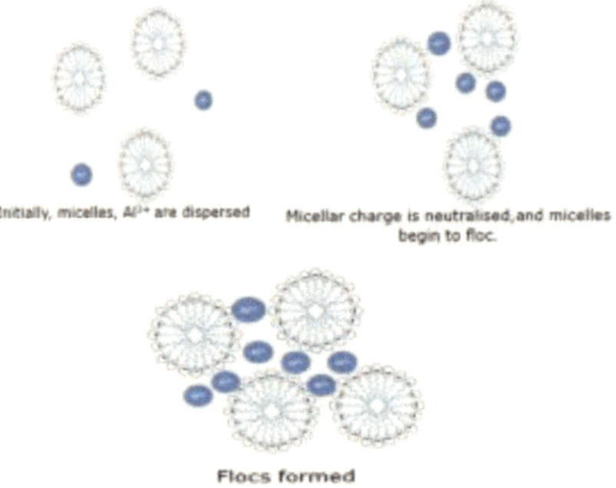

Figure 5. Cluster formation process [34]. Reprinted with permission. Copyright 2015 American Institute of Physics.

3.2. Application of SAXS technique

As discussed earlier in this chapter, SAXS is a powerful technique which allows for measuring particles in the size range of 1–100 nm. Particles must be presented in a dispersed form. Usually, amphiphilic molecules exhibit aggregation through self-assembly or organization. SAXS is an ideal technique to analyse the phenomenon of aggregation/self-assembly of these molecules. In this section, we discuss various research studies that were performed with SAXS and are relevant for corrosion science.

Jensen et al. [36] studied the formation of surfactant micelles under non-isothermal condition using synchrotron SAXS. The system was equipped with stop-flow mixing technique for kinetic investigations. The studied surfactant was dodecyl maltoside (DDM) in dimethyl formamide (DMF). Time-resolved SAXS was used with X-ray wavelength 0.995 A° and sample detector distance 1 m. Scattering vector was measured in the range of 0.01–0.6 $A^{\circ -1}$. DDM was applied in concentrations of 1.5, 1.8, 2, 2.6 and 3 wt.% in DMF. It was noticed that scattering from hydrocarbon tail caused oscillation in scattering spectrum and generated negative contrast, while maltoside head group offered positive contrast. The ellipsoidal core-corona particles were imagined for micelles and scattering form vector was calculated. SAXS pattern was fitted with a structural model mathematically expressed as:

$$I(q) = \frac{N}{V}\left[\frac{\psi}{P_{mic}} P_{mic}(q) + (1-\psi) P_{surf}(q)\right] \quad (4)$$

where N/V is the total concentration of surfactant molecules, ψ is the fraction of surfactant molecule in micelle, I (q) is the total scattering from solution, $P_{surf}(q)$ is the interference contribution from each point from surfactant and $P_{mic}(q)$ is the interference contribution from micelle (each point). This study revealed new information on the formation mechanism of micelles and suggested that insertion/expulsion of surfactant molecules was responsible for equilibration process of the surfactant [37].

The SAXS technique can be used to determine a critical micelle concentration (CMC) of a surfactant molecule. Surfactants are amphiphilic molecules with hydrophilic and hydrophobic parts (polar or non-polar) that are related to their behavior in solutions. This phenomenon takes place at a specific surfactant concentration known as critical micelle concentration (CMC). CMC depends on the surfactant structure (e.g. hydrocarbon chain length) and on environmental factors, such as pH, temperature and type of solvent.

Lucena et al. [38] studied non-ionic surfactant micelle structure through SAXS and determined critical micellar concentration (CMC), see **Figure 6**. The studied system was nonylphenolpolyethoxylated surfactant mixed with each solvent: octane, decane and dodecane. Ethylene glycol monobutyl ether was used as a polar additive. The authors used SAXS with X-ray wavelength of 0.1542 nm. Samples were analysed in quartz capillary with 1 mm diameter (external) and 10 μm thickness. Sample to detector distance was 700 mm. From the SAXS data, pair distance distribution function (P(r)) was derived using GIFT software. It was observed that at concentrations lower the CMC, surfactant interacts with organic solvent, whereas at the CMC, self-organization takes place through ejection of the ethoxylated part. Ethylene glycol monobutyl, which was used as a polar additive, enhances the micelle formation due to entropic effect. The driving force for the interaction between the polar additive and the polar region of the surfactant are the enthalphic changes that enhance the micelle formation through an increasing cohesion. This study is relevant to corrosion science as CMC reflects important properties (chemical and physical) of a surfactant in its application as a corrosion inhibitor. Interested readers can refer to related literature for broader understanding of the topic and for the development of experimental design [39, 40].

Shrestha et al. [41] studied non-ionic fluorinated micelles to describe the self-organization of perfluoroalkyl sulphonamide ethoxylate in aqueous solution. SAXS experiment was performed with X-ray wavelength of 0.1542 nm with quartz capillary (1 mm diameter (outer) and 10 μm thickness). This conventional SAXS operated at 40 kV and 50 mA. It was observed that rod-like structures form at lower temperature range (10 to 40°C) while planar formations are observed with temperature increase (up to 60°C). P(r) vs. r peak at low r regime exhibited a well-pronounced peak that was nearly flat towards high r regime. Changes in surfactant concentration did not affect the structure of particles. The authors also evaluated the effect of oil on the aggregation and phase transformation of the surfactant. This study is closely related to corrosion science, particularly to oil and gas industry where partitioning between oil and aqueous phase often affects the performance of as corrosion inhibitors.

4. Future prospects of SAXS and SANS applications in corrosion science

In situ monitoring of interactions between surfactant and other organic/inorganic compounds, including nanoparticles is highly important for corrosion applications, but only a limited number of techniques is currently available for these characterizations. Therefore, it is desirable to improve small angle scattering resolution and data capturing. There are number

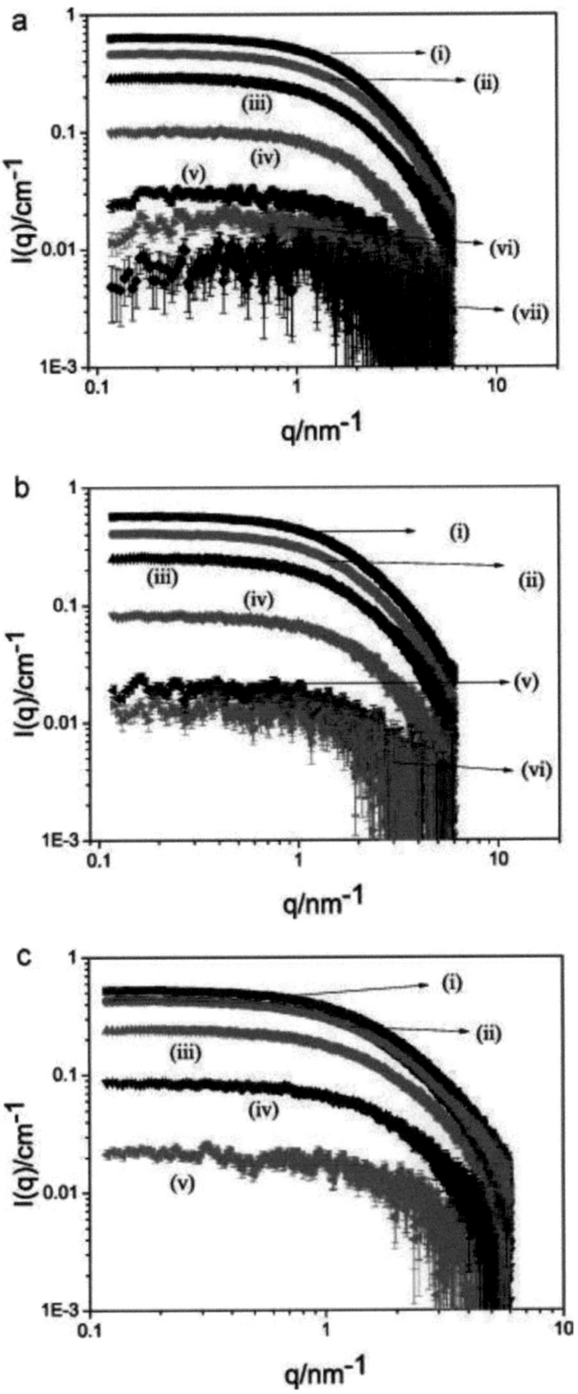

Figure 6. Scattering intensity curve I(q) for the solvents: octane (a), decane (b) and dodecane (c). (i) 0.2 mol/L; (ii) 0.15 mol/L; (iii) 0.1 mol/L; (iv) 0.05 mol/L; (v) 0.02 mol/L; (vi) 0.015 mol/L and (vii) 0.01 mol/L [38]. Reprinted with permission. Copyright 2012 Elsevier B. V.

of suitable applications of these techniques in corrosion investigations and we discuss below the eminent ones that we consider most relevant for investigating surfactants as corrosion inhibitors.

a. Determining critical micelle concentration (CMC): SAXS and SANS can be effectively utilized to determine CMC, which is an important factor in corrosion science as outlined earlier in this chapter. It is evident from earlier studies that CMC largely depends on various factors, such as temperature and pH of the studied system along with the ionic strength of the solution. The applications of the SANS and SAXS techniques to determine CMC for surfactants in corrosive media, such as for example production fluids, could provide mechanistic insights into the performance of the surfactants as corrosion inhibitors under simulated industrial conditions.

b. Surfactant aggregation study: The ability of effective corrosion inhibition by a surfactant compound has been reported to depend strongly on aggregation [42], which is related to properties of the studied system, such as pH and temperature. There are various corrosion inhibitor molecules that have been studied, for example using electrochemical impedance spectroscopy, but no studies applied both the SAXS and SANS techniques, leaving a gap in understanding of their inhibition mechanisms.

c. Monitoring (*in-situ*) of corrosion product formations: Corrosion phases at metal and alloy surfaces evolve in time and can be affected by alloy composition. SAXS and SANS could be used as in situ monitoring techniques of the corrosion-phase evolution. SANS is particularly suitable for such study, as it has the potential to characterize the bulk matrix without altering sample due to the beam exposure. It should be noted however that roughness of the steel substrate and multiple scattering from particles could possibly introduce error to the experiment.

d. Describing surfactant molecules at precursor state: Information at precursor level about interaction of surfactant molecules and other molecules in the corrosive media, such as for example corrosion-inhibitor enhancers or nanoparticles, is not abundantly available. This lack of knowledge presents significant gap in corrosion investigations as nucleation and growth in corrosive media are very important for predicting film formation kinetics and for understanding the dynamics of the corrosion process. Furthermore, the acts of nucleation, growth and interactions among molecules are important aspects of corrosion inhibition and could be elucidated through SAXS and SANS.

5. Conclusions and outlook

In this chapter, we discuss the use of small angle X-ray scattering (SAXS) and small angle neutron scattering (SANS) for analysis of organic surfactants that are utilized as effective corrosion inhibitors. We describe various examples of research studies performed with SAXS and SANS. These methods have emerged as popular techniques in biology, nanoscience and chemical sciences, but have not been extensively explored in the field of corrosion science. We believe that this chapter provides important information for future applications of SAXS and SANS in investigating surfactant molecules as corrosion inhibitors. This chapter includes principles of these techniques, lists available type of measurements, highlights errors that can be Encountered in research studies and briefly discusses data processing steps. This chapter also illustrates the ways for possible future experiments involving surfactants in corrosion science, such as determining critical micelle concentration or the effect of aggregation of the surfactant molecules.

Acknowledgements

The financial support from Curtin University for D. D. (Curtin International Postgraduate Research Scholarship) is greatly appreciated.

Author details

Deepak Dwivedi and Kateřina Lepková*

*Address all correspondence to: k.lepkova@curtin.edu.au

Department of Chemical Engineering, Curtin Corrosion Engineering Industry Centre, Curtin University, Western Australia, Australia

References

[1] Dwivedi D, Lepkova K, Becker T. Carbon steel corrosion: A review of key surface properties and characterization methods. RSC Advances. 2017;**7**(8):4580-4610

[2] Maqsood Ahmad Malik MA H, Firdosa N, AL-Thabaiti SA, Khan Z. Anti-corrosion ability of surfactants: A review. International Journal of Electrochemicl Sciene. 2011;**6**;1927-1948

[3] Ueno T, Saito K, Yano M, Ito M, Shoji T, Sakuma N, Kato A, Manabe A, Hashimoto A, Gilbert EP, Keiderling U, Ono K. Multiple magnetic scattering in small-angle neutron scattering of Nd-Fe-B nanocrystalline magnet. Scientific Reports. 2016;**6**:28167

[4] Kittel C. Introduction to Solid State Physics. USA: Willey; 2004. Vol. 8th Edition (printed in USA). p. 704

[5] Brulet A, Lairez D, Lapp A, Cotton JP. Improvement of data treatment in small-angle neutron scattering. Journal of Applied Crystallography. 2007;**40**(1):165-177

[6] Feigin LA & Svergun DI. Structure Analysis by Small-Angle X-Ray and Neutron Scattering. In: Taylor GW, editor. New York: Plenum Press; 1987

[7] Glatter O, Kratky O. Small Angle X-ray Scattering. New York: Academic Press; 1982. p. 515

[8] Kostorz G. Small-angle scattering studies of phase separation and defects in inorganic materials. Journal of Applied Crystallography. 1991;**24**(5):444-456

[9] Ohnuma M, Suzuki J, Ohtsuka S, Kim SW, Kaito T, Inoue M, Kitazawa H. A new method for the quantitative analysis of the scale and composition of nanosized oxide in 9Cr-ODS steel. Acta Materialia. 2009;**57**(18):5571-5581

[10] Tatchev D. Structure analysis of multiphase systems by anomalous small-angle X-ray scattering. Philosophical Magazine. 2008;**88**(12):1751-1772

[11] Wang ZW, Yan Y, Qiao LJ. Nanocrystalline layer on the bearing surfaces of artificial hip implants induced by biotribocorrosion processes. Biosurface and Biotribology. 2015;**1**(2):130-134

[12] Brady MP, Rother G, Anovitz LM, Littrell KC, Unocic KA, Elsentriecy HH, Song GL, Thomson JK, Gallego NC, Davis B. Film breakdown and nano-porous Mg(OH)2 formation from corrosion of magnesium alloys in salt solutions. Journal of Electrochemical Society. 2015;**162**(4);C140-C149

[13] Xu L, Zhang E, Yang K. Phosphating treatment and corrosion properties of Mg-Mn-Zn alloy for biomedical application. Journal of Material Science: Materials in Medicine. 2009;**20**(4):859-867

[14] Xu L, Zhang E, Yin D, Zeng S, Yang K. In vitro corrosion behaviour of Mg alloys in a phosphate buffered solution for bone implant application. Journal of Material Science: Materials in Medicine. 2008;**19**(3):1017-1025

[15] Ingham B, Ko M, Laycock N, Kirby NM, Williams DE. First stages of siderite crystallisation during CO2 corrosion of steel evaluated using in situ synchrotron small- and wide-angle X-ray scattering. Faraday Discuss. 2015;**180**:171-190

[16] Mori Y, Furukawa M, Hayashi T, Nakamura K. Size distribution of gold nanoparticles used by small angle X-ray scattering. Particulate Science and Technology. 2006;**24**(1):97-103

[17] Rosalie JM, Pauw BR. An in-depth TEM/SAXS study of aging induced precipitation in MgZn. arXiv.org, e-Print Arch., Condensed Matter. 2012;1-15. arXiv:1210.5366v1 [cond-mat.mtrl-sci]

[18] Schneidman-Duhovny D, Kim SJ, Sali A. Integrative structural modeling with Small Angle X-ray scattering profiles. BMC Structural Biology. 2012;**12**:17

[19] Schnablegger H, Singh Y. The SAXS Guide: Getting Acquainted with the Principles. 3rd ed. Austria: Anton Paar GmbH; 2013

[20] Bardhan J, Park S, Makowski L. SoftWAXS: A computational tool for modeling wide-angle X-ray solution scattering from biomolecules. Journal of Applied Crystallography. 2009;**42**(5):932-943

[21] Boldon L, Laliberte F, Liu L. Review of the fundamental theories behind small angle X-ray scattering, molecular dynamics simulations, and relevant integrated application. Nano Reviews. 2015;**6**(1):25661

[22] Brian Richard P. Everything SAXS: Small-angle scattering pattern collection and correction. Journal of Physics: Condensed Matter. 2013;**25**(38):383201

[23] Hassan PA, Fritz G, Kaler EW. Small angle neutron scattering study of sodium dodecyl sulfate micellar growth driven by addition of a hydrotropic salt. Journal of Colloid and Interface Science. 2003;**257**(1):154-162

[24] Hammouda B. The SANS Toolbox In http://www.ncnr.nist.gov/staff/hammouda/the_SANS_toolbox.pdf

[25] Jackson AJ. Introduction to Small-Angle Neutron Scattering and Neutron Reflectometry. Gaithersburg: NIST Center for Neutron Research; 2008. https://www.ncnr.nist.gov/summerschool/ss10/pdf/SANS_NR_Intro.pdf

[26] Mobin M, Parveen M, Rafiquee MZA. Synergistic effect of sodium dodecyl sulfate and cetyltrimethyl ammonium bromide on the corrosion inhibition behavior of l-methionine on mild steel in acidic medium. Arabian Journal of Chemistry. 2017;**10**(Supplement 1). S1364-S1372

[27] Roy P, Sukul D. Protein-surfactant aggregate as a potential corrosion inhibitor for mild steel in sulphuric acid: zein-SDS system. RSC Advances. 2015;**5**(2):1359-1365

[28] Khan JM, Chaturvedi SK, Khan RH. Elucidating the mode of action of urea on mammalian serum albumins and protective effect of sodium dodecyl sulfate. Biochemical and Biophysical Research Communications. 2013;**441**(3):681-688

[29] Mehan S, Chinchalikar AJ, Kumar S, Aswal VK, Schweins R. Small-angle neutron scattering study of structure and interaction of nanoparticle, protein, and surfactant complexes. Langmuir. 2013;**29**(36):11290-11299

[30] Magnus Bergström, L, Garamus, V. M., Structural behaviour of mixed cationic surfactant micelles: A small-angle neutron scattering study. Journal of Colloid and Interface Science 2012, **381**(1), 89-99.

[31] Singh PK, Kumbhakar M, Ganguly R, Aswal VK, Pal H, Nath S. Time-resolved fluorescence and small angle neutron scattering study in pluronics-surfactant supramolecular assemblies. The Journal of Physical Chemistry B. 2010;**114**(11):3818-3826

[32] Kang K-H, Lim K-H. A SANS study for structural transition of micelles of cationic octadecyl trimethyl ammonium chloride and anionic ammonium dodecyl sulfate surfactants in aqueous solutions. Colloids and Surfaces A: Physicochemical and Engineering Aspects. 2011;**391**(1-3):69-79

[33] Ray D, Aswal VK, Kohlbrecher J. Micelle-induced depletion interaction and resultant structure in charged colloidal nanoparticle system. Journal of Applied Physics. (Melville, NY, U. S.). 2015;**117**(16):164310/1-164310/9

[34] Nadaf R, Kutty B, Narayanan J, Ramagiri SV, Kumar S, Aswal VK, Bellare JR, Goyal PS. Flocculation of SDS micelles with Al3+: SANS, SAXS and TEM study. AIP Conference Proceedings. 2015;**1665**(1, Solid State Physics), 040013/1-040013/3

[35] Rother G, Keiser JR, Brady MP, Unocic KA, Anovitz LM, Littrell KC, Peascoe-Meisner RA, Santella ML, Wesolowski DJ, Cole DR. Small-angle neutron scattering study of the wet and dry high-temperature oxidation of alumina- and chromia-forming stainless steels. Corrosion Science. 2012;**58**:121-132

[36] Jensen GV, Lund R, Gummel J, Monkenbusch M, Narayanan T, Pedersen JS. Direct observation of the formation of surfactant micelles under nonisothermal conditions by synchrotron SAXS. Journal of American Chemical Society. 2013;**135**(19):7214-7222

[37] Zhang R, Somasundaran P. Advances in adsorption of surfactants and their mixtures at solid/solution interfaces. Advances in Colloid and Interface Science. 2006;123-126:213-229

[38] Lucena IL, Canuto JDS, Caroni ALPF, Fonseca JLC, NetoAAD, Dantas TNC. Characterization of nonionic surfactant micellar structures in organic solvents by small angle X-ray scattering (SAXS). Colloids and Surfaces A: Physicochemical and Engineering Aspects. 2012;**408**:48-56

[39] Paria S, Khilar KC. A review on experimental studies of surfactant adsorption at the hydrophilic solid–water interface. Advances in Colloid and Interface Science. 2004;**110**(3):75-95

[40] Atkin R, Craig VS J, Wanless EJ, Biggs S. Mechanism of cationic surfactant adsorption at the solid–aqueous interface. Advances in Colloid and Interface Science. 2003;**103**(3):219-304

[41] Shrestha LK, Sharma SC, SatoT, GlatterO, AramakiK. Small-angle X-ray scattering (SAXS) study on nonionic fluorinated micelles in aqueous system. Journal of Colloid and Interface Science. 2007;**316**(2):815-824

[42] Asefi D, Arami M, Mahmoodi NM. Electrochemical effect of cationic gemini surfactant and halide salts on corrosion inhibition of low carbon steel in acid medium. Corrosion Science. 2010;**52**(3):794-800